Lecture Notes in Computer Science 2643

Edited by G. Goos, J. Hartmanis, and J. van Leeuwen

Springer
Berlin
Heidelberg
New York
Barcelona
Hong Kong
London
Milan
Paris
Tokyo

Marc Fossorier   Tom Høholdt
Alain Poli (Eds.)

Applied Algebra,
Algebraic Algorithms and
Error-Correcting Codes

15th International Symposium, AAECC-15
Toulouse, France, May 12-16, 2003
Proceedings

 Springer

Series Editors

Gerhard Goos, Karlsruhe University, Germany
Juris Hartmanis, Cornell University, NY, USA
Jan van Leeuwen, Utrecht University, The Netherlands

Volume Editors

Marc Fossorier
University of Hawaii, Department of Electrical Engineering
2540 Dole Street, Holmes Hall 455, Honolulu, HI 96822, USA
E-mail: marc@spectra.eng.hawaii.edu

Tom Høholdt
The Technical University of Denmark, Department of Mathematics
Bldg. 303, DK-2800 Lyngby, Denmark
E-mail: T.Hoeholdt@mat.dtu.dk

Alain Poli
IRIT - Université Paul Sabatier
31602 Toulouse cédex, France
E-mail: poli@cict.fr

Cataloging-in-Publication Data applied for

A catalog record for this book is available from the Library of Congress.

Bibliographic information published by Die Deutsche Bibliothek.
Die Deutsche Bibliothek lists this publication in the Deutsche Nationalbibliografie;
detailed bibliographic data is available in the Internet at <http://dnb.ddb.de>.

CR Subject Classification (1998): E.4, I.1, E.3, G.2, F.2

ISSN 0302-9743
ISBN 3-540-40111-3 Springer-Verlag Berlin Heidelberg New York

Springer-Verlag Berlin Heidelberg New York
a member of BertelsmannSpringer Science+Business Media GmbH

http://www.springer.de

© Springer-Verlag Berlin Heidelberg 2003
Printed in Germany

Typesetting: Camera-ready by author, data conversion by PTP-Berlin GmbH
Printed on acid-free paper   SPIN: 10930380   06/3142   5 4 3 2 1 0

Preface

The AAECC symposium was started in June 1983 by Alain Poli (Toulouse), who, together with Roger Desq, Daniel Lazard, and Paul Camion, organized the first conference. The meaning of the acronym AAECC changed from "Applied Algebra and Error Correcting Codes" to "Applied Algebra, Algebraic Algorithms, and Error Correcting Codes." One reason for this was the increasing importance of complexity, particularly for decoding algorithms. During the AAECC-12 symposium the conference committee decided to enforce the theory and practice of the coding side as well as the cryptographic aspects. Algebra was conserved, as in the past, but was slightly more oriented to algebraic geometry codes, finite fields, complexity, polynomials, and graphs.

For AAECC-15 the main subjects covered were:

- Block codes.
- Algebra and codes: rings, fields, AG codes.
- Cryptography.
- Sequences.
- Algorithms, decoding algorithms.
- Algebra: constructions in algebra, Galois groups, differential algebra, polynomials.

The talks of the six invited speakers characterized the aims of AAECC-15:

- P. Sole ("Public Key Cryptosystems Based on Rings").
- S. Lin ("Combinatorics Low Density Parity Check Codes").
- J. Stern ("Cryptography and the Methodology of Provable Security").
- D. Costello ("Graph-Based Convolutional LDPC Codes").
- I. Shparlinsky ("Dynamical Systems Generated by Rational Functions").
- A. Lauder ("Algorithms for Multivariate Polynomials over Finite Fields").

Except for AAECC-1 (published in the journal Discrete Mathematics, 56, 1985) and AAECC-7 (Discrete Mathematics, 33, 1991), the proceedings of all the symposia have been published in Springer-Verlag's Lecture Notes in Computer Science series (vols. 228, 229, 307, 356, 357, 508, 673, 948, 1255, 1719, 2227). It is the policy of AAECC to maintain a high scientific standard. This has been made possible thanks to the many referees involved. Each submitted paper was evaluated by at least two international researchers.

AAECC-15 received 40 submissions; 25 were selected for publication in these proceedings while 8 additional works were contributed to the symposium as oral presentations.

The symposium was organized by Marc Fossorier, Tom Høholdt, and Alain Poli, with the help of the 'Centre Baudis' in Toulouse.

We express our thanks to Jinghu Chen and Juntan Zhang of the University of Hawaii for their dedicated work on these proccedings, to the Springer-Verlag staff, especially Alfred Hofmann and Anna Kramer, and to the referees.

February 2003   M. Fossorier, T. Høholdt, A. Poli

Organization

Steering Committee

Conference General Chairman: Alain Poli (Univ. of Toulouse, France)
Conference Co-chairman: Tom Høholdt (Technical Univ. of Denmark, Denmark)
Publication: Marc Fossorier, Jinghu Chen and Juntan Zhang (Univ. of Hawaii, USA)
Local Arrangements: Marie-Claude Gennero, Geneviève Cluzel (Univ. of Toulouse, France)

Conference Committee

J. Calmet
G. Cohen
S.D. Cohen
G.L. Feng
M. Giusti
J. Heintz
T. Høholdt
H. Imai
H. Janwa
J.M. Jensen

R. Kohno
H.W. Lenstra Jr.
S. Lin
O. Moreno
H. Niederreiter
A. Poli
T.R.N. Rao
S. Sakata
P. Sole

Program Committee

T. Berger
E. Biglieri
J. Calmet
C. Carlet
D. Costello
T. Ericson
P. Farrell
M. Fossorier
J. Hagenauer
S. Harari
T. Helleseth

E. Kaltofen
T. Kasami
L.R. Knudsen
S. Litsyn
R.J. McEliece
R. Morelos-Zaragoza
H. Niederreiter
P. Sole
H. Tilborg

Table of Contents

Cryptography and the Methodology of Provable Security 1
  Jacques Stern

Dynamical Systems Generated by Rational Functions 6
  Harald Niederreiter, Igor E. Shparlinski

Homotopy Methods for Equations over Finite Fields 18
  Alan G.B. Lauder

Three Constructions of Authentication/Secrecy Codes 24
  Cunsheng Ding, Arto Salomaa, Patrick Solé, Xiaojian Tian

The Jacobi Model of an Elliptic Curve and Side-Channel Analysis 34
  Olivier Billet, Marc Joye

Fast Point Multiplication on Elliptic Curves through Isogenies 43
  Eric Brier, Marc Joye

Interpolation of the Elliptic Curve Diffie–Hellman Mapping 51
  Tanja Lange, Arne Winterhof

An Optimized Algebraic Method for Higher Order Differential Attack . . . 61
  Yasuo Hatano, Hidema Tanaka, Toshinobu Kaneko

Fighting Two Pirates 71
  Hans Georg Schaathun

Copyright Control and Separating Systems 79
  Sylvia Encheva, Gérard Cohen

Unconditionally Secure Homomorphic Pre-distributed Commitments 87
  Anderson C.A. Nascimento, Akira Otsuka, Hideki Imai,
  Joern Mueller-Quade

A Class of Low-Density Parity-Check Codes Constructed Based
on Reed–Solomon Codes with Two Information Symbols 98
  Ivana Djurdjevic, Jun Xu, Khaled Abdel-Ghaffar, Shu Lin

Relative Duality in MacWilliams Identity 108
  L.S. Kazarin, V.M. Sidelnikov, Igor B. Gashkov

Good Expander Graphs and Expander Codes: Parameters and Decoding . 119
  H. Janwa

On the Covering Radius of Certain Cyclic Codes 129
  Oscar Moreno, Francis N. Castro

Unitary Error Bases: Constructions, Equivalence, and Applications 139
  Andreas Klappenecker, Martin Rötteler

Differentially 2-Uniform Cocycles – The Binary Case 150
  K.J. Horadam

The Second and Third Generalized Hamming Weights of Algebraic
Geometry Codes 158
  Domingo Ramirez-Alzola

Error Correcting Codes over Algebraic Surfaces 169
  Thanasis Bouganis

A Geometric View of Decoding AG Codes............................ 180
  Thanasis Bouganis, Drue Coles

Performance Analysis of M-PSK Signal Constellations in Riemannian
Varieties 191
  Rodrigo Gusmão Cavalcante, Reginaldo Palazzo Jr.

Improvements to Evaluation Codes and New Characterizations of Arf
Semigroups 204
  Maria Bras-Amorós

Optimal 2-Dimensional 3-Dispersion Lattices 216
  Moshe Schwartz, Tuvi Etzion

On g-th MDS Codes and Matroids 226
  Keisuke Shiromoto

On the Minimum Distance of Some Families of $\mathbb{Z}_{2^k}$-Linear Codes 235
  Fabien Galand

Quasicyclic Codes of Index ℓ over F_q Viewed as $F_q[x]$-Submodules of
$F_{q^\ell}[x]/\langle x^m - 1 \rangle$ 244
  Kristine Lally

Fast Decomposition of Polynomials with Known Galois Group 254
  Andreas Enge, François Morain

Author Index... 265

Cryptography and the Methodology of Provable Security

Jacques Stern

Dépt d'Informatique, ENS – CNRS, 45 rue d'Ulm, 75230 Paris Cedex 05, France
Jacques.Stern@ens.fr
http://www.di.ens.fr/users/stern

1 Introduction

Public key cryptography was proposed in the 1976 seminal article of Diffie and Hellman [6]. One year later, Rivest, Shamir and Adleman introduced the RSA cryptosystem as a first example. From an epistemological perspective, one can say that Diffie and Hellman have drawn the most extreme consequence of a principle stated by Auguste Kerckhoffs in the XIXth century: "le mécanisme dè chiffrement doit pouvoir tomber sans inconvénient aux mains de l'ennemi[1]". Indeed, Diffie and Hellman understood that only the deciphering operation has to be controlled by a secret key: the enciphering method may perfectly be executed by means of a publicly available key, provided it is virtually impossible to infer the secret deciphering key from the public data.

Today, algorithms have replaced mechanisms and the wording "virtually impossible" has been given a formal meaning using the theory of complexity. This allows a correct specification of the security requirements, which in turn can be established by means of a *security proof*.

2 The RSA Cryptosystem

In modern terms, a public-key encryption scheme on a message space $\mathcal{M}$ consists of three algorithms $(\mathcal{K}, \mathcal{E}, \mathcal{D})$:

- the key generation algorithm $\mathcal{K}(1^k)$ outputs a random pair of private/public keys $(\mathsf{sk}, \mathsf{pk})$, relatively to a security parameter k
- the encryption algorithm $\mathcal{E}_{\mathsf{pk}}(m; r)$ outputs a ciphertext c corresponding to the plaintext $m \in \mathcal{M}$, using random coins r
- the decryption algorithm $\mathcal{D}_{\mathsf{sk}}(c)$ outputs the plaintext m associated to the ciphertext c.

We will occasionnally omit the random coins and write $\mathcal{E}_{\mathsf{pk}}(m)$ in place of $\mathcal{E}_{\mathsf{pk}}(m; r)$. Note that the decryption algorithm is deterministic.

The famous RSA cryptosystem has been proposed by Rivest, Shamir and Adleman [14]. The key generation algorithm of RSA chooses two large primes p,

---

[1] The enciphering mechanism may fall into the enemy's hands without drawback

M. Fossorier, T. Hoeholdt, and A. Poli (Eds.): AAECC 2003, LNCS 2643, pp. 1–5, 2003.

q of equal size and issues the so-called modulus $n = 3Dpq$. The sizes of p, q are set in such a way that the binary length $|n|$ of n equals the security parameter k. Additionally, en exponent e, relatively prime to $\varphi(n) = 3D(p-1)(q-1)$ is chosen, so that the public key is the pair (n, e). Thè private key d is the inverse of e modulo $\varphi(n)$. Variants allow the use of more than two prime factors.

Encryption and decryption are defined as follows:

$$\mathcal{E}_{n,e}(m) = 3Dm^e \bmod n \qquad \mathcal{D}_{n,d}(c) = 3Dc^d \bmod n.$$

Note that both operations are deterministic and are mutually inverse to each other. Thus, the RSA encryption function is a permutation. It is termed a *trapdoor permutation* since decryption can only be applied given the private key.

The basic security assumption on which the RSA cryptosystem relies is its *one-wayness* (OW): using only public data, an attacker cannot recover the plaintext corresponding to a given ciphertext. In the general formal setting provided above, an encryption scheme is one-way if the success probability of any adversary $\mathcal{A}$ attempting to invert $\mathcal{E}$ (without the help of the private key), is negligible, i.e. asymptotically smaller than the inverse of any polynomial function of the security parameter. Probabilities are taken over the message space $\mathcal{M}$ and the randoin coins Ω. These include both the random coins r used for the encryption scheme, and the internal random coins of the adversary. In symbols:

$$\mathsf{Succ}^{\mathsf{ow}}(\mathcal{A}) = 3D\Pr[(\mathsf{pk}, \mathsf{sk}) \leftarrow \mathcal{K}(1^k), m \overset{R}{\leftarrow} \mathcal{M} : \mathcal{A}(\mathsf{pk}, \mathcal{E}_{\mathsf{pk}}(m)) = 3Dm].$$

Formally, the assumption that RSA is one-way is stronger than the hardness of factoring. Still, it is widely believed and the only method to assess the strength of RSA is to check whether the size of the modulus n outreaches the current performances of the various factoring algorithms.

3 From Naive RSA to OAEP

The "naive" RSA algorithm defined in the previous section cannot be used as it stands: in particular, it has algebraic multiplicative properties which are highly undesirable from a security perspective. Accordingly, it was found necessary to define formatting schemes adding some redundancy. For several years, this worked by trials and errors, as shown by the subtle attack against the PKCS #1 v1.5 encryption scheme devised by Bleichenbacher [4]. In this attack, the adversary discloses the secret key of an SSL server based on the information coming form the error messages received when an incorrectly formatted ciphertext is submitted to the server. Thus, a more formal approach appeared necessary.

The starting point of the new appraoch is semantic security, also called *polynomial security/indistinguishability of encryptions*, introduced by Goldwasser and Micali [9] : an encryption scheme is *semantically secure* if no polynomial-time attacker can learn any bit of information about the plaintext from the ciphertext, except its length. More formally, an encryption scheme is semantically secure if, for any two-stage adversary $\mathcal{A} = 3D(A_1, A_2)$ running in polynomial

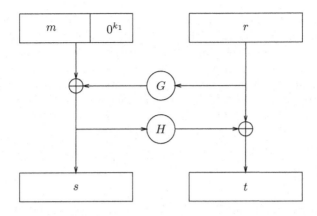

Fig. 1. Optimal Asymmetric Encryption Padding

time, the advantage $\mathsf{Adv}^{\mathsf{ind}}(\mathcal{A})$ is negligible, where $\mathsf{Adv}^{\mathsf{ind}}(\mathcal{A})$ is formally defined as

$$2 \times \Pr\left[\begin{array}{l}(\mathsf{pk},\mathsf{sk}) \leftarrow \mathcal{K}(1^k), (m_0, m_1, s) \leftarrow A_1(\mathsf{pk}), \\ b \xleftarrow{R} \{0,1\}, c = 3D\mathcal{E}_{\mathsf{pk}}(m_b) : A_2(m_0, m_1, s, c) = 3Db\end{array}\right] - 1,$$

where the probability space includes the internal random coins of the adversary, and m_0, m_1 are two equal length plaintexts chosen by A_1 in the message-space $\mathcal{M}$.

The above definition only covers passive adversaries. It is a *chosen–plaintext* or CPA attack since the attacker can only encrypt plaintext. In extended models, the adversary is given access to a decryption oracle which returns the decryption of any ciphertext c, with the only restriction that it should be different from the challenge ciphertext c. This scenario is is termed the *adaptive chosen-ciphertext attack* (CCA2) [13].

The OAEP padding scheme (optimal asymmetric encryption padding) was proposed by Bellare and Rogaway [3] in 1994. It is depicted on Fig. 1 . For a long time it was believed that OAEP achieved CCA2 security, based on an almost mathematical proof relying on the one-wayness of the RSA function. The word "almost" here refers to the use of the so-called *random oracle model* which models G and H in Fig. 1 as functions which return random independant values, which is not formally correct.

4   Rise, Fall, and Repair of OAEP

In 2001, Victor Shoup [16] showed by means of a subtle counter-example, the the proof of Bellare and Rogaway only applied in the restricted attack setting where the adversary can query the decryption oracle before it receives the challenge ciphertext c (sometimes referred as CCA1. It did not necessarily mean that OAEP was itself flawed. In any case, a new proof was needed.

Surprisingly, the repaired proof appeared shortly afterwards in [8]. Albeit based on the same methodology, it is significantly different and uses additional algebraic tools, notably two-dimensional lattices, which did not appear in the earlier proof. Thus, the multiplicative properties of RSA, which motivated the quest for formatting schemes, help for the security proof. It should also be noted that alternative formatting schemes, with a more direct security proof have been recently designed. However, OAEP is a widely used standard [15] and it is unclear whether it will be replaced by these challengers.

5   Conclusion

The lesson to learn from the above is that cryptography should proceed with care. Twenty-five centuries were needed before the discovery of public key cryptography by Diffie and Hellman. It took twenty-five more years to understand how RSA could be correctly practiced. No cryptographic algorithm can be designed and validated in twenty-five minutes or twenty-five hours, not even twenty-five days.

References

1. M. Bellare, A. Desai, D. Pointcheval, and P. Rogaway. Relations among Notions of Security for Public-Key Encryption Schemes. In *Crypto '98*, LNCS 1462, pages 26–45. Springer-Verlag, Berlin, 1998.
2. M. Bellare and P. Rogaway. Random Oracles Are Practical: a Paradigm for Designing Efficient Protocols. In *Proc. of the 1st CCS*, pages 62–73. ACM Press, New York, 1993.
3. M. Bellare and P. Rogaway. Optimal Asymmetric Encryption – How to Encrypt with RSA. In *Eurocrypt '94*, LNCS 950, pages 92–111. Springer-Verlag, Berlin, 1995.
4. D. Bleichenbacher. A Chosen Ciphertext Attack against Protocols based on the RSA Encryption Standard PKCS #1. In *Crypto '98*, LNCS 1462, pages 1–12. Springer-Verlag, Berlin, 1998.
5. D. Coppersmith. Finding a Small Root of a Univariate Modular Equation. In *Eurocrypt '96*, LNCS 1070, pages 155–165. Springer-Verlag, Berlin, 1996.
6. W. Diffie and M.E. Hellman. New Directions in Cryptography, IEEE Transactions on Information Theory, v. IT-22, 6, Nov 1976, pages 644–654.
7. D. Dolev, C. Dwork, and M. Naor. Non-Malleable Cryptography. *SIAM Journal on Computing*, 30(2):391–437, 2000.
8. E. Fujisaki, T. Okamoto, D. Pointcheval, and J. Stern. RSA–OAEP is Secure under the RSA Assumption. In *Crypto '2001*, LNCS 2139, pages 260–274. Springer-Verlag, Berlin, 2001.
9. S. Goldwasser and S. Micali. Probabilistic Encryption. *Journal of Computer and System Sciences*, 28:270–299, 1984.
10. C. Hall, I. Goldberg, and B. Schneier. Reaction Attacks Against Several Public-Key Cryptosystems. In *Proc. of ICICS'99*, LNCS, pages 2–12. Springer-Verlag, 1999.

11. M. Naor and M. Yung. Public-Key Cryptosystems Provably Secure against Chosen Ciphertext Attacks. In *Proc. of the 22nd STOC*, pages 427–437. ACM Press, New York, 1990.

12. T. Okamoto and D. Pointcheval. REACT: Rapid Enhanced-security Asymmetric Cryptosystem Transform. In *CT – RSA '2001*, LNCS 2020, pages 159–175. Springer-Verlag, Berlin, 2001.

13. C. Rackoff and D. R. Simon. Non-Interactive Zero-Knowledge Proof of Knowledge and Chosen Ciphertext Attack. In *Crypto '91*, LNCS 576, pages 433–444. Springer-Verlag, Berlin, 1992.

14. R. Rivest, A. Shamir, and L. Adleman. A Method for Obtaining Digital Signatures and Public Key Cryptosystems. *Communications of the ACM*, 21(2):120–126, February 1978.

15. RSA Data Security, Inc. Public Key Cryptography Standards – PKCS. Available from `http://www.rsa.com/rsalabs/pubs/PKCS/`.

16. V. Shoup. OAEP Reconsidered. In *Crypto '2001*, LNCS 2139, pages 239–259. Springer-Verlag, Berlin, 2001.

Dynamical Systems Generated by Rational Functions

Harald Niederreiter[1] and Igor E. Shparlinski[2]

[1] Department of Mathematics, National University of Singapore
2 Science Drive 2, Singapore 117543, Republic of Singapore
nied@math.nus.edu.sg
[2] Department of Computing, Macquarie University
NSW 2109, Australia
igor@comp.mq.edu.au

Abstract. We consider dynamical systems generated by iterations of rational functions over finite fields and residue class rings. We present a survey of recent developments and outline several open problem.

1 Introduction

Given an algebraic domain $\mathcal{D}$ and a map $\mathcal{F} : \mathcal{D} \to \mathcal{D}$, one can consider orbits

$$\mathbf{x}_{n+1} = \mathcal{F}\left(\mathbf{x}_n\right), \qquad n = 0, 1, \ldots, \tag{1}$$

generated by iterations of the map $\mathcal{F}$ starting with some initial point $\mathbf{x}_0 \in \mathcal{D}$. In particular, for any $n, k \geq 0$ we have

$$\mathbf{x}_{n+k} = \mathcal{F}^{(k)}\left(\mathbf{x}_n\right), \tag{2}$$

where $\mathcal{F}^{(k)}$ is the kth iteration of $\mathcal{F}$, with the usual convention that $\mathcal{F}^{(0)}$ is the identity map. Traditionally, such sequences have been considered over infinite domains $\mathcal{D}$, especially over the complex and real numbers, where an extensive and well-developed theory exists, see [1,30,64,66]. It is also known that number-theoretic transformations very often lead to very interesting dynamical systems. For example, one can associate dynamical systems with continued fractions, various number systems, the $3x + 1$ transformation and other number-theoretic constructions, see [2,3,14,17,18,34,35,36,40,41,42,43,44,45,47,63,65,68,69,71] and references therein.

However, in this paper we mainly concentrate on dynamical systems associated with transformations generated by rational functions, and polynomials in particular, over finite fields $\mathbb{F}_q$ of q elements and residue class rings $\mathbb{Z}_m$ modulo $m \geq 2$. Thus, typically we have $\mathcal{D} = \mathbb{F}_q^s$ or $\mathcal{D} = \mathbb{Z}_m^s$, in particular, the case $s = 1$ is of special interest.

Sequences of elements of $\mathbb{F}_q$ and $\mathbb{Z}_m$ generated by the recurrence relation (1) have been studied for a long time, mainly in the theory of pseudorandom number generators. Initially, only affine transformations have been considered [32,46,

M. Fossorier, T. Hoeholdt, and A. Poli (Eds.): AAECC 2003, LNCS 2643, pp. 6–17, 2003.

48], but after discovering several major disadvantages of such transformations, the emphasis in the research efforts has shifted to nonlinear constructions [13, 49,50,55]. Although most of the works in this direction use terminology from the area of pseudorandom numbers, the questions considered are the same as the basic questions in studying dynamical systems, that is, period length, fixed points, distribution and so on. Here we give a brief survey of several recently obtained results on the behaviour of such sequences and an outline of the recently discovered method in [51,54] which has led to a series of new results in this area [11,26,27,28,52,53,55,56,57,58,59,60,61].

Finally, we describe open problems and show some obstacles which prevent us from immediate application of the aforementioned method. It is important to remark that these obstacles are mostly of a purely algebraic nature and related to some algebraic properties of iterations of rational maps. Thus, we hope that the algebraic community may find a scope of new interesting problems which may succumb to their efforts.

2   Notation

As usual, we denote by $\mathbb{F}_q$ a finite field of q elements and by $\mathbb{Z}_m = \mathbb{Z}/m\mathbb{Z}$ the residue class ring modulo m. We identify $\mathbb{Z}_m$ with the set $\{0, 1, \ldots, m - 1\}$. In the case where $q = p$ prime we apply the same convention to $\mathbb{F}_p = \mathbb{Z}_p$.

We recall that for a given sequence of N points

$$\Gamma = (\gamma_{1,n}, \ldots, \gamma_{s,n})_{n=1}^{N}$$

of the s-dimensional unit cube, its *discrepancy* $\Delta(\Gamma)$ is defined as

$$\Delta(\Gamma) = \sup_{B \subseteq [0,1)^s} \left| \frac{T_\Gamma(B)}{N} - |B| \right|,$$

where $T_\Gamma(B)$ is the number of points of the sequence Γ which hit the box

$$B = [\alpha_1, \beta_1) \times \ldots \times [\alpha_s, \beta_s) \subseteq [0, 1)^s$$

and the supremum is taken over all such boxes.

Finally, given a sequence of N points

$$\mathcal{U} = (u_{1,n}, \ldots, u_{s,n})_{n=1}^{N}$$

of the s-fold Cartesian product $\mathbb{Z}_m^s$, its *modulo m discrepancy* $D_m(\mathcal{U})$ is defined as the ordinary discrepancy $\Delta(\Gamma)$ of the sequence

$$\Gamma = \left(\frac{u_{1,n}}{m}, \ldots, \frac{u_{s,n}}{m} \right)_{n=1}^{N}.$$

We apply this definition to points over $\mathbb{F}_p$ as well, where p is prime.

We finally note that if $\mathcal{D}$ is a finite domain, then any sequence generated by (1) becomes eventually periodic with least period $t \leq \#\mathcal{D}$. In fact, we always assume that the corresponding sequence is *purely periodic* (which can be achieved by a simple shift) and denote by t the corresponding period length.

Throughout the paper, the implied constants in the symbols 'O' may sometimes depend on the integer parameter $s \geq 1$ and on parameters in the recursion (1), but not on $\#\mathcal{D}$ or N.

3   Survey of the Results

The simplest and most natural specialisation of the general formula (1) is the case of the *polynomial generator* over a residue class ring $\mathbb{Z}_m$, that is,

$$x_{n+1} = f(x_n), \qquad n = 0, 1, \ldots, \tag{3}$$

where $f(X) \in \mathbb{Z}_m[X]$ is a polynomial of a fixed degree $d \geq 2$ and $x_0 \in \mathbb{Z}_m$.

The case $m = p$ was studied in [51] where it was shown that if $1 \leq N \leq t$, then for any $s \geq 1$ for the modulo p discrepancy of the set

$$\mathcal{X}_{s,N} = (x_n, \ldots, x_{n+s-1}), \qquad 0 \leq n \leq N - 1,$$

the bound

$$D_p(\mathcal{X}_{s,N}) = O\left(N^{-1/2}p^{1/2}(\log p)^{-1/2} (\log \log p)^s\right) \tag{4}$$

holds.

Because obviously $t \leq p$, this bound is nontrivial in a rather narrow (but not void) range $p \geq t \geq N \gg p(\log p)^{-1+\varepsilon}$ with some fixed $\varepsilon > 0$. In particular, the problem of finding polynomials f in (3) for which t is close to p arises. Clearly, the maximal value $t = p$ is achieved if and only if f is a permutation polynomial of $\mathbb{F}_p$ representing a permutation which is a cycle of length p. For random f the expected value of the maximal t that can be obtained with a suitable initial value x_0 is of the order of magnitude $p^{1/2}$ (see [16] for this and related results). This is in line with the *birthday paradox* which is underlying the famous *Pollard factorisation algorithm*, see [12].

It is natural to consider an analogue of (3) with rational functions instead of polynomials. Obviously, in this case one should take care of possible zero divisors occurring in the denominator. For example, if $m = p$ prime, then one has to decide how to compute x_{n+1} if x_n is a pole of $f(X) \in \mathbb{F}_p(X)$. For instance, one can "define" $0^{-1} = 0$. This introduces some additional complications, but generally does not change the approach. On the other hand, this case, although theoretically interesting, has never been worked out in detail, probably because it does not seem to give any advantages compared to the polynomial generator (but is computationally more expensive). However, there is a very special case which brings many nice surprises. Namely, the case of functions of the form

$$f(X) = \frac{\alpha X + \beta}{\gamma X + \delta} \tag{5}$$

with $\alpha\delta - \beta\gamma \neq 0$, which by substitution can be reduced to less symmetric but simpler functions of the form

$$f(X) = aX^{-1} + b$$

(with the above convention that $0^{-1} = 0$). This case is known as the *inversive generator*. Instead of the bound (4), a much stronger bound

$$D_p(\mathcal{X}_{s,N}) = O\left(N^{-1/2}p^{1/4}(\log p)^s\right) \tag{6}$$

has been obtained in [54] for $s = 1$ and then in [28] for arbitrary $s \geq 1$, which is nontrivial for $t \geq N \gg p^{1/2}(\log p)^{2s+\varepsilon}$ with some fixed $\varepsilon > 0$. Certainly, the same question of how to guarantee a large value of t immediately comes up (although the requirements are much more relaxed now). And here we come to one more advantage of the inversive generator – its inner structure is likely to beat the birthday paradox! For example, it is known when such sequences achieve the largest possible period, which is obviously $t = p$, see [15]. The corresponding condition is that the quotient of the roots of the polynomial $X^2 - bX - a \in \mathbb{F}_p[X]$ is an element of order $p+1$ in the multiplicative group of $\mathbb{F}_{p^2}$. In particular, finding a and b which produce sequences with $t = p$ is of about the same complexity as that of finding primitive roots (and also happens about as often).

Finally, the bounds (4) and (6) both rely on the *Weil bound* of exponential sums, see [38, Chapter 5], typically in the form given in [39]. One can also try to use similar arguments to study the distribution of these sequences modulo a composite number. In this case the Weil bound can be replaced by the *Hua bound* of rational exponential sums with polynomials and rational functions [5, 6,7,8,9,10,70]. Naturally, the strength of both bounds (4) and (6) deteriorates quite dramatically. However, and this is yet another attractive feature of the inversive generator, in the case $s = 1$ the bound (6) remains the same when p is replaced by a prime power (and probably by any modulus m), see [53].

Trying to increase the period length, one can consider multivariate generalizations of (3). For example, the sequences generated by the relation

$$x_{n+r} = f\left(x_{n+r-1}, \ldots, x_n\right), \qquad n = 0, 1, \ldots, \tag{7}$$

given r initial values $x_0, \ldots, x_{r-1}$ and $f(X_{r-1}, \ldots, X_0) \in \mathbb{F}_p[X_{r-1}, \ldots, X_0]$, have been considered in [26,27]. Also, using vector notation

$$\mathbf{x_n} = (x_{n+r-1}, \ldots, x_n)$$

and

$$\mathcal{F} = (f(X_{r-1}, \ldots, X_0), X_{r-1}, \ldots, X_1), \tag{8}$$

one can put (7) into the form (1).

Although an analogue of the bound (4) is given in [26], that result applies only to a rather special class of polynomials f. It has turned out that extending this result to general polynomials leads to a certain algebraic problem about linear independence of iterations of polynomials, which we discuss in Sect. 4. This obstacle has been partially removed in [27], but the general question still remains unsettled.

In the above results we have always assumed that the degrees of the involved polynomials and rational functions are bounded. There is, however, a very interesting and important special case where we can avoid this restriction and actually even get a much stronger result than (4). This is the case of monomials $f(x) = x^e$, that is, we have

$$x_{n+1} = x_n^e, \qquad n = 0, 1, \dots . \tag{9}$$

The corresponding sequence is known as the *power generator* and has been introduced for some cryptographic applications (especially when considered modulo an *RSA integer* $m = pl$ which is a product of two distinct primes). In the special case $\gcd(e, \varphi(m)) = 1$, where $\varphi(m)$ is the *Euler function*, it is called the *RSA generator* and in the special case $e = 2$ it is called the *Blum–Blum–Shub generator* (see [4,33]).

For the case of RSA moduli the first discrepancy bound was given in [22]. This bound has been improved in [24]. The method of [24] is quite sensitive to the arithmetic structure of the modulus. For example, for a prime modulus $m = p$ it reaches its maximal strength, producing a bound which is nontrivial for sequences (9) of period $t \gg p^{1/2+\varepsilon}$ with some fixed $\varepsilon > 0$, see [19]. For RSA moduli one already needs $t \gg p^{3/4+\varepsilon}$, see [24], and the method fails to produce any nontrivial bounds for general moduli. On the other hand, the approach of [22], although it loses the race for RSA moduli, can be extended to the general case, see [21]. Also, for moduli which are large powers of small primes, an alternative approach has been proposed in [20]. We remark that all these results apply only to the one-dimensional distribution (unless e is small, see [24]), thus studying the multidimensional distribution of the power generator (9) remains an important and challenging open problem which probably requires some new ideas.

Finally, motivated by a question posed in [31], a somewhat dual construction of dynamical systems from polynomials has been considered in [62]. Instead of iterating a fixed function of the form (5), starting from some given point, one can fix a point x_0 in an extension field $\mathbb{F}_{q^r}$ of $\mathbb{F}_q$ and study the distribution of the values $f(x_0)$ when f runs over all $q^3 - q$ distinct functions (5) with $\alpha, \beta, \gamma, \delta \in \mathbb{F}_q$ and $\alpha\delta - \beta\gamma \neq 0$.

4 How Does the Method Work?

Here we give a brief outline of the ideas underlying the proofs of the results described in Sect. 3.

It is a generic principle that to study the distribution and other "statistical" properties of any sequence $(\mathbf{x}_n)_{n=0}^{\infty}$ in any domain $\mathcal{D}$ being embedded in an abelian group structure, one usually considers character sums

$$S(\chi) = \sum_{n=0}^{N-1} \chi(\mathbf{x}_n), \tag{10}$$

where χ is a nonprincipal character of the corresponding group. For a recent survey of character sums and their applications we refer to [67].

Typically, individual single sums are hard to study, and our first observation helps us to create a double sum $W(\chi)$ which is closely associated with $S(\chi)$. We remark that for any integer $k \geq 0$ we have

$$S(\chi) = \sum_{n=k}^{N+k-1} \chi(\mathbf{x}_n) + O(k). \tag{11}$$

Indeed, the sums in (10) and (11) "disagree" for at most $2k$ values of $n = 0, 1, \ldots, N + k - 1$, and since $|\chi(\mathbf{x})| = 1$, $\mathbf{x} \in \mathcal{D}$, we obtain the above identity. Summing up these identities for $k = 0, \ldots, K - 1$, we obtain

$$KS(\chi) = W(\chi) + O(K^2),$$

where

$$W(\chi) = \sum_{k=0}^{K-1} \sum_{n=k}^{N+k-1} \chi(\mathbf{x}_n) = \sum_{k=0}^{K-1} \sum_{n=0}^{N-1} \chi(\mathbf{x}_{n+k}),$$

thus achieving our first goal.

Now assuming that the sequence $(\mathbf{x}_n)_{n=0}^{\infty}$ satisfies (1) and using the observation (2), we obtain

$$W(\chi) = \sum_{k=0}^{K-1} \sum_{n=0}^{N-1} \chi\left(\mathcal{F}^{(k)}(\mathbf{x}_n)\right).$$

Our next step is to reduce $W(\chi)$ to a sum which does not depend on the specific sequence $(\mathbf{x}_n)_{n=0}^{\infty}$ at all and to which therefore general-purpose bounds of character sums over $\mathcal{D}$ can be applied. We achieve this in two steps. First of all, we write

$$|W(\chi)| \leq \sum_{n=0}^{N-1} \left| \sum_{k=0}^{K-1} \chi\left(\mathcal{F}^{(k)}(\mathbf{x}_n)\right) \right|,$$

and then by applying the Cauchy-Schwarz inequality we get

$$|W(\chi)|^2 \leq N \sum_{n=0}^{N-1} \left| \sum_{k=0}^{K-1} \chi\left(\mathcal{F}^{(k)}(\mathbf{x}_n)\right) \right|^2.$$

If $1 \leq N \leq t$, which is of our principal interest, then (1) implies that the elements $\mathbf{x}_0, \ldots, \mathbf{x}_{N-1}$ are pairwise distinct. So we can only add more nonnegative terms to the last sum if we replace $\mathbf{x}_n$ with $\mathbf{x}$ running through the whole domain $\mathcal{D}$ (which we now assume to be finite), reaching our second goal:

$$|W(\chi)|^2 \leq N \sum_{\mathbf{x} \in \mathcal{D}} \left| \sum_{k=0}^{K-1} \chi\left(\mathcal{F}^{(k)}(\mathbf{x})\right) \right|^2.$$

Now we use that $|z|^2 = z\bar{z}$ for any complex z and $\overline{\chi(\mathbf{u})} = \chi(-\mathbf{u})$ for any $\mathbf{u} \in \mathcal{D}$ (where we write the corresponding group operation additively). After changing the order of summation, we derive

$$|W(\chi)|^2 \leq N \sum_{k,l=0}^{K-1} \sum_{\mathbf{x} \in \mathcal{D}} \chi\left(\mathcal{F}^{(k)}(\mathbf{x}) - \mathcal{F}^{(l)}(\mathbf{x})\right). \tag{12}$$

The contribution of the terms with $k = l$ is exactly $K\#\mathcal{D}$. To estimate the contribution of the terms with $k \neq l$, we need some additional information about the character sums with functions of the form $\mathcal{F}^{(k)}(\mathbf{x}) - \mathcal{F}^{(l)}(\mathbf{x})$.

Now, assuming that generally these functions fall into a class of functions for which nontrivial upper bounds on character sums are known, we still have one more problem to overcome. Namely, usually we have to show that the function $\chi\left(\mathcal{F}^{(k)}(\mathbf{x}) - \mathcal{F}^{(l)}(\mathbf{x})\right)$ is not constant for any nonprincipal character χ and $k \neq l$.

For example, if $\mathcal{D} = \mathbb{F}_p$ and $\mathcal{F}$ is a polynomial in one variable, that is, for the polynomial generator (3), the last question is almost trivial. Indeed, if $\deg(f) = d \geq 2$, then $f^{(k)}(X)$ is a polynomial of degree exactly d^k, thus $f^{(k)}(X) - f^{(l)}(X)$ is a nonconstant polynomial whenever $k \neq l$. This almost immediately implies that $\chi\left(f^{(k)}(X) - f^{(l)}(X)\right)$ is not constant either, unless k or l is large (assuming that p is large). Now, applying the Weil bound and optimising the choice of K, one obtains (4). On the other hand, it is easy to see that for maps generated by multivariate polynomials, say when $\mathcal{D} = \mathbb{F}_p^s$ with $s \geq 2$, the degree argument does not apply any more, see [26,27].

Similarly, this argument does not apply to the inversive generator either. However, a more detailed study of the function $f^{(k)}(X) - f^{(l)}(X)$ helps to establish that it is a nonconstant rational function for $k \neq l$ in a certain range, see [54].

To study the s-dimensional distribution, one can follow the same type of arguments and usually should prove that for any nonzero vector $\mathbf{a} = (a_0, \ldots, a_{s-1}) \in \mathcal{D}^s$ the linear combination

$$L_{\mathbf{a}}(X) = \sum_{j=0}^{s-1} a_j \left(f^{(k+j)}(X) - f^{(l+j)}(X)\right)$$

is not constant.

We are now able to explain the dramatic difference between the bounds (4) and (6). Indeed, both are based on the Weil bound, whose strength decreases when the degree of the polynomial or rational function to which it is applied grows. For the polynomial generator we have polynomials whose degree goes up to d^{K+s-2}. This forces us to consider only logarithmically small values of K and leads to a rather weak bound (4). To be more precise, the total contribution to $|W(\chi)|^2$ from "trivial" cases with $k = l$ is $O(KNp)$, while the total contribution from other cases is $O\left(K^2 d^{K+s-2} N p^{1/2}\right)$. Thus

$$|W(\chi)|^2 = O\left(KNp + K^2 d^{K+s-2} N p^{1/2}\right).$$

Choosing the optimal value of K which balances both terms (in fact even quite rough balancing suffices), we arrive at (4) after simple calculations. On the other hand, as straightforward calculations show, for the inversive generator $f^{(k)}(X)$ is still of the form (5), thus the degree of the numerator and denominator of $L_{\mathbf{a}}(X)$ remains bounded by $2s$. This enables us to work with quite large values of K and together with the Weil bound in the form given in [39] leads to (6).

Finally, we remark that the above approach does not directly apply to the power generator (9) because e is typically quite large. However, using the specific shape of the iterations

$$f^{(k)}(X) = X^{e^k},$$

in [19,24] this problem has been overcome by choosing only certain special values of k (rather than all values from an interval $[0, K - 1]$) and using several other tricks to reduce the degree of the resulting polynomial. On the other hand, the method of [21,22] is very different and is rather based on multiplicativity of the e^k. Unfortunately, neither of the above approaches applies to studying the multidimensional distribution.

5 Open Problems

The most difficult problem with studying sequences generated by multivariate polynomials (for example by iterations of the map (8)) is proving the linear independence of several consecutive iterations. In some cases such linear independence results have been proved in [26,27], but the general case remains unresolved. To be more precise, let

$$f^{(k)}(X_{r-1}, \ldots, X_0) = f\left(f^{(k-1)}(X_{r-1}, \ldots, X_0), \ldots, f^{(k-r)}(X_{r-1}, \ldots, X_0)\right)$$

for $k = 1, 2, \ldots$, where $f^{(k)}(X_{r-1}, \ldots, X_0) = X_{r+k-1}$, $k = -r + 1, \ldots, 0$. What are the most general (necessary and sufficient) conditions under which the s polynomials $f^{(k)}, \ldots, f^{(k+s-1)}$ over $\mathbb{F}_q$ are *linearly independent* over $\mathbb{F}_q$ for every $k = 1, 2, \ldots$? For applications to sums of multiplicative characters one needs to study when these polynomials are *multiplicatively independent*; the technique of [25] may be used for this problem. Although we are unaware of any applications in sight, the more general question about the *algebraic independence* over $\mathbb{F}_q$ of $f^{(k)}, \ldots, f^{(k+s-1)}$ is of ultimate intrinsic interest.

Typically, the final step of the method outlined in Sect. 4 is applying the Weil bound (or some other bound on exponential sums) to the off-diagonal (that is, with $k \neq l$) sums in (12). On the other hand, if for $0 \leq k < l \leq K - 1$ the difference $\mathcal{F}^{(k)}(\mathbf{x}) - \mathcal{F}^{(l)}(\mathbf{x})$ is a permutation of $\mathcal{D}$, then the off-diagonal sums vanish (which may lead to a better bound on discrepancy). This observation has been made in [11] together with some examples of such domains $\mathcal{D}$ and maps $\mathcal{F}$. Nevertheless, these examples have a rather small value of K and do not produce substantially better bounds. It is possibly natural to use the theory of permutation polynomials over finite fields (see [37,38]) in order to construct nontrivial examples with a large value of K. Generally this question remains open.

It has been shown in [56] that the method of Sect. 4 can be used to improve the "individual" bound (6) "on average" over all possible initial values of the inversive generator. On the other hand, it is not clear how to improve "on average" the much weaker bound (4) for the polynomial generator, which would be a very desirable result.

We have already mentioned that for "random" polynomials f one should expect the period t of the polynomial generator (3) to be of order $p^{1/2}$, which is very far below the nontriviality threshold of the bound (4). On the other hand, some specially chosen polynomials may produce rather large values of t. For example, it is shown in [23] that the power generator (9) is likely to have a very large period. It also seems that the period of sequences produced by the *Dickson polynomials* can be studied by the same method, which would be yet another useful feature of these celebrated polynomials (see [37]).

References

1. L. Arnold, *Random Dynamical Systems*, Springer-Verlag, Berlin, 1998.
2. G. Barat, T. Downarowicz and P. Liardet, Dynamiques associées à une échelle de numération, *Acta Arith.* **103**, 41–78 (2002).
3. V. Berthé and H. Nakada, On continued fraction expansions in positive characteristic: equivalence relations and some metric properties, *Expositiones Math.* **18**, 257–284 (2000).
4. L. Blum, M. Blum and M. Shub, A simple unpredictable pseudo-random number generator, *SIAM J. Computing* **15**, 364–383 (1986).
5. T. Cochrane, Exponential sums modulo prime powers, *Acta Arith.* **101**, 131–149 (2002).
6. T. Cochrane, C. Li and Z.Y. Zheng, Upper bounds on character sums with rational function entries, *Acta Math. Sinica (English Ser.)* **19**, 1–11 (2003).
7. T. Cochrane and Z.Y. Zheng, Pure and mixed exponential sums, *Acta Arith.* **91**, 249–278 (1999).
8. T. Cochrane and Z.Y. Zheng, Exponential sums with rational function entries, *Acta Arith.* **95**, 67–95 (2000).
9. T. Cochrane and Z.Y. Zheng, On upper bounds of Chalk and Hua for exponential sums, *Proc. Amer. Math. Soc.* **129**, 2505–2516 (2001).
10. T. Cochrane and Z.Y. Zheng, A survey on pure and mixed exponential sums modulo prime powers, *Number Theory for the Millennium* (M.A. Bennett et al., eds.), vol. 1, pp. 271–300, A.K. Peters, Natick, MA, 2002.
11. S.D. Cohen, H. Niederreiter, I.E. Shparlinski and M. Zieve, Incomplete character sums and a special class of permutations, *J. Théorie des Nombres Bordeaux* **13**, 53–63 (2001).
12. R. Crandall and C. Pomerance, *Prime Numbers: A Computational Perspective*, Springer-Verlag, New York, 2001.
13. J. Eichenauer-Herrmann, E. Herrmann and S. Wegenkittl, A survey of quadratic and inversive congruential pseudorandom numbers, *Monte Carlo and Quasi-Monte Carlo Methods 1996* (H. Niederreiter et al., eds.), Lect. Notes in Stat., vol. 127, pp. 66–97, Springer-Verlag, New York, 1998.

14. G. Everest and T. Ward, *Heights of Polynomials and Entropy in Algebraic Dynamics*, Springer-Verlag, London, 1999.
15. M. Flahive and H. Niederreiter, On inversive congruential generators for pseudorandom numbers, *Finite Fields, Coding Theory, and Advances in Communications and Computing* (G.L. Mullen and P.J.-S. Shiue, eds.), pp. 75–80, Marcel Dekker, New York, 1993.
16. P. Flajolet and A.M. Odlyzko, Random mapping statistics, *Advances in Cryptology – EUROCRYPT '89* (J.-J. Quisquater and J. Vandewalle, eds.), Lect. Notes in Comp. Sci., vol. 434, pp. 329–354, Springer-Verlag, Berlin, 1990.
17. L. Flatto, Z-numbers and β-transformations, *Symbolic Dynamics and Its Applications*, Contemporary Math., vol. 135, pp. 181–201, Amer. Math. Soc., Providence, RI, 1992.
18. L. Flatto, J.C. Lagarias and B. Poonen, The zeta function of the beta transformation, *Ergodic Theory Dynam. Systems* **14**, 237–266 (1994).
19. J.B. Friedlander, J. Hansen and I.E. Shparlinski, Character sums with exponential functions, *Mathematika* **47**, 75–85 (2000).
20. J.B. Friedlander, J. Hansen, and I.E. Shparlinski, On the distribution of the power generator modulo a prime power, *Proc. DIMACS Workshop on Unusual Applications of Number Theory 2000*, Amer. Math. Soc., Providence, RI (to appear).
21. J.B. Friedlander, S.V. Konyagin and I.E. Shparlinski, Some doubly exponential sums over $\mathbb{Z}_m$, *Acta Arith.* **105**, 349–370 (2002).
22. J.B. Friedlander, D. Lieman and I.E. Shparlinski, On the distribution of the RSA generator, *Sequences and Their Applications* (C. Ding, T. Helleseth and H. Niederreiter, eds.), pp. 205–212, Springer-Verlag, London, 1999.
23. J.B. Friedlander, C. Pomerance and I.E. Shparlinski, Period of the power generator and small values of Carmichael's function, *Math. Comp.* **70**, 1591–1605 (2001).
24. J.B. Friedlander and I.E. Shparlinski, On the distribution of the power generator, *Math. Comp.* **70**, 1575–1589 (2001).
25. S. Gao, Elements of provable high orders in finite fields, *Proc. Amer. Math. Soc.* **127**, 1615–1623 (1999).
26. F. Griffin, H. Niederreiter and I.E. Shparlinski, On the distribution of nonlinear recursive congruential pseudorandom numbers of higher orders, *Applied Algebra, Algebraic Algorithms and Error-Correcting Codes* (M. Fossorier *et al.*, eds.), Lect. Notes in Comp. Sci., vol. 1719, pp. 87–93, Springer-Verlag, Berlin, 1999.
27. J. Gutierrez and D. Gomez-Perez, Iterations of multivariate polynomials and discrepancy of pseudorandom numbers, *Applied Algebra, Algebraic Algorithms and Error-Correcting Codes* (S. Boztaş and I.E. Shparlinski, eds.), Lect. Notes in Comp. Sci., vol. 2227, pp. 192–199, Springer-Verlag, Berlin, 2001.
28. J. Gutierrez, H. Niederreiter and I.E. Shparlinski, On the multidimensional distribution of inversive congruential pseudorandom numbers in parts of the period, *Monatsh. Math.* **129**, 31–36 (2000).
29. K. Huber, On the period length of generalized inversive pseudorandom number generators, *Applicable Algebra Engrg. Comm. Comput.* **5**, 255–260 (1994).
30. A. Katok and B. Hasselblatt, *Introduction to the Modern Theory of Dynamical Systems*, Cambridge University Press, Cambridge, 1995.
31. A. Klapper, The distribution of points in orbits of $\mathrm{PGL}_2(\mathrm{GF}(q))$ acting on $\mathrm{GF}(q^n)$, *Finite Fields, Coding Theory, and Advances in Communications and Computing* (G.L. Mullen and P.J.-S. Shiue, eds.), pp. 430–431, Marcel Dekker, New York, 1993.
32. D.E. Knuth, *The Art of Computer Programming*, vol. 2, 3rd ed., Addison-Wesley, Reading, MA, 1998.

33. J.C. Lagarias, Pseudorandom number generators in cryptography and number theory, *Cryptology and Computational Number Theory* (C. Pomerance, ed.), Proc. Symp. in Appl. Math., vol. 42, pp. 115–143, Amer. Math. Soc., Providence, RI, 1990.

34. J.C. Lagarias, Number theory and dynamical systems, *The Unreasonable Effectiveness of Number Theory*, Proc. Symp. in Appl. Math., vol. 46, pp. 35–72, Amer. Math. Soc., Providence, RI, 1992.

35. J.C. Lagarias, H.A. Porta and K.B. Stolarsky, Asymmetric tent map expansions. I. Eventually periodic points, *J. London Math. Soc.* **47**, 542–556 (1993).

36. J.C. Lagarias, H.A. Porta and K.B. Stolarsky, Asymmetric tent map expansions. II. Purely periodic points, *Illinois J. Math.* **38**, 574–588 (1994).

37. R. Lidl, G.L. Mullen and G. Turnwald, *Dickson Polynomials*, Longman, Harlow, 1993.

38. R. Lidl and H. Niederreiter, *Finite Fields*, Cambridge University Press, Cambridge, 1997.

39. C.J. Moreno and O. Moreno, Exponential sums and Goppa codes: I, *Proc. Amer. Math. Soc.* **111**, 523–531 (1991).

40. P. Morton, Periods of maps on irreducible polynomials over finite fields, *Finite Fields Appl.* **3**, 11–24 (1997).

41. P. Morton, Arithmetic properties of periodic points of quadratic maps. II, *Acta Arith.* **87**, 89–102 (1998).

42. P. Morton and J.H. Silverman, Rational periodic points of rational functions, *Internat. Math. Res. Notices* **2**, 97–110 (1994).

43. P. Morton and J.H. Silverman, Periodic points, multiplicities, and dynamical units, *J. Reine Angew. Math.* **461**, 81–122 (1995).

44. P. Morton and F. Vivaldi, Bifurcations and discriminants for polynomial maps, *Nonlinearity* **8**, 571–584 (1995).

45. W. Narkiewicz, *Polynomial Mappings*, Lect. Notes in Math., vol. 1600, Springer-Verlag, Berlin, 1995.

46. H. Niederreiter, Quasi-Monte Carlo methods and pseudo-random numbers, *Bull. Amer. Math. Soc.* **84**, 957–1041 (1978).

47. H. Niederreiter, The probabilistic theory of linear complexity, *Advances in Cryptology – EUROCRYPT '88* (C.G. Günther, ed.), Lect. Notes in Comp. Sci., vol. 330, pp. 191–209, Springer-Verlag, Berlin, 1988.

48. H. Niederreiter, *Random Number Generation and Quasi-Monte Carlo Methods*, SIAM, Philadelphia, 1992.

49. H. Niederreiter, New developments in uniform pseudorandom number and vector generation, *Monte Carlo and Quasi-Monte Carlo Methods in Scientific Computing* (H. Niederreiter and P.J.-S. Shiue, eds.), Lect. Notes in Stat., vol. 106, pp. 87–120, Springer-Verlag, New York, 1995.

50. H. Niederreiter, Design and analysis of nonlinear pseudorandom number generators, *Monte Carlo Simulation* (G.I. Schuëller and P.D. Spanos, eds.), pp. 3–9, A.A. Balkema Publishers, Rotterdam, 2001.

51. H. Niederreiter and I.E. Shparlinski, On the distribution and lattice structure of nonlinear congruential pseudorandom numbers, *Finite Fields Appl.* **5**, 246–253 (1999).

52. H. Niederreiter and I.E. Shparlinski, On the distribution of pseudorandom numbers and vectors generated by inversive methods, *Applicable Algebra Engrg. Comm. Comput.* **10**, 189–202 (2000).

53. H. Niederreiter and I.E. Shparlinski, Exponential sums and the distribution of inversive congruential pseudorandom numbers with prime-power modulus, *Acta Arith.* **92**, 89–98 (2000).

54. H. Niederreiter and I.E. Shparlinski, On the distribution of inversive congruential pseudorandom numbers in parts of the period, *Math. Comp.* **70**, 1569–1574 (2001).

55. H. Niederreiter and I.E. Shparlinski, Recent advances in the theory of nonlinear pseudorandom number generators, *Monte Carlo and Quasi-Monte Carlo Methods 2000* (K.-T. Fang, F.J. Hickernell and H. Niederreiter, eds.), pp. 86–102, Springer-Verlag, Berlin, 2002.

56. H. Niederreiter and I.E. Shparlinski, On the average distribution of inversive pseudorandom numbers, *Finite Fields Appl.* **8**, 491–503 (2002).

57. H. Niederreiter and I.E. Shparlinski, On the distribution of power residues and primitive elements in some nonlinear recurring sequences, *Bull. Lond. Math. Soc.* (to appear).

58. H. Niederreiter and A. Winterhof, Incomplete exponential sums over finite fields and their applications to new inversive pseudorandom number generators, *Acta Arith.* **93**, 387–399 (2000).

59. H. Niederreiter and A. Winterhof, On the distribution of compound inversive congruential pseudorandom numbers, *Monatsh. Math.* **132**, 35–48 (2001).

60. H. Niederreiter and A. Winterhof, On the lattice structure of pseudorandom numbers generated over arbitrary finite fields, *Applicable Algebra Engrg. Comm. Comput.* **12**, 265–272 (2001).

61. H. Niederreiter and A. Winterhof, On a new class of inversive pseudorandom numbers for parallelized simulation methods, *Periodica Math. Hungarica* **42**, 77–87 (2001).

62. H. Niederreiter and A. Winterhof, On the distribution of points in orbits of $PGL(2, q)$ acting on $GF(q^n)$, preprint, 2002.

63. A.G. Postnikov, *Ergodic Problems in the Theory of Congruences and of Diophantine Approximations*, Proc. Steklov Inst. Math., vol. 82, Amer. Math. Soc., Providence, RI, 1967.

64. C. Robinson, *Dynamical Systems: Stability, Symbolic Dynamics, and Chaos*, CRC Press, Boca Raton, FL, 1999.

65. J. Schmeling, Symbolic dynamics for β-shifts and self-normal numbers, *Ergodic Theory Dynam. Systems* **17**, 675–694 (1997).

66. K. Schmidt, *Dynamical Systems of Algebraic Origin*, Progress in Math., vol. 128, Birkhäuser Verlag, Basel, 1995.

67. I.E. Shparlinski, Exponential sums in coding theory, cryptology and algorithms, *Coding Theory and Cryptology* (H. Niederreiter, ed.), pp. 323–383, World Scientific Publ., Singapore, 2002.

68. J.H. Silverman, The field of definition for dynamical systems on $\mathbb{P}_1$, *Compositio Math.* **98**, 269–304 (1995).

69. J.H. Silverman, The space of rational maps on $\mathbb{P}_1$, *Duke Math. J.* **94**, 41–77 (1998).

70. S.B. Stečkin, An estimate of a complete rational trigonometric sum (Russian), *Trudy Mat. Inst. Steklov.* **143**, 188–207 (1977).

71. G.J. Wirsching, *The Dynamical System Generated by the $3n + 1$ Function*, Lect. Notes in Math., vol. 1681, Springer-Verlag, Berlin, 1998.

Homotopy Methods for Equations over Finite Fields

Alan G.B. Lauder*

Computing Laboratory, Oxford University
Oxford OX1 3QD, U.K.
alan.lauder@comlab.ox.ac.uk

Abstract. This paper describes an application of some ideas from homotopy theory to the problem of computing the number of solutions to a multivariate polynomial equation over a finite field. The benefit of the homotopy approach over more direct methods is that the running-time is far less dependent on the number of variables. The method was introduced by the author in another paper, where specific complexity estimates were obtained for certain special cases. Some consequences of these estimates are stated in the present paper.

1 Introduction

A basic problem in computational mathematics is to solve a system of polynomial equations with coefficients in a field. By "solve" one typically means finding a solution, or perhaps all of the solutions if there are known to be only finitely many. When the field admits a non-trivial norm, such as the field of real numbers, then powerful analytic methods can be brought to bear. For example, Newton's method for locating zeros of a real polynomial, and its various generalisations [1]. However, when no such norm exists, one is forced to fall back on algebraic techniques. Consider, for example, a single polynomial equation in one variable over a finite prime field. Using a deterministic algorithm, Berlekamp's root-counting algorithm, one can compute in an efficient manner the number of solutions of the equation in the prime field, see [3, Chapter 14]. By an ingenious application of randomisation, one can actually find all the solutions efficiently. However, finding a solution efficiently using a deterministic algorithm seems a much more difficult, and as yet unsolved, problem. By "efficient" here, and elsewhere, I mean in polynomial-time in the size of the input.

The purpose of the present short paper is to describe a method I have recently been exploring for the related problem of counting, rather than finding, solutions to equations; specifically, for computing the number of rational solutions to a single multivariate polynomial equation over a finite field [6]. This technique involves deforming one polynomial into another. Such deformations

* The author is supported by the EPSRC (Grant GR/N35366/01) and St John's College, Oxford.

only make any intuitive sense over a field, such as the real numbers, which admits a non-trivial norm. However, by the application of some rather deep theory, based upon work of Dwork from the 1960s [2], these deformations also become useful over finite fields. My work is inspired by the paper of Dwork in which he uses deformations to prove the "functional equation of the zeta function of a smooth projective hypersurface". From the point of view of algorithms, the interesting aspect of Dwork's deformation theory is that it can lead to remarkable improvements in computational complexity. The algorithms I will describe bear a passing resemblance to the "homotopy methods" for locating zeros of a system of complex multivariate polynomial equations [1, Section 4.2], whence the title of this paper. In these methods, one begins with an approximate zero of a perturbed system, and then one gradually homes in on a zero of the original system by following a path in some appropriate space. In Dwork's theory, from knowledge of the number of solutions of some perturbed polynomial equation, one can recover the number of solutions of the original polynomial equation by studying the path taken from the original to the perturbed polynomial.

A considerable body of work has appear in the last few years on the problem of counting solutions to equations over finite fields, see for example the references in [7]. To my knowledge, the paper [6] describes the first explicit algorithmic application of a "homotopy method" to this problem. I would be very interested in learning of any other applications of "homotopy methods" to algorithmic, or theoretical, problems on equations over finite fields.

2   The Method

Let $\mathbb{F}_q$ denote the finite field with q elements, where q is a power of a prime p. Let $f \in \mathbb{F}_q[X_1, \dots, X_n]$ be a homogeneous polynomial of degree d in the variables $X_1, \dots, X_n$. Assume that p does not divide d, so that Dwork's theory can be applied below. Our aim will be to compute $N(f)$, the number of rational projective solutions to the equation $f = 0$. Specifically, the number $N(f)$ is defined via the equation

$$(q - 1)N(f) + 1 = \#\{(x_1, \dots, x_n) \in \mathbb{F}_q^n \mid f(x_1, \dots, x_n) = 0\}.$$

To compute $N(f)$ in a naive fashion, by substituting in f all rational projective points, would require $q^{n-1} + \dots + q + 1$ evaluations of f. This is certainly $\Omega(q^{n-1})$ bit operations (ignoring the dependence on d which is not such a concern when just counting rational solutions). However, the dense input size to this problem is about $\binom{n+d}{n} \log(q)$ bits, i.e., $\Theta(\log(q))$ bits. One needs to reduce the dependence on $\log(q)$ from $\Omega(q^{n-1})$ to $\log(q)^{O(1)}$ to get a more practical algorithm. Some progress towards this can be made using p-adic cohomology, as I will now describe.

Under the assumption that the zero set of f is "smooth and in general position" (see [4, Page 75]), there is a cohomological formula

$$N(f) = \frac{q^{n-1} - 1}{q - 1} + (-1)^n \frac{\text{Trace}(\alpha)}{q}. \tag{1}$$

Here α is a matrix whose entries lie in the p-adic field obtained by lifting $\mathbb{F}_q$ to characteristic zero and adjoining a primitive pth root of unity. (The primitive pth root is required since Dwork reduces everything to additive character sums, and these are only defined in fields with pth roots.) The matrix α has dimension

$$\frac{1}{d}\{(d-1)^n + (-1)^n(d-1)\} \in \mathbb{Z}.$$

It is the matrix of Frobenius on the primitive middle-dimensional p-adic cohomology space, which we will just call the *Frobenius matrix*. One wishes to construct this Frobenius matrix, or at least find its trace. Dwork's theory is constructive, and using a fairly direct approach the Frobenius matrix can be explicitly computed. This direct approach was first used by Kedlaya [5], and is extremely good for curves where $n = 3$. However, the computational complexity of this direct approach for an equation in n variables appears to be $(pd\log(q))^{\Theta(n)}$ bit operations, at least using the most straightforward generalisation. This is just the same complexity as in the theorem of the author and Wan [7, Theorem 1], but using Kedlaya's method the exponent can be roughly halved. The reason for this complexity is that the computations required involve n-variate polynomials of total degree around $pd\log(q)$, and such polynomials have roughly $(pd\log(q))^n$ terms. (Note that the factor p in the complexity is undesirable, but it is difficult to see how it could be replaced by $\log(p)$ using only p-adic cohomology. Grothendieck's l-adic cohomology theory might achieve this if it could be made constructive in general. This can be done for curves, using methods going back to Weil, but even in this case it only gives good algorithms at present for small genus.) The aim of the homotopy method is to remove the dependence in n from the exponent of $p\log(q)$. Specifically, I believe a complexity of $c_n(pd^n\log(q))^{O(1)}$ bit operations, where c_n depends only on n, should be possible. I have worked out this approach for a special family of equations – the precise results obtained are described in the next section. In the remainder of this section I will sketch the homotopy method and explain why it should be useful.

We write our polynomial f in the form

$$f = \sum_{i=1}^n a_i X_i^d + h(X_1, \ldots, X_n)$$

where h is a homogeneous polynomial of degree d with no diagonal terms. Generically $a_1 \ldots a_n \neq 0$, and we shall assume that this is the case. We wish to "deform" f to a diagonal form $\sum_{i=1}^n a_i X_i^d$ by making the remainder term h tend to zero. To this end, we introduce an extra variable Y which controls the deformation. Define

$$f_Y = \sum_{i=1}^n a_i X_i^d + Y h(X_1, \ldots, X_n) \in \mathbb{F}_q[Y][X_1, \ldots, X_n].$$

Setting $Y = 1$ gives our original polynomial $f = f_1$. Setting $Y = 0$ gives the diagonal form f_0. Intuitively, as Y moves from 1 to 0 our original polynomial is deformed into a diagonal form. Let $N(f_y)$ denote the number of rational projective zeros of the specialised polynomial f_y, where we take $Y = y \in \mathbb{F}_q$. For all but finitely many y in the algebraic closure of $\mathbb{F}_q$, the zero set of f_y will be "smooth and in general position". We will say that such a y defines a *smooth fibre*. If y defines a smooth fibre, a Frobenius matrix α_y is defined, and we have the formula

$$N(f_y) = \frac{q^{n-1} - 1}{q - 1} + (-1)^n \frac{\text{Trace}(\alpha_y)}{q}.$$

A *generic* Frobenius matrix $\alpha(Y)$ for the polynomial f_Y is also defined. Its entries are p-adic analytic functions in the variable Y which will converge on the Teichmüller lifting of any point $y \in \mathbb{F}_q$ which defines a smooth fibre. For those y at which $\alpha(Y)$ is defined, the generic Frobenius matrix $\alpha(Y)$ converges to the specialised Frobenius matrix α_y. Dwork's deformation theory yields the following factorisation

$$\alpha(Y) = C(Y^q)^{-1}\alpha(0)C(Y). \tag{2}$$

Here $C(Y)$ is a matrix of p-adic analytic functions which need not converge on the Teichmüller lifting of any non-zero point in $\mathbb{F}_q$. One may compute $\alpha(0)$ easily, as it is the Frobenius matrix α_0 of a diagonal form and has a nice Kronecker product decomposition. The matrix $C(Y)$ is the solution matrix of a system of linear differential equations: $dC(Y)/dY = C(Y)B(Y)$ and $C(0) = I$. Here the matrix $B(Y)$ contains entries which are rational functions in Y over the integers, and it can be computed using a method due to Dwork. (The matrix $B(Y)$ for the elliptic curve case is calculated in Dwork's original paper [2, Section 8].) All the theory is now in place to describe the deformation approach for computing the number of projective zeros of the original polynomial f: First, compute the rational matrix $B(Y)$ using Dwork's method. Second, compute an expansion of $C(Y)$ around the origin by solving the differential system numerically. Third, compute $\alpha(0) = \alpha_0$ directly from its Kronecker product decomposition. Fourth, recover an expansion of $\alpha(Y)$ around the origin from Equation (2). Fifth, use this expansion to compute $\alpha = \alpha_1 = \alpha(1)$. Finally, the trace of the matrix α yields the number of projective solutions, as in Equation (1).

One subtlety in this approach should be pointed out: because some y in the algebraic closure of $\mathbb{F}_q$ do not define smooth fibres, the expansion of $\alpha(Y)$ about the origin will not converge on the Teichmüller lifting of non-zero points of $\mathbb{F}_q$. Thus for the final step one needs some method of "continuing" $\alpha(Y)$ to the Teichmüller liftings of the points $y \in \mathbb{F}_q$ which define smooth fibres. In the next section I describe an easier, though non-generic, situation in which the singular fibres do not cause such a complication. I have implemented the algorithm in this simpler situation for some small examples with the help of Frederik Vercauteren.

The reason that the complexity is improved using the above method is that the deformation is always one-dimensional regardless of the number of variables n in the original problem. In practice, this means that one computes with univariate, rather than n-variate, polynomials. Even in the case of curves, in certain special cases the deformation method requires less space than that of Kedlaya although the time complexity is the same.

Note that throughout this section I have glossed over one important point: in all algorithms which exploit p-adic cohomology in some way, it is essential to first compute the "semi-linear Frobenius matrix" rather than the Frobenius matrix itself. This turns the "q" into a "$p \log_p(q)$" in the complexity estimates, but it does make the formulae a bit more involved.

3   Results

I will finish now by stating the results that have been obtained in my paper [6] based upon these ideas. They pertain to certain special equations, namely Artin-Schreier equations with a diagonal leading form. Let $f \in \mathbb{F}_q[X_1, \ldots, X_n]$ be of degree d where p does not divide d (f is not necessarily homogeneous). We will say that f has a diagonal leading form if it can be written as $f = \sum_{i=1}^{n} a_i X_i^d + h(X_1, \ldots, X_n)$ where $a_1 \ldots a_n \neq 0$ and h has degree strictly less than d. Let $N(f)$ be the number of affine solutions in $\mathbb{F}_q^{n+1}$ to the equation $Z^p - Z = f(X_1, \ldots, X_n)$.

Theorem 1. *There exists an explicit deterministic algorithm with the following input, output and complexity. The input is a polynomial $f \in \mathbb{F}_q[X_1, \ldots, X_n]$ with a diagonal leading form of degree d not divisible by the characteristic p, where $p > 2$. The output is the number $N(f)$ of rational solutions to the affine equation $Z^p - Z = f$. The running time is*

$$\tilde{\mathcal{O}}(c_n d^{\min(4n+1, 3n+3)} \log(q)^3 p^2)$$

bit operations, where c_n depends only on n.

Here the soft-Oh $\tilde{\mathcal{O}}$ notation suppresses logarithmic factors in the parameters d^n, $\log(q)$ and p. The restriction to diagonal leading form is useful as it allows us to avoid the difficulty of "crossing" over singular fibres in the family, i.e., in the fifth step the expansion of $\alpha(Y)$ will converge at the correct points in this case.

A curious corollary of the above is the following result.

Corollary 1. *Let $f \in \mathbb{Z}[X_1, \ldots, X_n]$ have the form*

$$f = \sum_{i=1}^{n} a_i X_i^d + h(X_1, \ldots, X_n)$$

where $a_1 \ldots a_n \neq 0$ and h has total degree less than d. There exists an explicit deterministic algorithm which takes as input a prime p and outputs the number of solutions to the equation $f = 0 \bmod p$, and runs in $\tilde{\mathcal{O}}(p^2)$ bit operations.

Here the algorithm itself depends upon f, and the hidden constant also depends upon f. The naive bound here would be $\tilde{\mathcal{O}}(p^{n-1})$ using Berlekamp's root-counting algorithm, as alluded to in the introduction.

Acknowledgements. This paper was written to accompany the author's invited talk at AAECC 15, Toulouse, 2003. He wishes to thank the organisers of the conference, most especially Professor Poli, and also Professors Richard Brent and Shuhong Gao for giving helpful comments on the paper.

References

1. J.-P. Dedieu, Newton's method and some complexity aspects of the zero-finding problem, in "Foundations of Computational Mathematics", (R.A. DeVore, A. Iserles, E. Suli), LMS Lecture Note Series 284, Cambridge University Press, 2001, 45–67.
2. B. Dwork, On the zeta function of a hypersurface II, Ann. Math. (2) 80, (1964), 227–299.
3. J. von zur Gathen and J. Gerhard, Modern Computer Algebra, Cambridge University Press, 1999.
4. N.M. Katz, On the differential equations satisfied by period matrices, Pub. Math. IHES 35, (1968), 71–106.
5. K. Kedlaya, Counting points on hyperelliptic curves using Monsky-Washnitzer cohomology, Journal of the Ramanujan Mathematical Society, 16 (2001), 323–338.
6. A.G.B. Lauder, Deformation theory and the computation of zeta functions, submitted. Preprint available at:
   `http://web.comlab.ox.ac.uk/oucl/work/alan.lauder/`
7. A.G.B. Lauder and D. Wan, Counting points on varieties over finite fields of small characteristic, to appear in Algorithmic Number Theory: Lattices, Number Fields, Curves and Cryptography (Mathematical Sciences Research Institute Publications), J.P. Buhler and P. Stevenhagen (eds), Cambridge University Press.

Three Constructions of Authentication/Secrecy Codes

Cunsheng Ding[1], Arto Salomaa[2], Patrick Solé[3], and Xiaojian Tian[1]

[1] Department of Computer Science, HKUST, Kowloon, Hong Kong, China
`cding@cs.ust.hk`
[2] Turku Centre for Computer Science
DataCity, Lemminkäisenkatu 14A, FIN-20520, Turku, Finland
`asalomaa@cs.utu.fi`
[3] CNRS-I3S, ESSI, Route des Colles, 06 903 Sophia Antipolis, France
`ps@essi.fr`

Abstract. In this paper, we present three algebraic constructions of authentication codes with secrecy. The codes have simple algebraic structures and are easy to implement. They are asymptotically optimal with respect to certain bounds.

1 Introduction

The authentication model introduced by Simmons involves three parties: a transmitter, a receiver, and an opponent. The transmitter wants to send a piece of information (called a *source state*) to the receiver through a public communication channel. A *source state* s is encoded to obtain a *message* $m = E_k(s)$ with an encoding rule E_k shared by the transmitter and receiver, and is sent to the receiver through the channel. When m is received, the receiver will recover the message and check the authenticity of the message using the encoding rule E_k. The encoding rule E_k is usually a family of mappings indexed by the parameter k, where k is from a space $\mathcal{K}$, which is called the *key space*. All the possible source states s form the *source state space* $\mathcal{S}$, and all possible messages m form the *message space* $\mathcal{M}$. Each of the three spaces $\mathcal{S}$, $\mathcal{M}$ and $\mathcal{K}$ is associated with a probability distribution. In this paper, we assume that both $\mathcal{S}$ and $\mathcal{K}$ have uniform probability distribution. We use $\mathcal{E}$ to denote the encoding rule space. Its probability distribution depends on that of $\mathcal{K}$ and the design of the encoding algorithm E_k. If the mapping $k \mapsto E_k$ is a one-to-one correspondence from $\mathcal{K}$ to $\mathcal{E}$ and all keys are equally likely, then all encoding rules are equally likely.

An *authentication code* is a four-tuple $(\mathcal{S}, \mathcal{K}, \mathcal{M}, E_k)$. There are two types of authentication codes: authentication codes with secrecy and those without secrecy. In an authentication code with secrecy, a source state s is sent to the receiver in an encrypted form. In this case, the secret key k shared by both the sender and receiver is used for both encryption and authentication purpose. In an authentication code without secrecy, a source state is sent to the receiver in plaintext. In this case, the secret key is used only for authentication purpose. In

M. Fossorier, T. Hoeholdt, and A. Poli (Eds.): AAECC 2003, LNCS 2643, pp. 24–33, 2003.

this paper we consider only authentication codes with secrecy. It is possible that an encoding rule may map a source state onto more than one message (this is called *splitting*). Here we consider only authentication codes without splitting.

Within this authentication model, we assume that an opponent can insert his message into the channel, and can substitute an observed message m with another message m'. We consider two kinds of attacks, the *impersonation* and *substitution* attacks. In an impersonation attack an opponent inserts his message into the channel and wishes to make the receiver accept it as authentic. In a substitution attack the opponent observed a message sent by the transmitter and will replace it with his message $m' \neq m$, hoping that the receiver accepts it as authentic. We use P_I and P_S to denote the success probabilities with respect to the two attacks.

Authentication codes with secrecy have been considered in [2,3,4,7,10,11,13, 14,15]. Most constructions are combinatorial. In this paper we present three algebraic constructions of authentication codes with secrecy. These codes are asymptotically optimal against both impersonation and substitution attacks.

2   Bounds on Authentication Codes

We summarize some of the known bounds needed in the sequel. We also use $\mathcal{M}$, $\mathcal{E}$, and $\mathcal{S}$ to denote the random variable of the messages, encoding rules, and source states. We use $\mathcal{M}^r$ to denote the random variables of the first r messages, and $H(\mathcal{E}|\mathcal{M}^r)$ the conditional entropy of $\mathcal{E}$ given that the first r messages have been observed.

The following is called the *information-theoretic bound* [8], [10], [12].

Lemma 1. *In any authentication code,*

$$P_I \geq 2^{H(\mathcal{E}|\mathcal{M})-H(\mathcal{E})}, \ \ P_S \geq 2^{H(\mathcal{E}|\mathcal{M}^2)-H(\mathcal{E}|\mathcal{M})}.$$

The following is called the *combinatorial bound* [9].

Lemma 2. **[6,9]** *In any authentication system without splitting,*

$$P_I \geq \frac{|\mathcal{S}|}{|\mathcal{M}|} \ \ and \ P_S \geq \frac{|\mathcal{S}| - 1}{|\mathcal{M}| - 1}.$$

If both equalities are achieved, then $|\mathcal{E}| \geq |\mathcal{M}|$.

3   Construction I

Let $q = p^h$, where p is an odd prime and $h \in \mathbb{N}$. Let $\mathrm{Tr}(x)$ be the trace function from $\mathrm{GF}(q^n)$ to $\mathrm{GF}(q)$. We use $\mathcal{S}$, $\mathcal{K}$, $\mathcal{M}$, and $\mathcal{E}$ to denote the source state space, key space, message space, and encoding rule space, respectively. Define

$$(\mathcal{S}, \mathcal{K}, \mathcal{M}, \mathcal{E}) = (\mathrm{GF}(q^n), \mathrm{GF}(q^n), \mathrm{GF}(q^n) \times \mathrm{GF}(q), \{E_k|k \in \mathcal{K}\}), \tag{1}$$

where for any $k \in \mathcal{K}$ and $s \in \mathcal{S}$,

$$E_k(s) = (s + k, \mathrm{Tr}(sk)).$$

We denote $m_1 = s+k$ and $m_2 = \mathrm{Tr}(sk)$. The first part is the encrypted message. The second part m_2 is the redundant part for authentication.

Theorem 1. *The authentication code of (1) provides at least*

$$\log_2\left(\frac{q^n - (q-1)q^{n/2}}{q}\right)$$

bits of secrecy protection if n is even, and at least

$$\log_2\left(\frac{q^n - q^{(n+1)/2}}{q}\right)$$

bits of secrecy protection if n is odd.
   Furthermore, we have

$$P_I = \begin{cases} \frac{1}{q} + \frac{1}{q^{1+n/2}} & n \text{ even, } p \equiv 1 \bmod 4 \text{ or } (p \equiv 3 \bmod 4 \text{ and } nh \equiv 0 \bmod 4), \\ \frac{1}{q} + \frac{q-1}{q^{1+n/2}} & n \text{ even, } p \equiv 3 \bmod 4 \text{ and } nh \equiv 2 \bmod 4, \\ \frac{1}{q} + \frac{1}{q^{(1+n)/2}} & n \text{ odd,} \end{cases}$$

and

$$\frac{q^{n-1} - (q-1)q^{n/2}}{q^n + (q-1)q^{n/2}} \le P_S \le \frac{q^{n-1} + (q-1)q^{n/2}}{q^n - (q-1)q^{n/2}}. \tag{2}$$

Optimality of the Codes

We now prove that P_I and P_S of the authentication code of (1) meet the lower bound of Lemma 2 asymptotically.
   In the authentication code of (1),

$$|\mathcal{S}| = q^n, \quad |\mathcal{M}| = q^{n+1}$$

so the bound on P_I given in Lemma 2 is

$$P = \frac{1}{q}.$$

Since

$$\lim_{n\to\infty}\left[\frac{1}{q} + \frac{1}{q^{1+n/2}}\right] = \lim_{n\to\infty}\left[\frac{1}{q} + \frac{q-1}{q^{1+n/2}}\right] = \lim_{n\to\infty}\left[\frac{1}{q} + \frac{1}{q^{(1+n)/2}}\right] = \frac{1}{q},$$

from (2), We obtain

$$\lim_{n\to\infty} P_I = \frac{1}{q}.$$

It follows that

$$\lim_{n \to \infty} \frac{P}{P_I} = \frac{\lim_{n \to \infty} P}{\lim_{n \to \infty} P_I} = 1.$$

The bound on P_S given in Lemma 2 is

$$Q = \frac{q^n - 1}{q^{n+1} - 1}.$$

Because

$$\lim_{n \to \infty} \left[\frac{q^{n-1} - (q-1)q^{n/2}}{q^n + (q-1)q^{n/2}} \right] = \lim_{n \to \infty} \left[\frac{q^{n-1} + (q-1)q^{n/2}}{q^n - (q-1)q^{n/2}} \right] = \frac{1}{q},$$

from (2), We obtain

$$\lim_{n \to \infty} P_S = \frac{1}{q}.$$

On the other hand, we have

$$\lim_{n \to \infty} Q = \lim_{n \to \infty} \frac{q^n - 1}{q^{n+1} - 1} = \frac{1}{q}.$$

It follows that

$$\lim_{n \to \infty} \frac{Q}{P_S} = \frac{\lim_{n \to \infty} Q}{\lim_{n \to \infty} P_S} = 1.$$

4   Construction II

Let n be a positive integer. Let $\text{Tr}(x)$ be the trace function from $\text{GF}(q^n)$ to $\text{GF}(q)$. Define

$$(\mathcal{S}, \mathcal{K}, \mathcal{M}, \mathcal{E}) = (\text{GF}(q^n)^*, \text{GF}(q^n)^*, \text{GF}(q^n)^* \times \text{GF}(q), \{E_k | k \in \mathcal{K}\}), \qquad (3)$$

where for any $k \in \mathcal{K}$ and $s \in \mathcal{S}$,

$$E_k(s) = (sk, \text{Tr}(s + k)).$$

We denote $m_1 = sk$ and $m_2 = \text{Tr}(s+k)$. The first part is the encrypted message. The second part m_2 is the redundant part for authentication.

Theorem 2. *The authentication code of (3) provides at least*

$$\log_2 \left(\frac{q^n - 1 - 2(q-1)q^{n/2}}{q} \right)$$

bits of secrecy protection.
    Furthermore we have

$$\frac{1}{q} - \frac{2(q-1)q^{n/2}}{q(q^n - 1)} \le P_I \le \frac{1}{q} + \frac{2(q-1)q^{n/2}}{q(q^n - 1)}$$

and

$$\frac{-2(q^2-1)q^{n/2} + q^n - 1}{q[2(q-1)q^{n/2} + q^n - 1]} \le P_S \le \frac{2(q^2-1)q^{n/2} + q^n - 1}{q[-2(q-1)q^{n/2} + q^n - 1]}$$

Optimality of the Codes

We now prove that P_I and P_S of authentication code (3) meet the lower bound of Lemma 2 asymptotically.

In the authentication code of (3),

$$|\mathcal{S}| = q^n - 1, \quad |\mathcal{M}| = q(q^n - 1)$$

So the bound on P_I given in Lemma 2 is

$$P = \frac{1}{q}.$$

Since

$$\lim_{n \to \infty} \left[\frac{1}{q} - \frac{2(q-1)q^{n/2}}{q(q^n-1)} \right] = \lim_{n \to \infty} \left[\frac{1}{q} + \frac{2(q-1)q^{n/2}}{q(q^n-1)} \right] = \frac{1}{q},$$

from Theorem 2, We obtain

$$\lim_{n \to \infty} P_I = \frac{1}{q}.$$

It follows that

$$\lim_{n \to \infty} \frac{P}{P_I} = \frac{\lim_{n \to \infty} P}{\lim_{n \to \infty} P_I} = 1.$$

The bound on P_S given in Lemma 2 is

$$Q = \frac{q^n - 2}{q(q^n - 1) - 1}.$$

Because

$$\lim_{n \to \infty} \left[\frac{-2(q^2-1)q^{n/2} + q^n - 1}{q[2(q-1)q^{n/2} + q^n - 1]} \right] = \lim_{n \to \infty} \left[\frac{2(q^2-1)q^{n/2} + q^n - 1}{q[-2(q-1)q^{n/2} + q^n - 1]} \right] = \frac{1}{q},$$

from Theorem 2, We obtain

$$\lim_{n \to \infty} P_S = \frac{1}{q}.$$

On the other hand, we have

$$\lim_{n \to \infty} Q = \lim_{n \to \infty} \frac{q^n - 2}{q(q^n - 1) - 1} = \frac{1}{q}.$$

It follows that

$$\lim_{n \to \infty} \frac{Q}{P_S} = \frac{\lim_{n \to \infty} Q}{\lim_{n \to \infty} P_S} = 1.$$

5   Construction III

In this section, we first describe the general construction and then present two specific constructions of authentication codes with secrecy using perfect nonlinear mappings.

5.1   The General Construction

Let $(A, +)$ and $(B, +)$ be two finite abelian groups, and let Π be a mapping from A to B such that $\Pi(-x) = \Pi(x)$ for all $x \in A$. We construct an authentication code $(\mathcal{S}, \mathcal{K}, \mathcal{M}, \mathcal{E})$ by defining

$$(\mathcal{S}, \mathcal{K}, \mathcal{M}, \mathcal{E}) = (A, A, A \times B, \{E_k | k \in \mathcal{K}\}), \tag{4}$$

where for any $k \in \mathcal{K}$ and $s \in \mathcal{S}$,

$$E_k(s) = (s + k, \Pi(s) + \Pi(k)).$$

We denote $m_1 = s + k$, and $m_2 = \Pi(s) + \Pi(k)$. The first part is the encrypted message. The second part m_2 is the redundant part for authentication.

Impersonation Attack

We assume that an opponent knows the structure of the system except the secret key k or equivalently the corresponding encoding rule E_k. We now discuss the security of this system with respect to impersonation attack.

The impersonation attack is as follows. The opponent picks up an element $m = (m_1, m_2) \in \mathcal{M}$, and sends it to the receiver. The receiver will compute $s = m_1 - k$ and $\Pi(s) + \Pi(k)$. Then he will check whether $\Pi(m_1 - k) + \Pi(k) = m_2$. Hence

$$P_I = \max_{m_1, m_2} \Pr[\Pi(m_1 - k) + \Pi(k) = m_2]. \tag{5}$$

Substitution Attack

An opponent has observed one message $m = (m_1, m_2)$, where

$$m_1 = s + k, \quad m_2 = \Pi(s) + \Pi(k). \tag{6}$$

He wants to replace m with another message $m' = (m_1', m_2')$, where $m_1 \neq m_1'$.

The substitution attack is to modify the first part m_1 of the message, and possibly the second part m_2. This is equivalent to adding an element $\delta_1 \neq 0$ to m_1, and an element δ_2 to m_2. This is successful if and only if

$$\Pi(s) + \Pi(k) + \delta_2 = \Pi(s + \delta_1) + \Pi(k)$$

which is equivalent to $\Pi(s + \delta_1) - \Pi(s) = \delta_2$. Whence

$$P_S = \max_{\delta_1 \neq 0, \delta_2} \Pr[\Pi(s + \delta_1) - \Pi(s) = \delta_2]. \tag{7}$$

By (5) and (7), the probabilities P_I and P_S depend totally on the mapping Π. In the sequel, we construct codes by choosing proper mappings Π within this general framework. To this end, we need optimal nonlinear functions Π.

5.2   First Specific Construction of Codes in This Family

Let f be a function from an abelian group $(A, +)$ to another abelian group $(B, +)$. The derivatives are defined as $D_a f(x) = f(x + a) - f(x)$. A robust measure of the nonlinearity of functions is given by

$$P_f = \max_{0 \neq a \in A} \max_{b \in B} \Pr(D_a f(x) = b), \tag{8}$$

where $\Pr(E)$ denotes the probability of the occurrence of event E.

It is straightforward to see that $P_f \geq \frac{1}{|B|}$. If the equality is achieved, we say that function $f : A \to B$ has perfect nonlinearity. In this case $|B|$ must divide $|A|$.

Define a function $\Pi(x)$ from $\mathrm{GF}(q)^{2t}$ to $\mathrm{GF}(q)$ as

$$\Pi(x_1, x_2, \cdots, x_{2t}) = x_1 x_2 + x_3 x_4 + \cdots + x_{2t-1} x_{2t}. \tag{9}$$

Then Π has perfect nonlinearity.

Theorem 3. *Let Π be the function of (9), and let $(A, +) = (\mathrm{GF}(q)^{2t}, +)$ and $(B, +) = (\mathrm{GF}(q), +)$. Then the authentication code of (4) provides at least $\log_2(q^{2t-1} - q^{t-1})$ bits of secrecy protection. Furthermore, we have*

$$P_I = \frac{1}{q} + \frac{q-1}{q^{t+1}}, \quad P_S = \frac{1}{q}.$$

Optimality of the Codes of Theorem 3

Clearly, $P_I = \frac{1}{q} + \frac{q-1}{q^{t+1}}$ does not meet the lower bound on P_I given in Lemma 2. We now prove that it meets the bound on P_I given in Lemma 1 asymptotically.

Clearly, $H(\mathcal{E}) = \log_2 q^{2t}$. We now compute $H(\mathcal{E}|\mathcal{M})$. Suppose that a message $m = (m_1, m_2)$ has been observed. Since all encoding rules and all source states are used with equal probability, we can get the probability distribution of the messages. We distinguish even q and odd q. When q is even, we have

m	uncertainty of e	probability of m	number of m
$(0, 0)$	$\log_2[q^{2t-1} + (q-1)q^{t-1}]$	$[q^{2t-1} + (q-1)q^{t-1}]/q^{4t}$	1
$(0, m_2 \neq 0)$	$\log_2[q^{2t-1} - q^{t-1}]$	$[q^{2t-1} - q^{t-1}]/q^{4t}$	$q - 1$
$(m_1 \neq 0, m_2)$	$\log_2[q^{2t-1}]$	$[q^{2t-1}]/q^{4t}$	$(q^{2t} - 1)q$

So we obtain

$$H(\mathcal{E}|\mathcal{M}) = \frac{q^{2t-1} + (q-1)q^{t-1}}{q^{4t}} \log_2[q^{2t-1} + (q-1)q^{t-1}]$$

$$+ \frac{(q-1)(q^{2t-1} - q^{t-1})}{q^{4t}} \log_2[q^{2t-1} - q^{t-1}]$$

$$+ \frac{(q^{2t} - 1)q q^{2t-1}}{q^{4t}} \log_2[q^{2t-1}].$$

Thus the bound on P_I given in Lemma 1 is

$$Q := 2^{H(\mathcal{E}|\mathcal{M})-H(\mathcal{E})}$$

$$= \frac{[q^{2t-1} + (q-1)q^{t-1}]^{\frac{q^{2t-1}+(q-1)q^{t-1}}{q^{4t}}} [q^{2t-1} - q^{t-1}]^{\frac{(q-1)(q^{2t-1}-q^{t-1})}{q^{4t}}}}{q^{2t}} \times$$

$$[q^{2t-1}]^{\frac{(q^{2t}-1)qq^{2t-1}}{q^{4t}}}$$

It follows that

$$\lim_{t\to\infty} \frac{Q}{P_I} = 1.$$

Similarly, when q is odd, we have

m	uncertainty of e	probability of m	number of m
if $\Pi(m_1) = 2m_2$	$\log_2[q^{2t-1} + (q-1)q^{t-1}]$	$[q^{2t-1} + (q-1)q^{t-1}]/q^{4t}$	q^{2t}
if $\Pi(m_1) \neq 2m_2$	$\log_2[q^{2t-1} - q^{t-1}]$	$[q^{2t-1} - q^{t-1}]/q^{4t}$	$q^{2t}(q-1)$

In this case, we can also prove that P_I asymptotically achieves the lower bound given in Lemma 1.

Note that $P_S = \frac{1}{q}$ is almost optimal compared with the bound P_S given in Lemma 2.

Secrecy Protection

When a message $m = (m_1 = s + k, m_2 = \Pi(s) + \Pi(k))$ is observed. We have

$$\Pi(s) + \Pi(m_1 - s) = m_2.$$

When $m_1 \neq 0$ and q is even, the number of solutions to this equation is q^{2t-1}. In other cases, the minimum number of solutions to this equation is $q^{2t-1} - q^{t-1}$. Hence the secrecy protection for the source state is at least $\log_2(q^{2t-1} - q^{t-1})$ bits.

5.3   Second Specific Construction of Codes in This Family

Define $\Pi(x) = \mathrm{Tr}_{\mathrm{GF}(q^n)/\mathrm{GF}(q)}(x^2)$, where n is a positive integer, q is an odd prime, and $\mathrm{Tr}_{\mathrm{GF}(q^n)/\mathrm{GF}(q)}$ is the trace function. Since x^2 is a perfect nonlinear mapping from $\mathrm{GF}(q^n)$ to itself, Π is a perfect nonlinear mapping from $\mathrm{GF}(q^n)$ to $\mathrm{GF}(q)$.

Theorem 4. *Let $\Pi = \mathrm{Tr}_{\mathrm{GF}(q^n)/\mathrm{GF}(q)}(x^2)$, and let $(A, +) = (\mathrm{GF}(q^n), +)$ and $(B, +) = (\mathrm{GF}(q), +)$ Then the authentication code of (4) provides at least $\log_2(q^{n-1} - (q-1)q^{n/2-1})$ bits of secrecy protection when n is even, and at least $\log_2(q^{n-1} - q^{(n-1)/2})$ bits of secrecy protection when n is odd. Furthermore, we have $P_S = \frac{1}{q}$, and*

$$P_I = \begin{cases} \frac{1}{q} + \frac{q-1}{q^{n/2+1}} \ \text{or} \ \frac{1}{q} + \frac{1}{q^{n/2+1}}, & \text{if } n \text{ even} \\ \frac{1}{q} + \frac{1}{q^{(n+1)/2}}, & \text{if } n \text{ odd} \end{cases}$$

Similarly we can prove that the code of Theorem 4 is asymptotically optimal with respect to the bounds of Lemma 1.

More Specific Constructions of Codes in This Family

Define $\Pi(x) = \mathrm{Tr}_{\mathrm{GF}(p^m)/\mathrm{GF}(p^h)}(x^s)$, where m and h are integers with $1 \leq h|m$, p is an odd prime, and $\mathrm{Tr}_{\mathrm{GF}(p^m)/\mathrm{GF}(p^h)}$ is the trace function. If

- $s = p^k + 1$, where $m/\gcd(m,k)$ is odd, or
- $s = (3^k + 1)/2$, where $p = 3, k$ is odd, and $\gcd(m,k) = 1$,

then Π has perfect nonlinearity. These mappings give more authentication codes with secrecy. However, computing the two probabilities P_I and P_S may be hard.

6   Concluding Remarks

The first two constructions of this paper are specific, while the third construction is generic in the sense that any perfect nonlinear mapping may be employed to obtain authentication codes with secrecy. Thus new functions with perfect nonlinearity give new authentication/secrecy codes. Note that authentication codes with secrecy have six parameters. It is in general hard to compare two classes of them. We have not found any existing class of authentication codes that could be compared with those presented in this paper.

There are optimal authentication codes with perfect secrecy constructed using both combinatorial and algebraic methods. Compared with these codes, the authentication codes presented in this paper have the advantage of using less keys (encoding rules), and the disadvantage of not being optimal. It is not practical if the key space of an authentication code is too large. The objective of this paper is to present three classes of authentication/secrecy codes which use less secret keys (encoding rules) but are asymptotically optimal.

References

1. E.F. Brickel, A few results in message authentication, *Congressus Numerantium* 43 (1984), 141–154.
2. L.R.A. Casse, K.M. martin, and P R. Wild, Bounds and characterizations of authentication/secrecy schemes, *Designs, Codes and Cryptography* 13 (1998), 107–129.
3. M. De Soete, Some constructions for authentication-secrecy codes, *Advances in Cryptology – Eurocrypt' 88, Lecture Notes in Comput. Sci.,* Vol. 330, Springer-Verlag, 1988, 57–76.
4. E. Gilbert, F.J. MacWilliams and N.J.A. Sloane, Codes which detect deception, *Bell Systems Tech. J.* 53 (1974), 405–424.
5. R. Lidl and H. Niederreiter, Finite Fields, Encyclopedia of Mathematics and Its Application 20, Cambridge University Press, Cambridge, 1997.
6. J. L. Massey, Cryptography – A selective survey, in *Digital Communications* (1986), 3–21.

7. C. Mitchell, M. Walker and P. Wild, The combinatorics of perfect authentication schemes, *SIAM J. Disc. Math.* 7 (1994), 102–107.
8. D. Pei, Information-theoretic bounds for authentication codes and block designs, *J. Cryptography* 8 (1995), 177–188.
9. R.S. Rees and D.R. Stinson, Combinatorial characterizations of authentication codes, *Designs, Codes and Cryptography* 7 (1996), 239–259.
10. U. Rosenbaum, A lower bound on authentication after having observed a sequence of messages, *J. Cryptography* 6 (1993), 135–156.
11. R, Safavi-Naini and L. Tombak, Authentication codes in plaintext and chosen-content attacks, *Advances in Cryptology – Eurocrypt' 94, Lecture Notes in Comput. Sci.,* Vol. 950 (1995), 254–265.
12. A. Sgarro, Information-theoretic bounds for authentication frauds, *J. Computer Security* 2 (1993), 53–63.
13. G.J. Simmons, Authentication theory/coding theory, *Advances in Cryptology – Crypto' 84, Lecture Notes in Comput. Sci.,* Vol. 196 (1984), 411–431.
14. D.R. Stinson, A construction for authentication/secrecy codes from certain combinatorial designs, *J. Cryptography* 1 (1988), 119–127.
15. D.R. Stinson and L. Teirlinck, A construction for authentication/secrecy codes from 3-homogeneous permutation groups, *European J. Combin.* 11 (1990), 73–79.

The Jacobi Model of an Elliptic Curve and Side-Channel Analysis

Olivier Billet[1,2] and Marc Joye[1]

[1] Gemplus Card International, Card Security Group
La Vigie, Av. du Jujubier, ZI Athélia IV, 13705 La Ciotat Cedex, France
marc.joye@gemplus.com, billet@eurecom.fr
http://ecwww.eurecom.fr/~billet/
http://www.geocities.com/MarcJoye/
http://www.gemplus.com/smart/
[2] Télécom Paris (ENST), 46 rue Barrault, 75634 Paris Cedex 13, France
http://www.enst.fr
UNSA, Laboratoire Dieudonné, Parc Valrose, 06108 Nice Cedex 02, France
http://math.unice.fr

Abstract. A way for preventing SPA-like attacks on elliptic curve systems is to use the same formula for the doubling and the general addition of points on the curve. Various proposals have been made in this direction with different results. This paper re-investigates the Jacobi form suggested by Liardet and Smart (CHES 2001). Rather than considering the Jacobi form as the intersection of two quadrics, the addition law is directly derived from the underlying quartic. As a result, this leads to substantial memory savings and produces the fastest unified addition formula for curves of order a multiple of 2, as those required for OK-ECDH or OK-ECDSA.

Keywords: elliptic curve cryptosystems, unified addition formula, side-channel analysis, SPA-like attacks, smart cards.

1 Introduction

In 1996, Kocher introduced the so-called *side-channel attacks*. By monitoring some side-channel information (e.g., timing [9] or power consumption [10]), an attacker tries to retrieve secret data involved in a cryptographic computation. For elliptic curve cryptosystems, a naïve implementation of the point multiplication is particularly susceptible to such attacks as the classical formulæ for doubling a point and for adding two (distinct) points are different. Hence, according to the implemented crypto-algorithm, a *simple power analysis* (SPA)[1] can yield the value of multiplier d used in the computation of $\boldsymbol{Q} = d\boldsymbol{P}$ on an elliptic curve.

A promising way for preventing SPA-like attacks consists in re-writing the addition formula so that the doubling and the general addition become indistinguishable from some side-channel information. In particular, an addition formula valid for both the doubling and the addition would be helpful.

---

[1] There is another class of side-channel attacks, the *differential attacks*, but we do not consider them as efficient protections are known (e.g., see [6]).

M. Fossorier, T. Hoeholdt, and A. Poli (Eds.): AAECC 2003, LNCS 2643, pp. 34–42, 2003.

1.1   Related Work

The use of a unified formula for the addition of points on an elliptic curve as a means for preventing SPA-like attacks has been independently suggested in [8] and [12]. In [8], the authors suggest the Hessian form while in [12], the intersection of two quadrics is considered. Unified addition formulæ for the general Weierstraß parameterization are given in [3].

1.2   Our Results

Building on [12], we consider the Jacobi form not as the intersection of two quadrics but as a quartic. This allows points to be represented with triplets instead of quadruplets. Furthermore, this considerably reduces the number of required (field) multiplications. As a result, we obtain the most efficient (memory-wise and computation-wise) unified formula for adding points on an elliptic curve whose order is a multiple of 2 (as required for OK-ECDH [1] or OK-ECDSA [2]). In particular, compared to [12], we get a 23% speed-up improvement with fewer memory requirements.

1.3   Organization of the Paper

The rest of this paper is organized as follows. In the next section, we respectively review the Jacobi form of an elliptic curve as the intersection of two quadrics and as a quartic. Then, in Sect. 3, we show how the quartic model helps to prevent SPA-like attacks. Finally, we conclude in Sect. 4.

2   Jacobi Models

Throughout this paper, we assume that $\mathbb{K}$ represents a field of characteristic $\operatorname{Char} \mathbb{K} \neq 2$. Furthermore, since our ultimate goal is the study over large prime fields, we also assume that $\operatorname{Char} \mathbb{K} \neq 3$.

2.1   Intersection of Two Quadrics

It is well known that any elliptic curve over $\mathbb{K}$ can be embedded as the intersection of two quadrics in $\mathbb{P}^3$ [4, Chapter 7]. Indeed, a point (x, y) on a Weierstraß elliptic curve

$$y^2 = x^3 + ax + b$$

corresponds to the point (X, Y, Z, T) on the intersection

$$\begin{cases} X^2 - TZ = 0 \\ Y^2 - aXZ - bZ^2 - TX = 0 \end{cases} \tag{1}$$

via the map $(x, y) \longmapsto (X, Y, Z, T) = (x, y, 1, x^2)$.

As for the Weierstraß parameterization, the group law on the intersection of two quadrics has a nice geometrical interpretation [13]. Let $P_1 = (X_1, Y_1, Z_1, T_1)$ and $P_2 = (X_2, Y_2, Z_2, T_2)$ be two points on the intersection given by Eq. (1). Then, the sum $P_3 = P_1 + P_2 = (X_3, Y_3, Z_3, T_3)$ is given by

$$X_3 = \mathsf{F}(P_1, P_2)\,\mathsf{H}(P_1, P_2)\,, \qquad\qquad Z_3 = \mathsf{H}(P_1, P_2)^2\,,$$
$$Y_3 = Y_1\,\mathsf{G}(P_2, P_1) + Y_2\,\mathsf{G}(P_1, P_2)\,, \qquad T_3 = \mathsf{F}(P_1, P_2)^2\,,$$

where

$$\mathsf{F}(P_1, P_2) = T_1 T_2 - 2aX_1 X_2 - 4b(X_1 Z_2 + Z_1 X_2) + a^2 Z_1 Z_2\,,$$
$$\begin{aligned}\mathsf{G}(P_1, P_2) = {}&T_1^2 T_2 + 2aX_1 T_1 X_2 + 4bX_1 T_1 Z_2 + 3aZ_1 T_1 T_2\\ &+ 12bZ_1 T_1 X_2 - 3a^2 Z_1 T_1 Z_2 + 4bZ_1 X_1 T_2 - 2a^2 X_1 Z_1 X_2\\ &- 4abX_1 Z_1 Z_2 - a^3 Z_1^2 Z_2 - 8b^2 Z_1^2 Z_2\,,\end{aligned}$$
$$\mathsf{H}(P_1, P_2) = 2Y_1 Y_2 + X_1 T_2 + T_1 X_2 + a(X_1 Z_2 + Z_1 X_2) + 2bZ_1 Z_2\ .$$

The above formulæ present the particularity of being valid for *both* the doubling and the general addition. This was therefore considered in [12] as a means for preventing side-channel attacks. However, as already noticed there, these addition formulæ are overly involved to be of any use in real-life implementations.

Following [5], attention was therefore restricted to the particular case given by

$$\begin{cases} X^2 + Y^2 - T^2 = 0 \\ (1 - \lambda)\,X^2 + Z^2 - T^2 = 0 \end{cases}. \tag{2}$$

As shown in [12], this corresponds to the Weiertraß equation

$$y^2 = x(x + 1)(x + \lambda) \qquad (\cup\{O\}) \tag{3}$$

via the inverse maps

$$(x, y) \longmapsto (-2y, x^2 - \lambda, x^2 + 2x\lambda + \lambda, x^2 + 2x + \lambda), \quad O \longmapsto (0, 1, 1, 1)$$

and

$$(X, Y, Z, T) \longmapsto \begin{cases} O & \text{if } X = 0 \text{ and } Y = Z = T, \\ \left(\dfrac{\lambda(Z - T)}{(1 - \lambda)Y - Z + \lambda T}, \dfrac{\lambda(1 - \lambda)X}{(1 - \lambda)Y - Z + \lambda T}\right) & \text{otherwise}. \end{cases}$$

From the Weierstraß form (Eq. (3)), it clearly appears that this elliptic curve has three points of order 2. This implies that this elliptic curve contains a copy of $\mathbb{Z}_2 \times \mathbb{Z}_2$ and so its group order is a multiple of 4.

The addition law, $P_3 = P_1 + P_2$, on the specialized intersection given by Eq. (2) becomes

$$X_3 = T_1 Y_2 \cdot X_1 Z_2 + Z_1 X_2 \cdot Y_1 T_2\,, \qquad Z_3 = T_1 Z_1 T_2 Z_2 - k^2 X_1 Y_1 X_2 Y_2\,,$$
$$Y_3 = T_1 Y_2 \cdot Y_1 T_2 - Z_1 X_2 \cdot X_1 Z_2\,, \qquad T_3 = (T_1 Y_2)^2 + (Z_1 X_2)^2\,, \tag{4}$$

so that 16 (field) multiplications (plus 1 multiplication by constant k^2) are required. The inverse of a point $P_1 = (X_1, Y_1, Z_1, T_1)$ is $-P_1 = (-X_1, Y_1, Z_1, T_1)$. Again, we see that (excluding the neutral element $(0, 1, 1, 1)$) there are three points of order 2, namely, $(0, -1, 1, 1)$, $(0, 1, -1, 1)$ and $(0, 1, 1, -1)$.

2.2 Jacobi Quartic

Jacobi also studied quartics of the form

$$y^2 = (1 - x^2)(1 - k^2 x^2) \tag{5}$$

with $k \neq 0, \pm 1$. This type of elliptic curve can be parameterized with Jacobi's elliptic functions, namely the "*sinus amplitudinus*" and its derivative. So, a point $P = (x, y)$ on the Jacobi quartic given by Eq. (5) is represented as $(\mathrm{sn}(u), \mathrm{cn}(u)\,\mathrm{dn}(u))$. From the rich body of addition formulæ for elliptic functions (see [15] for instance), we directly derive the addition law on the quartic. We have

$$\mathrm{sn}(u_1 + u_2) = \frac{\mathrm{sn}(u_1)\,\mathrm{cn}(u_2)\,\mathrm{dn}(u_2) + \mathrm{sn}(u_2)\,\mathrm{cn}(u_1)\,\mathrm{dn}(u_1)}{1 - k^2\,\mathrm{sn}(u_1)^2\,\mathrm{sn}(u_2)^2} ,$$

$$\mathrm{cn}(u_1 + u_2) = \frac{\mathrm{cn}(u_1)\,\mathrm{cn}(u_2) - \mathrm{sn}(u_1)\,\mathrm{sn}(u_2)\,\mathrm{dn}(u_1)\,\mathrm{dn}(u_2)}{1 - k^2\,\mathrm{sn}(u_1)^2\,\mathrm{sn}(u_2)^2} , \tag{6}$$

$$\mathrm{dn}(u_1 + u_2) = \frac{\mathrm{dn}(u_1)\,\mathrm{dn}(u_2) - k^2\,\mathrm{sn}(u_1)\,\mathrm{sn}(u_2)\,\mathrm{cn}(u_1)\,\mathrm{cn}(u_2)}{1 - k^2\,\mathrm{sn}(u_1)^2\,\mathrm{sn}(u_2)^2} .$$

Hence, if $(x_i, y_i) = (\mathrm{sn}(u_i), \mathrm{cn}(u_i)\,\mathrm{dn}(u_i))$ for $i = 1, 2$ are two generic points on the Jacobi quartic, the sum $(x_3, y_3) = (x_1, y_1) + (x_2, y_2)$ is given by

$$(x_3, y_3) = (\mathrm{sn}(u_1 + u_2), \mathrm{cn}(u_1 + u_2)\,\mathrm{dn}(u_1 + u_2)) .$$

Here again, the formulæ present the tremendous advantage of remaining valid for the doubling (i.e., when $u_1 = u_2$).

3 Preventing SPA-Like Attacks

We consider slightly more general quartics than those originally considered by Jacobi. Namely, we investigate quartics given by

$$y^2 = \epsilon x^4 - 2\delta x^2 + 1 . \tag{7}$$

(Jacobi quartics correspond to the case $\epsilon = k^2$ and $\delta = (1 + k^2)/2$.)

Remarkably, all elliptic curves with a point of order 2 can be expressed by a quartic equation of the form of Eq. (7). Let E denote an elliptic curve (over $\mathbb{K}$)[2] given by a Weierstraß equation

$$y^2 = x^3 + ax + b$$

with its point 'at infinity' O. Suppose that E has a point of order 2, say, $(\theta, 0) \in E(\mathbb{K})$. Then, the above Weierstraß elliptic curve is birationnally equivalent to the (extended) Jacobi quartic

$$Y^2 = \epsilon X^4 - 2\delta X^2 Z^2 + Z^4 \tag{8}$$

---

[2] Remember that we assume $\mathrm{Char}\,\mathbb{K} \neq 2, 3$.

with $\epsilon = -(3\theta^2 + 4a)/16$ and $\delta = 3\theta/4$, under the transformations

$$\begin{cases} (\theta, 0) \longmapsto (0 : -1 : 1), \\ (x, y) \longmapsto \left(2(x - \theta) : (2x + \theta)(x - \theta)^2 - y^2 : y\right), \\ O \longmapsto (0 : 1 : 1), \end{cases} \qquad (9)$$

and

$$\begin{cases} (0 : 1 : 1) \longmapsto O, \\ (0 : -1 : 1) \longmapsto (\theta, 0), \\ (X : Y : Z) \longmapsto \left(2\dfrac{(Y + Z^2)}{X^2} - \dfrac{\theta}{2}, \ Z\dfrac{4(Y + Z^2) - 3\theta X^2}{X^3}\right). \end{cases} \qquad (10)$$

In Eqs. (9) and (10), the notation $(X : Y : Z)$ stands for equivalence classes. Two triplets $(X_1 : Y_1 : Z_1)$ and $(X_2 : Y_2 : Z_2)$ represent the same point if and only if there exists $t \in \mathbb{K} \setminus \{0\}$ such that $X_1 = tX_2$, $Y_1 = t^2Y_2$ and $Z_1 = tZ_2$.

We now give the group law on the elliptic curve given by Eq. (8). From [11] (see also [7]), the sum of two points $(X_1 : Y_1 : Z_1)$ and $(X_2 : Y_2 : Z_2)$ is given by $(X_3 : Y_3 : Z_3)$ where

$$X_3 = X_1 Z_1 Y_2 + Y_1 X_2 Z_2,$$

$$\begin{aligned} Y_3 = &\left[(Z_1 Z_2)^2 + \epsilon(X_1 X_2)^2\right]\left[Y_1 Y_2 - 2\delta X_1 X_2 Z_1 Z_2\right] \\ &+ 2\epsilon X_1 X_2 Z_1 Z_2 (X_1^2 Z_2^2 + Z_1^2 X_2^2), \end{aligned} \qquad (11)$$

$$Z_3 = (Z_1 Z_2)^2 - \epsilon(X_1 X_2)^2.$$

The negation of a point $(X : Y : Z)$ on the (extended) Jacobi quartic (Eq. (8)) is given by $(-X : Y : Z)$.

To sum up, our unified method for adding points on an elliptic curve with a point of order 2 goes as follows. We first represent the given Weierstraß curve as a quartic given by Eq. (8) and transform points P_1 and P_2 according to Eq. (9). Next, given the two points on the corresponding quartic, $(X_1 : Y_1 : Z_1)$ and $(X_2 : Y_2 : Z_2)$, we apply the addition formula given by Eq. (11) and obtain $(X_3 : Y_3 : Z_3) = (X_1 : Y_1 : Z_1) + (X_2 : Y_2 : Z_2)$. Finally, we recover the sum as a point on the initial Weierstraß curve, $P_3 = P_1 + P_2$, by the transformation given by Eq. (10).

Figure 1 details the procedure for adding points $(X_1 : Y_1 : Z_1)$ and $(X_2 : Y_2 : Z_2)$. It appears that only **13** (field) **multiplications** (plus 3 multiplications by constants) are required. We insist that the same procedure equally applies for doubling a point. We further note that the neutral element is $(0 : 1 : 1)$ and that the procedure remains valid for it too.

When constant δ (resp. ϵ) is small, the cost of a multiplication by δ (resp. ϵ) can be neglected. A good choice consists in imposing a small value for ϵ since this removes two multiplications by constants – as shown in Fig. 1, there are 2 multiplications by ϵ and 1 multiplication by δ. In particular, most elliptic curves over the prime field $\mathbb{K} = \mathbb{F}_p$, with three points of order 2, can be rescaled

$$
\begin{array}{ll}
T_1 \leftarrow X_1; T_2 \leftarrow Y_1; T_3 \leftarrow Z_1; T_4 \leftarrow X_2; T_5 \leftarrow Y_2; T_6 \leftarrow Z_2 & \\
T_7 \leftarrow T_1 \cdot T_3 & (= X_1 Z_1) \\
T_7 \leftarrow T_2 + T_7 & (= X_1 Z_1 + Y_1) \\
T_8 \leftarrow T_4 \cdot T_6 & (= X_2 Z_2) \\
T_8 \leftarrow T_5 + T_8 & (= X_2 Z_2 + Y_2) \\
T_2 \leftarrow T_2 \cdot T_5 & (= Y_1 Y_2) \\
T_7 \leftarrow T_7 \cdot T_8 & (= X_3 + Y_1 Y_2 + X_1 X_2 Z_1 Z_2) \\
T_7 \leftarrow T_7 - T_2 & (= X_3 + X_1 X_2 Z_1 Z_2) \\
T_5 \leftarrow T_1 \cdot T_4 & (= X_1 X_2) \\
T_1 \leftarrow T_1 + T_3 & (= X_1 + Z_1) \\
T_8 \leftarrow T_3 \cdot T_6 & (= Z_1 Z_2) \\
T_4 \leftarrow T_4 + T_6 & (= X_2 + Z_2) \\
T_6 \leftarrow T_5 \cdot T_8 & (= X_1 X_2 Z_1 Z_2) \\
T_7 \leftarrow T_7 - T_6 & (= X_3) \\
T_1 \leftarrow T_1 \cdot T_4 & (= X_1 Z_2 + X_2 Z_1 + X_1 X_2 + Z_1 Z_2) \\
T_1 \leftarrow T_1 - T_5 & (= X_1 Z_2 + X_2 Z_1 + Z_1 Z_2) \\
T_1 \leftarrow T_1 - T_8 & (= X_1 Z_2 + X_2 Z_1) \\
T_3 \leftarrow T_1 \cdot T_1 & (= X_1^2 Z_2^2 + X_2^2 Z_1^2 + 2 X_1 X_2 Z_1 Z_2) \\
T_6 \leftarrow T_6 + T_6 & (= 2 X_1 X_2 Z_1 Z_2) \\
T_3 \leftarrow T_3 - T_6 & (= X_1^2 Z_2^2 + X_2^2 Z_1^2) \\
T_4 \leftarrow \epsilon \cdot T_6 & (= 2\epsilon X_1 X_2 Z_1 Z_2) \\
T_3 \leftarrow T_3 \cdot T_4 & (= 2\epsilon X_1 X_2 Z_1 Z_2 (X_1^2 Z_2^2 + X_2^2 Z_1^2)) \\
T_4 \leftarrow \delta \cdot T_6 & (= 2\delta X_1 X_2 Z_1 Z_2) \\
T_2 \leftarrow T_2 - T_4 & (= Y_1 Y_2 - 2\delta X_1 X_2 Z_1 Z_2) \\
T_4 \leftarrow T_8 \cdot T_8 & (= Z_1^2 Z_2^2) \\
T_8 \leftarrow T_5 \cdot T_5 & (= X_1^2 X_2^2) \\
T_8 \leftarrow \epsilon \cdot T_8 & (= \epsilon X_1^2 X_2^2) \\
T_5 \leftarrow T_4 + T_8 & (= Z_1^2 Z_2^2 + \epsilon X_1^2 X_2^2) \\
T_2 \leftarrow T_2 \cdot T_5 & (= (Z_1^2 Z_2^2 + \epsilon X_1^2 X_2^2)(Y_1 Y_2 - 2\delta X_1 X_2 Z_1 Z_2)) \\
T_2 \leftarrow T_2 + T_3 & (= Y_3) \\
T_5 \leftarrow T_4 - T_8 & (= Z_3) \\
X_3 \leftarrow T_7; Y_3 \leftarrow T_2; Z_3 \leftarrow T_5 &
\end{array}
$$

Fig. 1. Unified addition on an (extended) Jacobi quartic

to the case $\epsilon = 1$ as follows. Let $(\theta_1, 0)$, $(\theta_2, 0)$ and $(\theta_3, 0)$ denote the three points of order 2 on the Weierstraß curve, i.e., $y^2 = x^3 + ax + b = (x - \theta_1)(x - \theta_2)(x - \theta_3)$. Then, an application of map given by Eq. (9) with $\theta = \theta_1$ transforms the Weierstraß curve into the (extended) Jacobi curve

$$Y^2 = \epsilon X^4 - 2\delta X^2 Z^2 + Z^4 \qquad (12)$$

with $\delta = 3\theta_1/4$ and $\epsilon = -(3\theta_1^2 + 4a)/16 = (\theta_2 - \theta_3)^2/16$. If $p \equiv 3 \pmod 4$ then -1 is not a square modulo p and consequently either $(\theta_2 - \theta_3)$ or $-(\theta_2 - \theta_3)$ is a square modulo p. Letting ξ a square-root of the corresponding square $\pm(\theta_2 - \theta_3)$, it follows that $\epsilon = \xi^4/16 = (\xi/2)^4$. If $p \equiv 1 \pmod 4$ then there is a $1/8$ chance that we cannot find a pair of indices such that $(\theta_i - \theta_j)$ is a square modulo p. If such a pair exists, we let ξ denote the corresponding square-root and again, after a possible re-arrangement, we get $\epsilon = (\xi/2)^4$. The change of variable $X \leftarrow 2X/\xi$ then transforms the previous quartic into

$$Y^2 = X^4 - 2\rho\,X^2Z^2 + Z^4 \tag{13}$$

where $\rho = 4\delta/\xi^2$.

This latter case corresponds exactly to the curves considered by Liardet and Smart ([12]). The first advantage of using the quartic representation is that only 13 multiplications (in $\mathbb{F}_p$) plus 1 multiplication by constant ρ are required – the representation as the intersection of two quadrics requires 16 multiplications (in $\mathbb{F}_p$) plus 1 multiplication by constant k^2 (cf. § 2.1) – resulting in a $(\frac{16}{13}-1) \simeq 23\%$ speed improvement. The second advantage is that fewer memory resources are required since points are represented with triplets instead of quadruplets.

4   Conclusion

This paper revisited the Jacobi model initially suggested in [12] as a means for preventing side-channel attacks. Using an (extended) form of the Jacobi quartic, we derived a unified addition formula for adding or doubling points with only 13 field multiplications. This is the fastest known unified addition law for elliptic curves whose order is a multiple of 2.

References

1. Key Agreement Scheme OK-ECDH. Hitachi Ltd., 2001.
2. Digital Signature Scheme OK-ECDSA. Hitachi Ltd., 2001.
3. Éric Brier and Marc Joye. Weierstraß elliptic curves and side-channel attacks. In D. Naccache, editor, *Public Key Cryptography*, volume 2274 of *Lecture Notes in Computer Science*, pages 335–345. Springer-Verlag, 2002.
4. J.W.S. Cassels and E.V. Flynn. *Prolegomena to a middlebrow arithmetic of curves of genus 2*. Number 230 in London Mathematical Society, Lecture Notes Series. Cambridge Univ. Press, 2000.
5. D.V. Chudnovsky and G.V. Chudnovsky. Sequences of numbers generated by addition in formal groups and new primality and factorization tests. *Adv. Appl. Math.*, 7:385–434, 1986/87.
6. Jean-Sébastien Coron. Resistance against differential power analysis for elliptic curve cryptosystems. In Ç.K. Koç and C. Paar, editors, *Cryptographic Hardware and Embedded Systems (CHES '99)*, volume 1717 of *Lecture Notes in Computer Science*, pages 292–302. Springer-Verlag, 1999.
7. Jun-ichi Igusa. On the transformation theory of elliptic functions. *Amer. J. Math.*, 81:436–452, 1959.
8. Marc Joye and Jean-Jacques Quisquater. Hessian elliptic curves and side-channel attacks. In Ç.K. Koç, D. Naccache, and C. Paar, editors, *Cryptographic Hardware and Embedded Systems – CHES 2001*, volume 2162 of *Lecture Notes in Computer Science*, pages 402–410. Springer-Verlag, 2001.

9. Paul Kocher. Timing attacks on implementations of Diffie-Hellman, RSA, DSS, and other systems. In N. Koblitz, editor, *Advances in Cryptology – CRYPTO '96*, volume 1109 of *Lecture Notes in Computer Science*, pages 104–113. Springer-Verlag, 1996.

10. Paul Kocher, Joshua Jaffe, and Benjamin Jun. Differential power analysis. In M. Wiener, editor, *Advances in Cryptology – CRYPTO '99*, volume 1666 of *Lecture Notes in Computer Science*, pages 388–397. Springer-Verlag, 1999.

11. Peter S. Landweber. Supersingular elliptic curves and congruences for Legendre polynomials. In P.S. Landweber, editor, *Elliptic Curves and Modular Forms in Algebraic Topology*, volume 1326 of *Lecture Notes in Mathematics*, Springer-Verlag, 1988.

12. Pierre-Yvan Liardet and Nigel P. Smart. Preventing SPA/DPA in ECC systems using the Jacobi form. In Ç.K. Koç, D. Naccache, and C. Paar, editors, *Cryptographic Hardware and Embedded Systems – CHES 2001*, volume 2162 of *Lecture Notes in Computer Science*, pages 391–401. Springer-Verlag, 2001.

13. J.R. Merriman, S. Siksek, and N.P. Smart. Explicit 4-descents on an elliptic curve. *Acta Arith.*, 77(4):385–404, 1996.

14. Joseph H. Silverman. *The arithmetic of elliptic curves*, volume 106 of *Graduate Texts in Mathematics*. Springer-Verlag, 1986.

15. E.T. Whittaker and G.N. Watson. *A course of modern analysis*. Cambridge University Press, 4th edition, 1927.

A   Illustration

Here is an example of a cryptographic elliptic curve (i.e., the group order has a cofactor ≤ 4) over $\mathbb{F}_p$. This example is adapted from [12].

Let $p = 2^{192} - 2^{64} - 1$. Over $\mathbb{F}_p$, consider the Weierstraß elliptic curve E given by

$$E_{/\mathbb{F}_p} : y^2 = x^3 - 3x + b \quad (\cup \{O\}) \tag{14}$$

where $b = 5785156510951660859948362664535565676137370865272662811849$. The order of E is four times a prime:

$$\#E = 4 \cdot 1569275433846670190958947355830249374250393459078477724241 .$$

The three points of order 2 on the Weierstraß curve are $(\theta_i, 0)$ with

$$\begin{cases} \theta_1 = 3931134103214925937592361744683965239873651308020133387956 \\ \theta_2 = 3722240065524459449962883383651126589463273788373166826730 \\ \theta_3 = 2161748259540728720113669865088143302633269781215144746593 \end{cases} .$$

An application of transformation given by Eq. (9) shows that the Weierstraß curve E is equivalent to the Jacobi curve

$$Y^2 = \epsilon X^4 - 2\delta X^2 Z^2 + Z^4$$

with $\epsilon = 4392384375834284450995086699732976092557230326145055776522$ and $\delta = 2948350577411194453194271308512973929905238481015100409667$. Further, since there are three points of order 2, ϵ can be rescaled to the case $\epsilon = 1$ via the additional transformation $X \leftarrow 2X/\xi$ with

$$\xi = \sqrt{\theta_2 - \theta_3}$$
$$= 23623242405095704049612218239456174797431133842158 29517748 \,.$$

Consequently, the Weierstraß curve E (Eq. (14)) is also equivalent to the Jacobi curve

$$Y^2 = X^4 - 2\rho\, X^2 Z^2 + Z^4 \qquad (15)$$

with $\rho = 45135350573494704539962104900207506134698581607568 52710254$ via the map

$$\begin{cases} (\theta_1, 0) \longmapsto (0 : -1 : 1) \\ (x, y) \longmapsto \big(\xi(x - \theta_1) : (2x + \theta_1)(x - \theta_1)^2 - y^2 : y\big) \\ \boldsymbol{O} \longmapsto (0 : 1 : 1) \end{cases}$$

and conversely, the Jacobi curve (Eq. (15)) is equivalent to the initial Weierstraß curve (Eq. (14)) via the map

$$\begin{cases} (0 : 1 : 1) \longmapsto \boldsymbol{O} \\ (0 : -1 : 1) \longmapsto (\theta_1, 0) \\ (X : Y : Z) \longmapsto \left(\dfrac{\xi^2(Y + Z^2)}{2X^2} - \dfrac{\theta_1}{2}, Z\xi\, \dfrac{\xi^2(Y + Z^2) - 3\theta_1\, X^2}{2X^3} \right) \end{cases} \,.$$

Fast Point Multiplication on Elliptic Curves through Isogenies

Eric Brier and Marc Joye

Gemplus Card International, Card Security Group
La Vigie, Av. du Jujubier, ZI Athélia IV, 13705 La Ciotat Cedex, France
{eric.brier,marc.joye}@gemplus.com
http://www.gemplus.com/smart/
http://www.geocities.com/MarcJoye/

Abstract. Elliptic curve cryptosystems are usually implemented over fields of characteristic two or over (large) prime fields. For large prime fields, projective coordinates are more suitable as they reduce the computational workload in a point multiplication. In this case, choosing for parameter a the value -3 further reduces the workload. Over $\mathbb{F}_p$, not all elliptic curves can be rescaled through isomorphisms to the case $a = -3$. This paper suggests the use of the more general notion of isogenies to rescale the curve. As a side result, this also illustrates that selecting elliptic curves with $a = -3$ (as those recommended in most standards) is not restrictive.

Keywords: elliptic curves, scalar multiplication, isogenies, cryptography.

1 Introduction

Elliptic curves are plane curves defined by a polynomial equation having strong algebraic properties. In particular, it is possible to define an addition on points which yields a group structure. Furthermore, no sub-exponential algorithm is known to solve the Discrete Logarithm in the induced group.

From a practical viewpoint, fast addition formulæ are to be defined for efficient protocols using elliptic curve cryptography. As will be described in the next section, the case where elliptic curve parameter a is equal to -3 allows faster computation. Unfortunately, one cannot always obtain this value using the classical notion of isomorphism (though, according to the definition field, the probability is $1/2$ or $1/4$).

The aim of this paper is to show that it is possible to obtain the desired value $a = -3$ for an *isogenous* elliptic curve and to perform computations on this curve rather than on the original one. Being isogenous, both curves have the same number of rational points and mappings between curves (called *isogenies*) allow to relate point multiplication in both groups.

The rest of this paper is organized as follows. The next section reviews the addition formulæ on elliptic curves. Section 3 explains how, in some cases, isomorphisms of curves may speed up the scalar multiplication. This idea is generalized and extended in Sect. 4 through the use of isogenies. A direct application

M. Fossorier, T. Hoeholdt, and A. Poli (Eds.): AAECC 2003, LNCS 2643, pp. 43–50, 2003.

to elliptic curve cryptography is given in Sect. 5. Finally, Sect. 6 concludes the paper. (A concrete example of our technique is given in Appendix A.)

2   Elliptic Curve Arithmetic

Let $\mathbb{K}$ be a field with $\operatorname{Char} \mathbb{K} \neq 2, 3$. An *elliptic curve E over $\mathbb{K}$* is the set of points $(x, y) \in \mathbb{K} \times \mathbb{K}$ satisfying the Weierstraß equation

$$E_{/\mathbb{K}} : y^2 = x^3 + ax + b \tag{1}$$

along with the *point at infinity $\mathbf{O}$*. If this set is equipped with the so-called "chord-and-tangent" rule, it becomes an abelian group [Sil86, III.2].

We use the additive notation. The point at infinity is the neutral element, $\mathbf{P} + \mathbf{O} = \mathbf{O} + \mathbf{P} = \mathbf{P}$. For two points $\mathbf{P} = (x_0, y_0)$ and $\mathbf{Q} = (x_1, y_1)$ with $\mathbf{P} \neq -\mathbf{Q}$, their sum $\mathbf{R} = \mathbf{P} + \mathbf{Q} = (x_2, y_2)$ is given by

$$x_2 = \lambda^2 - x_0 - x_1 \quad \text{and} \quad y_2 = (x_1 - x_2)\lambda - y_1$$

where $\lambda = (y_0 - y_1)/(x_0 - x_1)$ when $\mathbf{P} \neq \mathbf{Q}$ and $\lambda = (3x_1^2 + a)/(2y_1)$ otherwise.

The above formulæ require an inversion (in $\mathbb{K}$), a usually costly operation, especially when $\mathbb{K}$ is a large prime field. For this reason, projective coordinates may be preferred. Within (Jacobian) projective coordinates [IEEE], the representation of a point is not unique, the triplets $(v^2 X : v^3 Y : vZ)$ for any $v \in \mathbb{K} \setminus \{0\}$ all represent the same point. The correspondence of $\mathbf{P} = (X_0 : Y_0 : Z_0)$ with its affine coordinates is given by $\mathbf{P} = (x_0, y_0)$ where $x_0 = X_0/Z_0^2$ and $y_0 = Y_0/Z_0^3$ if $Z_0 \neq 0$, and $\mathbf{P} = \mathbf{O}$ if $Z_0 = 0$. The addition formulæ of points $\mathbf{P} = (X_0 : Y_0 : Z_0)$ and $\mathbf{Q} = (X_1 : Y_1 : Z_1)$ (with $\mathbf{P} \neq -\mathbf{Q}$ and $Z_0, Z_1 \neq 0$) then become $\mathbf{R} = (X_2 : Y_2 : Z_2)$ where

$$\begin{cases} X_2 = R^2 - TW^2 \\ 2Y_2 = VR - MW^3 \quad \text{when } \mathbf{P} \neq \mathbf{Q} \\ Z_2 = Z_0 Z_1 W \end{cases} \tag{2}$$

with $U_0 = X_0 Z_1^2$, $U_1 = X_1 Z_0^2$, $S_0 = Y_0 Z_1^3$, $S_1 = Y_1 Z_0^3$, $W = U_0 - U_1$, $R = S_0 - S_1$, $T = U_0 + U_1$, $M = S_0 + S_1$, and $V = TW^2 - 2X_2$, and

$$\begin{cases} X_2 = M^2 - 2S \\ Y_2 = M(S - X_2) - T \quad \text{when } \mathbf{P} = \mathbf{Q} \\ Z_2 = 2Y_1 Z_1 \end{cases} \tag{3}$$

with $M = 3X_1^2 + a Z_1^4$, $S = 4X_1 Y_1^2$, and $T = 8Y_1^4$.

We see that the addition of two (different) points requires 16 multiplications and only 11 when $Z_1 = 1$. The doubling of a point requires 10 multiplications, including the multiplication by the parameter a. When a is small, this latter multiplication can be neglected. The value $a = -3$ is particularly attractive since then $M = 3(X_1 - Z_1{}^2)(X_1 + Z_1{}^2)$ and so only 8 multiplications are required to double a point. The same conclusion holds when $a = -3c^2$ for a small c (e.g., $a = -12$).

Another useful value for a is $a = 0$ since then the number of required multiplications decreases to 7. This case is not studied here because choosing $a = 0$ has too many implications on the endomorphism ring of the curve, which could decrease security (though no algorithm using this property is known today).

3   Isomorphisms

Two elliptic curves E and E', respectively given by the Weierstraß equations $E_{/\mathbb{K}} : y^2 = x^3 + ax + b$ and $E'_{/\mathbb{K}} : y^2 = x^3 + a'x + b'$, are *isomorphic over* $\mathbb{K}$ if and only if there exists a nonzero element $u \in \mathbb{K}$ such that $u^4 a' = a$ and $u^6 b' = b$. Moreover, the isomorphism is given by

$$\phi : E \xrightarrow{\sim} E', \begin{cases} (x, y) \longmapsto (u^{-2}x, u^{-3}y) \\ O \longmapsto O \end{cases} .$$

The elliptic curve $E_{/\mathbb{K}} : y^2 = x^3 + ax + b$ can thus be made isomorphic to the elliptic curve $E'_{/\mathbb{K}} : y^2 = x^3 - 3x + b'$ if and only if $a = -3u^4$ for some $u \in \mathbb{K} \setminus \{0\}$. When $\mathbb{K} = \mathbb{F}_p$, a (large) prime field, this occurs roughly with probability $1/2$ when $p \equiv 3 \pmod{4}$ and with probability $1/4$ when $p \equiv 1 \pmod{4}$. Consequently, there is a non-negligible probability that a *random* elliptic curve cannot be rescaled to the interesting[1] case $a = -3$. The next section investigates an alternative solution to overcome this limitation through the use of isogenies.

Note that we do not consider the more general case $a = -3c^2$ with c small. In order to make easier point compression/decompression, elliptic curve cryptosystems are usually defined over $\mathbb{F}_p$ with $p \equiv 3 \pmod{4}$. In that case, all elliptic curves that can be rescaled to $a = -3c^2$ can also be rescaled to $a = -3$, independently of the value of c.

4   Isogenies

An *isogeny* between two elliptic curves E and E' defined over $\mathbb{K}$ is a non-constant[2] morphism $\phi : E \to E'$. The *degree of isogeny* ϕ is defined to be

$$\deg \phi = [\overline{\mathbb{K}}(E) : \phi^* \overline{\mathbb{K}}(E')]$$

where $\phi^* : \overline{\mathbb{K}}(E') \to \overline{\mathbb{K}}(E), f \mapsto \phi^*(f) = f \circ \phi$ denotes the map induced by ϕ. (Remark that an isogeny of degree 1 is an isomorphism.)

---

[1] i.e., suitable for fast implementations; see Sect. 2.
[2] We do not consider the zero isogeny $\phi = [0]$.

A useful result is that for every isogeny $\phi : E \to E'$, there exists a unique isogeny $\hat{\phi} : E' \to E$, called the *dual isogeny* [Sil86, III.4], such that

$$\hat{\phi} \circ \phi = [m] \quad \text{and} \quad \phi \circ \hat{\phi} = [m]'$$

where $m = \deg \phi$ and $[m]$ (resp. $[m]'$) is the multiplication-by-m isogeny on E (resp. E'). Interestingly, this leads to a different way for computing $\boldsymbol{Q} = [rm]\boldsymbol{P}$ as $\boldsymbol{Q} = \hat{\phi}([r]'\phi(\boldsymbol{P}))$.

$$
\begin{array}{ccc}
\boldsymbol{P} \in E(\mathbb{K}) & \xrightarrow{\ [rm]\ } & \boldsymbol{Q} = [rm]\boldsymbol{P} \in E(\mathbb{K}) \\
\phi \downarrow & & \uparrow \hat{\phi} \\
\boldsymbol{P}' \in E'(\mathbb{K}) & \xrightarrow{\ [r]'\ } & \boldsymbol{Q}' = [r]'\boldsymbol{P}' \in E'(\mathbb{K})
\end{array}
$$

Fig. 1. Computing $\boldsymbol{Q} = [rm]\boldsymbol{P}$ through isogenies

Isogenies have been intensively studied in order to improve point counting algorithms. What we are interested in is to build an isogeny of small degree. We know that we can find an isogeny ϕ of degree m from E to a curve E' if and only if the equation

$$\Phi_m(j, X) = 0$$

where Φ_m is the m-th modular polynomial and j is the j-invariant of the curve, has a rational solution. If so, we can follow the method described in [BSS99, pp. 126–130] to find the isogenous curve equation. We check that the new curve is isomorphic to a curve with parameter $a = -3$. It then remains to compute the isogeny itself. An algorithm for producing the isogeny is presented in [CM94].

5   Application to Cryptography

The basic operation of elliptic curve cryptosystems is the point multiplication: given a point $\boldsymbol{P} = (x_1, y_1) \in E(\mathbb{K})$, one has to compute $\boldsymbol{Q} = [k]\boldsymbol{P} = (x_k, y_k)$ for some $1 \le k < \mathrm{ord}_E \, \boldsymbol{P}$. Assume that the definition field is $\mathbb{F}_p$ where p is a large prime. We have seen in Sect. 2 that in this case an elliptic curve with parameter $a = -3$ yields a point multiplication substantially faster when working within projective coordinates. We compute $\boldsymbol{Q} = [k]\boldsymbol{P}$ as $(X_k : Y_k : Z_k) = [k](x_1 : y_1 : 1)$ and then $(x_k, y_k) = (X_k/Z_k{}^2, Y_k/Z_k{}^3)$ with only 8 multiplications (in $\mathbb{F}_p$) per doubling. When E has not parameter $a = -3$ (or cannot be reduced to this case through isomorphism) then we can apply the above methodology.

Let ϕ denote an isogeny of degree m between the elliptic curves $E_{/\mathbb{F}_p} : y^2 = x^3 + ax + b$ and $E'_{/\mathbb{F}_p} : y^2 = x^3 - 3x + b'$. Since, for security reasons, point $\boldsymbol{P}$ has large prime order, we may assume w.l.o.g. that $\gcd(m, \mathrm{ord}_E \, \boldsymbol{P}) = 1$ and so m is invertible modulo $\mathrm{ord}_E \, \boldsymbol{P}$. We define $k_m \equiv k/m \pmod{\mathrm{ord}_E \, \boldsymbol{P}}$. Hence, we can obtain $\boldsymbol{Q} = [k]\boldsymbol{P}$ according to

$$\boldsymbol{Q} = \hat{\phi}([k_m]'\phi(\boldsymbol{P})) \ . \tag{4}$$

Example 1. We give a "toy" example to illustrate the technique. A concrete example (i.e., with cryptographic size) can be found in appendix.

Over the field $\mathbb{F}_{149}$, we define the elliptic curves

$$E_{/\mathbb{F}_{149}} : y^2 = x^3 + x + 113$$

and

$$E'_{/\mathbb{F}_{149}} : y^2 = x^3 - 3x - 14$$

which are isogenous via the maps

$$\varphi : E \longrightarrow E'$$
$$(x, y) \longmapsto \left(\frac{x^5+4x^4+99x^3+42x^2+99x+49}{(17x^2+34x+50)^2}, y \cdot \frac{x^6+6x^5+107x^4+126x^3+112x^2+116x+139}{(17x^2+34x+50)^3} \right)$$

$$\hat{\varphi} : E' \longrightarrow E$$
$$(x, y) \longmapsto \left(\frac{x^5+85x^4+60x^3+137x^2+26x+95}{(123x^2+87x+86)^2}, y \cdot \frac{x^6+53x^5+134x^4+74x^3+106x^2+50x+34}{(123x^2+87x+86)^3} \right)$$

Choosing a *random* point $\boldsymbol{P} = (24, 36)$ on the curve E, we have

$$\varphi(\boldsymbol{P}) \ \ = (115, 57) \in E'$$
$$\hat{\varphi}(\varphi(\boldsymbol{P})) = (13, 104) \in E$$

whereby it is easily checked that

$$\hat{\varphi}(\varphi(\boldsymbol{P})) = [5]\boldsymbol{P} \ .$$

6 Concluding Remarks

The first consequence of our work on isogenies is that computing a point multiplication can in most of cases be made using a curve with $a = -3$ even when such a value cannot be rescaled directly through isomorphism. This leads to a faster point multiplication.

The second consequence is that when choosing a random curve, one can restrict oneself to curves with parameter $a = -3$ and a random value for parameter b. This follows from the observation that most curves are mapped to a curve with $a = -3$ by an isogeny of small degree. The Discrete Logarithm Problem on the isogenous curve is then as hard as on the original curve.

This paper can be seen as a justification to the fact that most curves recommended in cryptographic standards use for parameter a the value -3.

References

[BSS99] Ian Blake, Gadiel Seroussi, and Nigel Smart. *Elliptic Curves in Cryptography*, volume 265 of *London Mathematical Society Lecture Note Series*. Cambridge University Press, 1999.

[CM94] Jean-Marc Couveignes and François Morain. Schoof's algorithm and isogeny cycles. *Algorithmic Number Theory (ANTS-I)*, volume 877 of *Lecture Notes in Computer Science*, pages 43–58. Springer-Verlag, 1994.

[Coh93] Henri Cohen. *A course in computational algebraic number theory*, volume 138 of *Graduate Texts in Mathematics*. Springer-Verlag, 1993.

[IEEE] IEEE Computer Society. IEEE standard specifications for public-key cryptography. IEEE Std 1363-2000, 2000.

[NIST] National Institute of Standards and Technology (NIST). Digital signature standard (DSS). FIPS PUB 186-2, 2000.

[SECG] Certicom Research. Standards for efficient cryptography. Version 1.0, 2000. Available at url http://www.secg.org/.

[Sil86] Joseph H. Silverman. *The arithmetic of elliptic curves*, volume 106 of *Graduate Texts in Mathematics*. Springer-Verlag, 1986.

A   A Concrete Example

The following example is taken from [BSS99]. The curve equation is

$$E_{/\mathbb{F}_p} : y^2 = x^3 + ax + b$$

with $\begin{cases} p = 2^{160} + 7 \\ a = 1 \\ b = 1010685925500572430206879608558642904226772615919 \\ \#E(\mathbb{F}_p) = 1461501637330902918203683038630093524408650319587 \end{cases}$.

[It should be noted that this curve is not isomorphic to a curve with parameter $a = -3$.]

This curve is isogenous to

$$E'_{/\mathbb{F}_p} : y^2 = x^3 - 3x + b'$$

with $b' = 632739926637917759594186681013274896520567575517$. The isogeny, whose degree is $m = 11$, is given by

$$\phi(x, y) = \left(\frac{u(x)}{w(x)^2}, y \cdot \frac{v(x)}{w(x)^3} \right)$$

where

$u(x) =$ 370690178134646041774135216324801714060106373562 +
86125631288803937543929609098811288659745650 3254 $x-$
710816068955948441247024728782103926686793729401 x^2-
660949999632736745607557761155749317154796231190 x^3+
104899805475825460399317568666533314730964676 7733 x^4-
682987121612996351004314311578456905260387427374 x^5-
540119027692075395022771129540214285068704207535 x^6+
796617542631289284086426144833751968835133978328 x^7+
563704518387054717437229862610850421868057177503 x^8-
424589103609629859400080142081400291591066256566 x^9+
876425480231036355075553669797171429349630674 30 $x^{10} + x^{11}$

$$
\begin{aligned}
v(x) = \;& 452495052944959116984585216749653914270910889527 \;+ \\
& 10693805604210560392041543215707500924556555960036 \; x+ \\
& 49633669527287210061590288590396941307155703319 \; x^2+ \\
& 7187917972982414851544661602618376504731328059 18 \; x^3+ \\
& 89347021676196991929966264352461222669954368759 4 \; x^4+ \\
& 95360109982590787180111885348013732359702675720 5 \; x^5+ \\
& 18886633622622298256790036603277695633740632445 0 \; x^6+ \\
& 43273774401193542010990857967967390374744414016 5 \; x^7+ \\
& 71042908863420436180409854193276857166599029914 8 \; x^8+ \\
& 12541374448563679881807060652928411368799099742 00 \; x^9+ \\
& 13619316715238473832049787660552099206554388341 06 \; x^{10}+ \\
& 22417483812474951414444834184963605685267920210 5 \; x^{11}+ \\
& 82258018006544282440002513632388213536827521732 \; x^{12}+ \\
& 86813934947230318852637374208533468404004929431 0 \; x^{13}+ \\
& 13146382203465545326133330504695757144024446011 45 \; x^{14} + x^{15}
\end{aligned}
$$

$$
\begin{aligned}
w(x) = \;& 32501987297990211150224018758786187104987287944 8 \;+ \\
& 13639370876451546179627903630561137999113723953 17 \; x+ \\
& 14119210182702613277972456182143505842794942380 51 \; x^2+ \\
& 70237665636693548622727735807424896198713030751 \; x^3+ \\
& 54041111397077468645890073148187670715073826739 0 \; x^4+ \\
& 18569674346155743905322181970978683599028319327 4 \; x^5
\end{aligned}
$$

and the dual isogeny is given by

$$
\hat{\phi}(x,y) = \left(\frac{\hat{u}(x)}{\hat{w}(x)^2}, \, y \cdot \frac{\hat{v}(x)}{\hat{w}(x)^3} \right)
$$

where

$$
\begin{aligned}
\hat{u}(x) = \;& 753112556937953823969300906862974140977871919143 \;+ \\
& 65809795177082413448982034865179786648563343729 4 \; x- \\
& 35183501384647667442846092389667749196892404625 7 \; x^2+ \\
& 25545246903672634894298841107332021021490857074 6 \; x^3- \\
& 18334420650014871229322739744326227560447380518 3 \; x^4+ \\
& 12575220850464771436402576110126009544585842516 34 \; x^5- \\
& 69024973193875784615274350969816794845898829972 4 \; x^6- \\
& 25965441700829881061329929468284859568810501028 \; x^7- \\
& 18567603955952416315987761166471180706572838123 6 \; x^8- \\
& 71626255671962842060388116547325332721478308415 8 \; x^9+ \\
& 10603481441742315326224603282062454269712931689 00 \; x^{10} + x^{11}
\end{aligned}
$$

$$\hat{v}(x) = 10670650273334049173710210327636668506972469630529 +$$
$$88336147979691530262079848987066256632327220 9634 \; x+$$
$$14735431332801354055617376300872602776018 6777184 \; x^2+$$
$$36807803650978825404154106770624195670055714497 \; x^3+$$
$$1165987233874930893186423892075380266512707088792 \; x^4+$$
$$13100365377373768479504240645795992122635 4724875 \; x^5+$$
$$12619205599792254278982225209950786887046 37834874 \; x^6+$$
$$17572904999035091818652865125825560159465 3396176 \; x^7+$$
$$8974057283078431908061484250040246497450 1364756 \; x^8+$$
$$13997158606865578615711722314515170921022 43078837 \; x^9+$$
$$12461925894789259757038393598014144929255 49285355 \; x^{10}+$$
$$6160267099924948800624474674093262102741 52875856 \; x^{11}+$$
$$428814934564835486109414410124450806255556418100 \; x^{12}+$$
$$24129716622610568822976745072532539144603 5097070 \; x^{13}+$$
$$12902057893044438073000565959308512080100 7210367 \; x^{14} + x^{15}$$

$$\hat{w}(x) = \; 135333262963423915607144923995277271284729908723 +$$
$$85973534686132758165705716586297948633659 6776258 \; x+$$
$$12939537751897638699542295601641576491598 24538936 \; x^2+$$
$$87954883075573755420722759559033630949224 1550000 \; x^3+$$
$$84018748623545236142884208513422519397400 3541259 \; x^4+$$
$$78457029243813111560519225078811093846232 2411742 \; x^5$$

Interpolation of the Elliptic Curve
Diffie–Hellman Mapping

Tanja Lange[1] and Arne Winterhof[2]

[1] Information Security and Cryptology, Ruhr-University of Bochum
Universitätsstr. 150, 44780 Bochum, Germany
`lange@itsc.ruhr-uni-bochum.de`
[2] Temasek Laboratories, National University of Singapore
10 Kent Ridge Crescent, Singapore 119260, Republic of Singapore
`tslwa@nus.edu.sg`

Abstract. We prove lower bounds on the degree of polynomials interpolating the Diffie–Hellman mapping for elliptic curves over finite fields and some related mappings including the discrete logarithm. Our results support the assumption that the elliptic curve Diffie–Hellman key exchange and related cryptosystems are secure.

1 Introduction

Computationally difficult problems as the discrete logarithm problem or the Diffie–Hellman problem play an important role in cryptology. For example the Diffie–Hellman key exchange and the ElGamal encryption (see e. g. [11]) are based on the fact that no representation of the discrete logarithm or the Diffie–Hellman mapping as an easily computable function is known. Evidently, solving the discrete logarithm problem suffices to solve the Diffie–Hellman problem. On the other hand for many groups Maurer and Wolf [8] presented a technique to reduce efficiently the discrete logarithm problem to the Diffie–Hellman problem. For a recent survey on the Diffie–Hellman problem see [9] and for surveys on the discrete logarithm see e. g. [3,11]. Initially, the discrete logarithm and the Diffie–Hellman mapping were suggested for use in practice for the multiplicative group of a finite field. Recently, several lower bounds on the complexity of functions representing the discrete logarithm or the Diffie–Hellman mapping in finite fields have been obtained (e. g. [4,15]), including lower bounds on the degree of interpolation polynomials of functions related to the discrete logarithm or Diffie–Hellman mapping. Subexponential algorithms for solving the discrete logarithm problem in finite fields are known (see e. g. [11]) which motivates considering other groups. An alternative used in practice is the group of points on an elliptic curve over a finite field suggested independently by Koblitz [5] and Miller [12]. The present article deals with the Diffie–Hellman problem and related mappings for elliptic curves.

Let E be an elliptic curve over the finite field $\mathbb{F}_q$ defined by the Weierstraß equation

$$E : Y^2 + h(X)Y = f(X)$$

M. Fossorier, T. Hoeholdt, and A. Poli (Eds.): AAECC 2003, LNCS 2643, pp. 51–60, 2003.

with a linear polynomial

$$h(X) = a_1 X + a_3, \quad a_1, a_3 \in \mathbb{F}_q,$$

and a cubic polynomial

$$f(X) = X^3 + a_2 X^2 + a_4 X + a_6, \quad a_2, a_4, a_6 \in \mathbb{F}_q,$$

such that over the algebraic closure $\overline{\mathbb{F}_q}$ there are no solutions $(x, y) \in \overline{\mathbb{F}_q}^2$ simultaneously satisfying the equations

$$y^2 + h(x)y = f(x), \quad 2y + h(x) = 0, \quad \text{and} \quad h'(x)y = f'(x).$$

In odd characteristic we may assume $Y^2 = f(X)$ and in even characteristic $h(X) = X$ or $h(X) = 1$. The latter case corresponds to supersingular curves for which the discrete logarithm problem, and thus the Diffie–Hellman problem, is easier due to the attack of Menezes, Okamoto, and Vanstone [10].

We denote by $\mathcal{O}$ the point at infinity. Let $P \neq \mathcal{O}$ be a point of order l on E. The *Diffie–Hellman problem* for the group G generated by P is the following: Given points nP and mP on E for some $1 \leq n, m \leq l - 1$ find the point nmP without knowing n and m. Since

$$2nmP = (n + m)^2 P - n^2 P - m^2 P$$

and bisecting (i. e. finding square roots in G if G is multiplicatively written) can be efficiently managed (see e. g. [1, Chapter 7.1]), we may consider the mapping on E defined by

$$F_0(nP) = n^2 P, \quad 1 \leq n \leq l - 1.$$

For given x the second coordinate y of a point $(x, y) \in \mathbb{F}_q^2$ on E can be easily determined up to two possibilities, y and $-y - h(x)$. Hence, we may consider a univariate mapping with the property

$$F_1(x_n) = x_{n^2}, \quad n \not\equiv 0 \bmod l,$$

where $nP = (x_n, y_n)$. Note that $x_n = x_k$ if $n \equiv \pm k \bmod l$. We prove lower bounds on the degree of interpolation polynomials of F_1 in Sect. 3. Moreover, we investigate interpolation polynomials of the bivariate mapping

$$F_2(x_n, x_m) = x_{nm}, \quad n, m \not\equiv 0 \bmod l,$$

in Sect. 4. Finally, we consider the following mapping, which is closely related to the discrete logarithm, in Sect. 5:

$$F_3(x_n) = \xi_n, \quad 1 \leq n \leq \lfloor l/2 \rfloor,$$

where

$$\xi_n = n_1 \beta_1 + n_2 \beta_2 + \ldots + n_r \beta_r \tag{1}$$

if

$$n = n_1 + n_2 p + \ldots + n_r p^{r-1}, \quad 0 \leq n_1, n_2, \ldots, n_r \leq p - 1,$$

for some fixed basis $\{\beta_1, \beta_2, \ldots, \beta_r\}$ of $\mathbb{F}_q$ over $\mathbb{F}_p$, $q = p^r$, and p is a prime. Similar results on F_3 were already proven in [7]. However, in this paper we consider much sparser sets of values of n, as well.

2   Preliminaries on Division Polynomials

In this section we recall some basic facts on division polynomials (see e.g. [2, 6,14,16]). We put $b_2 = a_1^2 + 4a_2$, $b_4 = a_1a_3 + 2a_4$, $b_6 = a_3^2 + 4a_6$, and $b_8 = a_1^2 a_6 + 4a_2 a_6 - a_1 a_3 a_4 + a_2 a_3^2 - a_4^2$.

The *division polynomials* $\psi_m(X, Y) \in \mathbb{F}_q[X, Y]/(Y^2 + h(X)Y - f(X))$, $m \geq 0$, are recursively defined by

$$\psi_0 = 0,$$
$$\psi_1 = 1,$$
$$\psi_2 = 2Y + h(X),$$
$$\psi_3 = 3X^4 + b_2 X^3 + 3b_4 X^2 + 3b_6 X + b_8,$$
$$\psi_4 = (2X^6 + b_2 X^5 + 5b_4 X^4 + 10b_6 X^3 + 10b_8 X^2 + (b_2 b_8 - b_4 b_6)X$$
$$+ (b_4 b_8 - b_6^2))\psi_2,$$
$$\psi_{2k+1} = \psi_{k+2}\psi_k^3 - \psi_{k-1}\psi_{k+1}^3, \quad k \geq 2,$$
$$\psi_{2k} = \psi_k(\psi_{k+2}\psi_{k-1}^2 - \psi_{k-2}\psi_{k+1}^2)/\psi_2, \quad k \geq 3,$$

where ψ_m is an abbreviation for $\psi_m(X,Y)$. If q is even or both q and m are odd then $\psi_m(X,Y) \in \mathbb{F}_q[X]$ is univariate and $\psi_m(X,Y) \in \psi_2(X,Y)\mathbb{F}_q[X]$ if q is odd and m is even. Therefore, as $\psi_2^2(X,Y) = 4f(X) + h^2(X)$, we have $\psi_m^2(X,Y), \psi_{m-1}(X,Y)\psi_{m+1}(X,Y) \in \mathbb{F}_q[X]$. In particular, we may write $\psi_{2k+1}(X)$ and $\psi_m^2(X)$.

The division polynomials can be used to state multiples of a point. Let $P = (x, y) \neq \mathcal{O}$, then the first coordinate of mP is given by

$$\frac{\theta_m(x)}{\psi_m^2(x)}, \quad \text{where } \theta_m(X) = X\psi_m^2 - \psi_{m-1}\psi_{m+1}.$$

The zeros of the denominator $\psi_m^2(X)$ are exactly the first coordinates of the nontrivial m-torsion points, i.e., the points $Q = (x, y) \in \overline{\mathbb{F}_q}^2$ on E with $mQ = \mathcal{O}$. Note, that these points occur in pairs $Q = (x, y)$ and $-Q = (x, -h(x) - y)$, which coincide only if $2Q = \mathcal{O}$, i.e., if x is a zero of $\psi_2^2(X)$.

We recall that the group of m-torsion points $E[m]$, for an elliptic curve E defined over a field of characteristic p, is isomorphic to $(\mathbb{Z}/m\mathbb{Z})^2$ if $p \nmid m$ and to a proper subgroup of $(\mathbb{Z}/m\mathbb{Z})^2$ if $p \mid m$. If m is a power of p then $E[m]$ is either isomorphic to $(\mathbb{Z}/m\mathbb{Z})$ or to $\{\mathcal{O}\}$. Accordingly, the degree of $\psi_m^2(X)$ is $m^2 - 1$ if $p \nmid m$ and strictly less than $m^2 - 1$ otherwise. In particular, for $p = 2$ and m a power of 2 we have $\deg(\psi_m^2) = m - 1$ if E is not supersingular and $\deg(\psi_m^2) = 0$ otherwise. By induction one can show that $\theta_m(X) \in \mathbb{F}_q[X]$ is monic of degree m^2.

3   Univariate Interpolation of the Diffie–Hellman Mapping

In this section we prove results on interpolation polynomials of F_1. First we consider large sets of given data.

Theorem 1. *Let E be an elliptic curve over $\mathbb{F}_q$ and $P \neq \mathcal{O}$ a point on E of order l divisible by a prime at least 11. For $1 \leq n \leq l-1$ let the first coordinate of nP be denoted by x_n and let S be a subset of $\{1 \leq n \leq l-1 : n^2 \not\equiv 0 \bmod l\}$ of cardinality $l-1-s$. Let $F(X) \in \mathbb{F}_q[X]$ satisfy*

$$F(x_n) = x_{n^2}, \quad n \in S.$$

Then we have

$$\deg(F) \geq \frac{l - 3s - 2}{38}.$$

Proof. Since otherwise the result is trivial we may assume that $s \leq l/3 - 1$. Then there exist $n, m \in S$ with $n^2 \not\equiv \pm m^2 \bmod l$, i.e., $x_{n^2} \neq x_{m^2}$ and $F(X)$ is not constant. The subset R of S defined by

$$R = \{n \in S : (2n \bmod l) \in S\}$$

has cardinality at least

$$|R| \geq |S| - 2s - 1 = l - 3s - 2.$$

Put $d = \deg(F)$. Using the polynomials $\psi_n^2(X)$ and $\theta_n(X)$ defined in the previous section we get for $n \in R$,

$$F\left(\frac{\theta_2(x_n)}{\psi_2^2(x_n)}\right) = F(x_{2n}) = x_{4n^2} = \frac{\theta_4(x_{n^2})}{\psi_4^2(x_{n^2})} = \frac{\theta_4(F(x_n))}{\psi_4^2(F(x_n))}.$$

Finally, we consider the polynomial

$$U(X) = H(X)(U_1(X) - U_2(X)),$$

with

$$H(X) = \psi_4^2(F(X))\psi_2^{2d}(X),$$

$$U_1(X) = F\left(\frac{\theta_2(X)}{\psi_2^2(X)}\right), \quad \text{and} \quad U_2(X) = \frac{\theta_4(F(X))}{\psi_4^2(F(X))}.$$

For even characteristic the degrees of $H(X)U_1(X)$ and $H(X)U_2(X)$ are different and $U(X)$ is not the zero polynomial. The degree of $U(X)$ is $18d$ if $a_1 \neq 0$, i.e., if E is non-supersingular, and $16d$ otherwise.

Now we consider the case of odd characteristic. There are exactly three distinct zeros $\gamma_1, \gamma_2, \gamma_3 \in \overline{\mathbb{F}_q}$ of $\psi_2^2(X)$. The polynomial $\psi_4^2(X)$ has nine distinct zeros. For a zero β of $\psi_4^2(X)$ let α be a solution of $F(\alpha) = \beta$. Since two α belonging to different β are different, at least six of them are not zeros of $\psi_2^2(X)$ and there exists a zero α of $\psi_4^2(F(X))$ with $\psi_2^2(\alpha) \neq 0$. Then

$$U(\alpha) = -\psi_2^{2d}(\alpha)\theta_4(F(\alpha)) = -\psi_2^{2d}(\alpha)(\alpha\psi_4^2(F(\alpha)) - \psi_3(F(\alpha))\psi_5(F(\alpha)))$$
$$= \psi_2^{2d}(\alpha)\psi_3(F(\alpha))\psi_5(F(\alpha)) \neq 0$$

since the zeros of $\psi_3(X)$ and $\psi_5(X)$ are the first coordinates of the finite 3- and 5-torsion points but $F(\alpha)$ is the first coordinate of a 4-torsion point. Thus $U(X)$ is not identical to zero and $\deg(U) \leq 19d$. Since $U(X)$ has at least $|R|/2$ zeros (we have $x_n = x_{l-n}$) we get $d \geq \deg(U)/19 \geq |R|/38$ and thus the result. $\square$

For odd l the bound on R (and therefore on $\deg(F)$) to $R \geq |S| - s = l - 2s - 1$.

If l is not divisible by a large prime then the Pohlig-Hellman algorithm [11,13] can solve efficiently the discrete logarithm problem and thus the Diffie–Hellman problem. Our result covers the interesting case of prime order groups where the Pohlig-Hellman algorithm is inefficient.

Theorem 1 is only nontrivial if $|S| \geq (2l + 37)/3$. However, under the assumption that $F(X)$ satisfies a property suggested by the proof of Theorem 1 we obtain a lower bound for much smaller S. The proof of the following lemma follows on the same lines as the proof of Theorem 1 by replacing $2n$ by kn.

Lemma 1. *Let $2 \leq k \leq l - 1$. If a nonconstant polynomial $F(X) \in \mathbb{F}_q[X]$ satisfies*

$$F(x_n) = x_{n^2} \quad and \quad F(x_{kn}) = x_{(kn)^2}, \quad n \in S,$$

for a subset $S \subseteq \{1 \leq n \leq l - 1\}$, then we have

$$\deg(F) \geq |S|/(2(k^4 + k^2 - 1)).$$

Now we restrict ourselves to the case that the order l is an odd prime. The general case can be reduced to this case to some extent by considering the subgroup of largest prime order.

Theorem 2. *Let E be an elliptic curve over $\mathbb{F}_q$ and $P \neq \mathcal{O}$ a point of prime order l. Let S be a subset of $\{1, \ldots, l - 1\}$ and $F(X) \in \mathbb{F}_q[X]$ satisfy*

$$F(x_n) = x_{n^2}, \quad n \in S.$$

Then for any $\varepsilon > 0$ and $|S| \geq \max(2l/(\varepsilon \log_2(l)), 5)$ there exists a constant $c = c(\varepsilon)$ such that

$$\deg(F) \geq c|S|/(l^{4\varepsilon} \log_2(l)).$$

Proof. Since $|S| \geq 5$ the polynomial $F(X)$ is not constant. Put $K = \lceil \varepsilon \log_2(l) \rceil - 2$, $S_i = \{1 \leq a \leq l - 1 : 2^i m \equiv a \bmod l, \; m \in S\}$, $0 \leq i \leq K$, and $R_{i,j} = S_i \cap S_j$, $0 \leq i < j \leq K$. Then adopting the proof of Theorem 8.2 of [15] we get,

$$(K + 1)|S| - \sum_{0 \leq i < j \leq K} |R_{i,j}| \leq \left| \bigcup_{i=0}^{K} S_i \right| \leq l - 1.$$

Therefore, there is a pair $0 \leq i < j \leq K$ such that

$$|R_{0,j-i}| = |R_{i,j}| \geq \frac{2|S|}{K} - \frac{2(l - 1)}{K(K + 1)} \geq \frac{|S|}{\varepsilon \log_2(l)}.$$

Putting $k = 2^{j-i} \leq 2^K$ and using Lemma 1 we get the result.   □

4 Bivariate Interpolation of the Diffie–Hellman Mapping

In this section we consider interpolation polynomials of the bivariate mapping F_2. Again we restrict ourselves to the case that the order l is an odd prime. First the given data needs no special structure.

Theorem 3. *Let E be an elliptic curve over $\mathbb{F}_q$ and $P \neq \mathcal{O}$ a point of prime order l. Let S_1 and S_2 be subsets of $\{1, 2, \ldots, l-1\}$ of cardinality $|S_1| = l-1-s$ and $|S_2| = u$. If $F(X, Y) \in \mathbb{F}_q[X, Y]$ satisfies*

$$F(x_n, x_m) = x_{nm}, \quad (n, m) \in S_1 \times S_2, \tag{2}$$

then we have

$$\deg(F) \geq \begin{cases} \min\left\{\left\lceil \frac{l-s-1}{s} \right\rceil - 1, \lceil (u-1)^{1/3} \rceil - 2 \right\}, & s > 0, \\ \lceil (u-1)^{1/3} \rceil - 2, & s = 0. \end{cases}$$

Proof. We may assume $l \geq u \geq 10$ and $s < (l-1)/2$ since otherwise the result is trivial. Put

$$d = \begin{cases} \min\left\{\left\lceil \frac{l-s-1}{s} \right\rceil - 1, \lceil (u-1)^{1/3} \rceil - 2 \right\}, & s > 0, \\ \lceil (u-1)^{1/3} \rceil - 2, & s = 0, \end{cases}$$

and let R be the set of $a \in S_1$ such that

$$(ia \bmod l) \in S_1, \quad i = 1, \ldots, d+1.$$

The cardinality of R is at least $l - 1 - (d+1)s > 0$. We fix $a \in R$. Then we have

$$F(x_{(i+1)a}, x_{b_j}) = x_{(i+1)ab_j}, \quad 0 \leq i, j \leq d,$$

where $b_0, \ldots, b_d$ are any distinct elements of S_2. Since otherwise the result is trivial we may suppose that $\deg_X(F)$, $\deg_Y(F) \leq d$, i.e.,

$$F(X, Y) = \sum_{i,j=0}^{d} c_{i,j} X^i Y^j.$$

The coefficients $c_{i,j}$ are uniquely determined by the following matrix equation,

$$C = \begin{pmatrix} c_{0,0} & \cdots & c_{0,d} \\ \vdots & & \vdots \\ c_{d,0} & \cdots & c_{d,d} \end{pmatrix}$$

$$= \begin{pmatrix} 1 & x_a & \cdots & x_a^d \\ \vdots & \vdots & & \vdots \\ 1 & x_{(d+1)a} & \cdots & x_{(d+1)a}^d \end{pmatrix}^{-1} \begin{pmatrix} x_{ab_0} & \cdots & x_{ab_d} \\ \vdots & & \vdots \\ x_{(d+1)ab_0} & \cdots & x_{(d+1)ab_d} \end{pmatrix} \begin{pmatrix} 1 & \cdots & 1 \\ x_{b_0} & \cdots & x_{b_d} \\ \vdots & & \vdots \\ x_{b_0}^d & \cdots & x_{b_d}^d \end{pmatrix}^{-1}.$$

The matrix C is regular if and only if the middle matrix on the right hand side is regular. A subset $\{b_0, \ldots, b_d\}$ of S_2 with this property exists if and only if the vectors $\underline{x}_k = (x_{kab})_{b \in S_2}$, $k = 1, \ldots, d+1$, are linearly independent. If $\underline{x}_k$, $k = 1, \ldots, d+1$, were linearly dependent, then there would exist an integer w with $1 \leq w \leq d+1$ and coefficients $d_1, \ldots, d_w \in \mathbb{F}_q$, $d_w \neq 0$, such that

$$\sum_{k=1}^{w} d_k x_{kab} = 0, \quad b \in S_2.$$

As at most 2 points with first coordinate equal to 0 exist and $u \geq 10$, we get $w > 1$. Since $x_{kab} = \theta_k(x_{ab})/\psi_k^2(x_{ab})$ the polynomial

$$G(X) = \sum_{k=1}^{w} d_k \theta_k(X) \prod_{\substack{j=1 \\ j \neq k}}^{w} \psi_j^2(X)$$

has at least $\lfloor u/2 \rfloor$ zeros and degree at most

$$1 + \sum_{k=1}^{w}(k^2 - 1) = (2w^3 + 3w^2 - 5w + 6)/6 \leq w^3/2 \leq (d+1)^3/2.$$

If $p \nmid w$ then points of order w on E exist over $\overline{\mathbb{F}_q}$. Let $\alpha \in \overline{\mathbb{F}_q}$ be the first coordinate of a point of order w. Then we have $\psi_w^2(\alpha) = 0$ and

$$G(\alpha) = d_w \theta_w(\alpha) \prod_{j=1}^{w-1} \psi_j^2(\alpha) \neq 0.$$

If $p \mid w$ then the degree of $\psi_w^2(X)$ is strictly less than $w^2 - 1$. Therefore the summand including $\theta_w(X)$ has largest degree and $G(X)$ has nonzero leading coefficient showing that $G(X)$ is not the zero polynomial. Hence, $(d+1)^3/2 \geq \lfloor u/2 \rfloor$ in contradiction to the definition of d. This shows that C is not singular and in particular each row of C has at least one nonzero entry and we have $\deg(F) \geq \deg_X(F) \geq d$. $\qquad\square$

Theorem 3 is only nontrivial if $|S_1| > (l-1)/2$. However, under restrictions on S_1 as used in the proof of Theorem 3 we can prove similar nontrivial results for very small sets.

Proposition 1. *Let S_1 and S_2 be subsets of $\{1, 2, \ldots, l-1\}$ with $|S_1| = v$ and $|S_2| = u$. Assume that there exists an integer a with $1 \leq a \leq l-1$ such that $S_1 \subseteq \{ia, (i+1)a, \ldots, (i+j)a\} \setminus \{0\}$ for some integers $i, j \geq 1$, where we understand the entries modulo l. If $F(X,Y) \in \mathbb{F}_q[X,Y]$ satisfies (2), then we have*

$$\deg(F) \geq \min\left\{v - 1, \lceil (u-1)^{1/3} \rceil - i - j + v - 2\right\}.$$

5 Interpolation of the Discrete Logarithm

In this section we turn our attention to interpolation polynomials of F_3 that reveal information about the discrete logarithm ind_P, i.e.,

$$\mathrm{ind}_P(x, y) = n \quad \text{if and only if} \quad (x, y) = nP, \quad 1 \leq n \leq l-1.$$

To make the choice of n unique we need to restrict the sets. Here we assume without loss of generality that it holds for $S \subseteq \{1, \ldots, \lfloor l/2 \rfloor\}$ but the same considerations work for sets $\tilde{S} \subseteq \{\lfloor l/2 \rfloor + 1, \ldots, l-1\}$. Throughout this section we restrict ourselves to the case $q \geq 7$. Then we have $l \leq q + 1 + 2\sqrt{q} < 2q$ by Hasse's Theorem (see e. g. [2]). Let the polynomial $F(X) \in \mathbb{F}_q[X]$ satisfy

$$F(x_n) = \xi_n, \quad n \in S, \tag{3}$$

for a subset $S \subseteq \{1, \ldots, \lfloor l/2 \rfloor\}$, where the elements ξ_n are defined by (1). Given ξ_n it is easy to determine the discrete logarithm.

First we recall a result of [7].

Theorem 4. *Let* $F(X) \in \mathbb{F}_q[X]$ *satisfy* (3) *for a subset* $S \subseteq \{N+1, \ldots, N + H\} \subseteq \{1, \ldots, \lfloor l/2 \rfloor\}$ *of cardinality* $|S| = H - s$. *Then we have*

$$\deg(F) \geq \left(1 - \frac{2}{p}\right)\frac{H-2}{3} - s - \frac{2}{3}.$$

A similar result for q even can also be proven. Theorem 4 is nontrivial only if the set of given data is relatively dense. By the properties used in the proof of the theorem we can obtain nontrivial results for less dense sets, again.

Corollary 1. *Let* $F(X) \in \mathbb{F}_q[X]$ *satisfy*

$$F(x_{n-1}) = \xi_{n-1}, \quad F(x_n) = \xi_n, \quad and \quad F(x_{n+1}) = \xi_{n+1}, \quad n \in S,$$

for a subset $S \subseteq \{2, \ldots, \lfloor l/2 \rfloor - 1\}$. *Then we have*

$$\deg(F) \geq |S|/3.$$

We can also use other properties of $F(X)$. For a finite prime field $\mathbb{F}_p$ we identify ξ_n with n by choosing $\beta_1 = 1$ and get the following result.

Proposition 2. *Let* $F(X) \in \mathbb{F}_p[X]$, $p \geq 7$, *satisfy*

$$F(x_n) = n \quad and \quad F(x_{2n}) = 2n, \quad n \in S,$$

for a subset $S \subseteq \{1, \ldots, \lfloor l/4 \rfloor\}$. *Then we have*

$$\deg(F) \geq \frac{|S|}{4}.$$

Proof. Let a be the leading coefficient of $F(X)$. For $n \in S$ we have

$$F\left(\frac{\theta_2(x_n)}{\psi_2^2(x_n)}\right) = F(x_{2n}) = 2n = 2F(x_n)$$

and the polynomial $U(X) = \psi_2^{2d}(X)(F(\theta_2(X)/\psi_2^2(X)) - 2F(X))$ has at least $|S|$ zeros and it is not the zero polynomial, since for the first coordinate α of a point of order 2 we have $U(\alpha) = a\theta_2(\alpha)^d \neq 0$. Hence, $4d \geq \deg(U) \geq |S|$. $\qquad\square$

The next result is nontrivial for finite fields $\mathbb{F}_q$, $q = p^r$, with $r \geq 3$.

Proposition 3. *Choose a polynomial basis* $\{1, \beta, \beta^2, \ldots, \beta^{r-1}\}$ *of* $\mathbb{F}_q$ *over* $\mathbb{F}_p$. *Let* $F(X) \in \mathbb{F}_q[X]$ *satisfy*

$$F(x_n) = \xi_n \quad and \quad F(x_{pn}) = \xi_{pn}, \quad n \in S,$$

for a subset $S \subseteq \{1, \ldots, \lfloor l/(2p) \rfloor\}$. *Then we have*

$$\deg(F) \geq |S|/p^2.$$

Proof. Put $d = \deg(F)$. The polynomial

$$U(X) = \psi_p^{2d}(X)(F(\theta_p(X)/\psi_p^2(X)) - \beta F(X))$$

of degree $p^2 d$ has at least $|S|$ zeros.   □

Now we allow arbitrary sets, use no special restrictions on $F(X)$, and choose an arbitrary basis of $\mathbb{F}_q$ over $\mathbb{F}_p$.

Theorem 5. *Let $F(X) \in \mathbb{F}_q[X]$ satisfy (3) for a subset $S \subseteq \{1, \ldots, \lfloor l/2 \rfloor\}$. Then with $t = \lceil \log_p(\lfloor l/2 \rfloor) \rceil$ we have*

$$\deg(F) \geq \frac{|S|(|S| - 2)}{3 \cdot 2^t(l - 2)}.$$

Proof. Since otherwise the result is trivial we may assume that $|S| \geq 3$. Put $d = \deg(F)$. Consider

$$D = \{1 \leq a \leq l - 1 : a \equiv m - n \bmod l, \; n, m \in S, n \neq a\}.$$

Obviously, there exists $a \in D$ such that there are at least

$$\frac{|S|(|S| - 2)}{|D|} \geq \frac{|S|(|S| - 2)}{l - 2}$$

representations $a \equiv m - n \bmod l$, $n, m \in S, n \neq a$. Choose this a and put

$$R = \{n \in S : n \neq a, \; a + n \equiv m \bmod l \text{ with } m \in S\}.$$

We have $|R| \geq |S|(|S| - 2)/(l - 2)$. If $a \leq l/2$ then for all $n \in R$ we have $n + a \in S$, i.e. $n + a \leq l/2$. Therefore, in this case there are at most 2^{t-1} possible elements $\omega \in \mathbb{F}_q$ such that

$$F(x_{n+a}) = F(x_n) + \omega,$$

namely the 2^{t-1} elements $\omega = \xi_a + \sum_{i \in I} \beta_i$ with $I \subseteq \{2, \ldots, t\}$. In the other case where $a > l/2$ the 2^{t-1} elements $\omega = -\xi_{l-a} - \sum_{i \in I} \beta_i$ work. As the cases are disjoint at least one of the mappings

$$H_\omega(x_n) = F(x_{n+a}) - F(x_n) - \omega$$

has at least $|R|/2^{t-1}$ zeros in $\mathbb{F}_q$. We choose this ω. Now by the usual addition formula we have

$$x_{n \pm a} = \frac{x_n^2 x_a + x_n x_a^2 + 2a_2 x_n x_a + a_4(x_n + x_a) + 2a_6 \mp 2y_n y_a}{(x_n - x_a)^2}.$$

The expression

$$H_\omega(x_n) = (F(x_{n+a}) - F(x_n) - \omega)(F(x_{n-a}) - F(x_n) - \omega)(x_n - x_a)^{4d}$$

is a polynomial expression in x_n and x_a, as it is the norm of $(F(x_{n+a}) - F(x_n) - \omega)(x_n - x_a)^{2d}$. Consider x_n as variable, then $H_\omega(X)$ is a polynomial of degree $6d$. Since

$$6d \geq \frac{|R|}{2^{t-1}} \geq \frac{|S|(|S| - 2)}{2^{t-1}(l - 2)}$$

the claim follows.   □

The bound is only nontrivial if $p \geq 3$ and $|S|$ is at least of the order of magnitude $l^{1/2}$. It achieves its strength for fields of large characteristic.

Acknowledgment. Parts of this paper were written during a visit of the second author to the Ruhr-University of Bochum. He wishes to thank Prof. Hans Dobbertin and the Institute of Information Security and Cryptology for hospitality and financial support. The second author is also supported by DSTA research grant R-394-000-011-422.

References

1. E. Bach and J.O. Shallit. *Algorithmic number theory, Vol.1: Efficient algorithms.* MIT Press, Cambridge, 1996.
2. I.F. Blake, G. Seroussi, and N.P. Smart. *Elliptic curves in cryptography*, volume 265 of *London Mathematical Society Lecture Note Series*. Cambridge University Press, 1999.
3. J. Buchmann and D. Weber. Discrete logarithms: recent progress. In *Coding theory, cryptography and related areas (Guanajuato, 1998)*, pages 42–56, Berlin, 2000. Springer.
4. D. Coppersmith and I.E. Shparlinski. On polynomial approximation of the discrete logarithm and the Diffie–Hellman mapping. *J. Cryptology*, 13:339–360, 2000.
5. N. Koblitz. Elliptic cryptosystems. *Math. Comp.*, 48:203–209, 1987.
6. S. Lang. *Elliptic curves: Diophantine analysis.* Springer, Berlin, 1978.
7. T. Lange and A. Winterhof. Polynomial Interpolation of the Elliptic Curve and XTR Discrete Logarithm. In *8th Annual International Computing and Combinatorics Conference (COCOON'02) (Singapore, 2002)*, volume 2387 of *Lect. Notes Comp. Sci.*, pages 137–143, Berlin, 2002. Springer.
8. U.M. Maurer and S. Wolf. The Relationship Between Breaking the Diffie–Hellman Protocol and Computing Discrete Logarithms. *SIAM Journal on Computing*, 28(5):1689–1721, 1999.
9. U.M. Maurer and S. Wolf. The Diffie–Hellman Protocol. *Designs, Codes, and Cryptography*, 19:147–171, 2000.
10. A.J. Menezes, T. Okamoto, and S.A. Vanstone. Reducing ellipitc curve logarithms to a finite field. *IEEE Trans. on Inform. Theory*, 39:1639–1646, 1993.
11. A.J. Menezes, P.C. van Oorschot, and S.A. Vanstone. *Handbook of Applied Cryptography*. CRC Press Series on Discrete Mathematics and its Applications. CRC Press, Boca Raton, FL, 1996.
12. V. Miller. Use of elliptic curves in cryptography. In *Advances in Cryptology - Crypto '85*, volume 263 of *Lect. Notes Comput. Sci.*, pages 417–426, Berlin, 1986. Springer.
13. S. Pohlig and M. Hellman. An improved algorithm for computing logarithms over $GF(p)$ and its cryptographic significance. *IEEE Trans. on Inform. Theory*, 24:106–110, 1978.
14. R. Schoof. Elliptic curves over finite fields and the computation of square roots mod p. *Math. Comp.*, 44:483–494, 1985.
15. I.E. Shparlinski. *Number theoretic methods in cryptography. Complexity lower bounds.* Birkhäuser, Basel, 1999.
16. J.H. Silverman. *The arithmetic of elliptic curves*, volume 106 of *Graduate texts in mathematics.* Springer, 1986.

An Optimized Algebraic Method for Higher Order Differential Attack

Yasuo Hatano[1], Hidema Tanaka[2], and Toshinobu Kaneko[1]

[1] Tokyo University of Science, 2641, Yamazaki, Noda, Chiba, 278-8510, Japan
j7302656@ed.noda.tus.ac.jp
kaneko@kaneko01.ee.noda.tus.ac.jp
[2] Communications Research Laboratory, 4-2-1, Nukui-Kitamachi, Koganei, Tokyo,
184-8795, Japan
hidema@crl.go.jp

Abstract. We show an optimization technique of algebraic method for higher order differential attack. Our technique is based on linear dependency and makes a small coefficient matrix for algebraic method using redefined unknown variables. We also show a technique of algebraic method for an attack equation which holds probabilistically. We demonstrate our method by attacking five-round MISTY1. Our method needs $2^{21.6}$ chosen plaintexts and $2^{32.4}$ computational cost. The computer simulation took about 6 minutes.

1 Introduction

Higher order differential attack is a famous attack of block ciphers. It uses a property of higher order differential of discrete functions [5]. A higher order differential attack is consisted of two phases; 1) searching for the necessary order of higher order differential and 2) solving an attack equation. The searching for the necessary order of higher order differential depends on a target crypto-algorithm. Some method exists to solve an attack equation; for example, an exhaustive search, an algebraic method and so on. The first attack using higher order differential, proposed by Jakobsen et.al. has been applied to $\mathcal{KN}$ cipher. In their attack, they used an exhaustive search to solve an attack equation to find the true key. An algebraic method, which is proposed by Shimoyama et. al. [7], transforms higher order equations into linear equations and determines the value of unknowns using Gauss-Jordan elimination method. The part of calculation of a coefficient matrix requires most computational cost. Moriai et.al. have generalized it on an attack of a CAST cipher [8].

In this paper, we show an optimization technique for algebraic method. We redefine variables using linear dependency and make a small coefficient matrix. Thus, our method requires smaller computational cost than the original. We use the calculation of rank of a coefficient matrix to redefine unknowns. We also show a technique to calculate the rank with a little complexity. Furthermore, we show a technique to solve an attack equation which holds probabilistically. We demonstrate the effectiveness of our technique by attacking five-round MISTY1.

M. Fossorier, T. Hoeholdt, and A. Poli (Eds.): AAECC 2003, LNCS 2643, pp. 61–70, 2003.

Our method needs $2^{21.6}$ chosen plaintexts and $2^{32.4}$ computational cost and the computer simulation took about 6 minutes.

2   Preliminaries

2.1   Higher Order Differential [5]

Let $E(\cdot)$ be a function as follows.

$$Y = E(X; K), \tag{1}$$

where $X \in \mathrm{GF}(2)^n$, $Y \in \mathrm{GF}(2)^m$, and $K \in \mathrm{GF}(2)^s$. Let $\{A_1, \cdots, A_i\}$ be a set of linear independent vectors in $\mathrm{GF}(2)^n$ and $V^{(i)}$ be the sub-space spanned by these vectors. Then, i-th order differential is defined as follows.

$$\Delta_{V^{(i)}}^{(i)} E(X; K) = \sum_{A \in V^{(i)}} E(X + A; K), \tag{2}$$

In the following, $\Delta^{(i)}$ denotes $\Delta_{V^{(i)}}^{(i)}$, when it is clearly understood.

In this paper, we use the following property of higher order differential.

Property: If the degree of $E(X; K)$ with respect to X equals to D, then

$$deg_X\{E(X; K)\} = \begin{cases} \Delta^{(D+1)} E(X; K) = 0 \\ \Delta^{(D)} E(X; K) = const. \end{cases}$$

2.2   Attack of a Block Cipher

Let us consider an R-round block cipher. Let $E_i(\cdot)$ be the i-round encryption function: $\mathrm{GF}(2)^n \times \mathrm{GF}(2)^{i \times s} \to \mathrm{GF}(2)^m$ and $H_{(R-1)}(X)$ be the $(R-1)$-th round output of that from an input X.

$$H_{(R-1)}(X) = E_{(R-1)}(X; \mathrm{K}_1, \ldots, \mathrm{K}_{(R-1)}), \tag{3}$$

where K_i denotes the key inputted to the i-th round.

If the degree of $E_{(R-1)}(\cdot)$ with respect to X is $D-1$, we have

$$\Delta_{V^{(D)}}^{(D)} H_{(R-1)}(X) = 0, \tag{4}$$

Let $\tilde{E}(\cdot)$ be the function, which outputs $H_{(R-1)}$ from a ciphertext $C \in \mathrm{GF}(2)^n$.

$$H_{(R-1)}(X) = \tilde{E}(C(X); \mathrm{K}_R) \tag{5}$$

From Eq. (4),(5) and Property, the following equation holds.

$$0 = \sum_{A \in V^{(D)}} \tilde{E}(C(X + A); \mathrm{K}_R) \tag{6}$$

In the following, we call Eq. (6) the attack equation.

2.3   Algebraic Method

Shimoyama et. al. have shown an effective method to solve the attack equation (6)[7,8]. The method, called *algebraic method* in this paper, is a linearization method for the attack equation, which replaces higher order variables like $x_i x_j$ by a new independent variable like y_{ij}. In the following, we call an equation, regarded as a linear equation, the linearized equation. The reader is referred to [7,8] for the details.

Let L be the number of independent unknowns in the linearized equation of Eq. (6). As mentioned above, since we derive the attack equation using an m-bit sub-block, we can rewrite equation Eq. (6) as follows.

$$\mathcal{A}k = b' \ , \ k = {}^t(k^{(1)}, k^{(2)}, \ldots, k^{(s)}, k^{(1)}k^{(2)}, \ldots, k^{(1)} \cdots k^{(s)}). \tag{7}$$

where $\mathcal{A}$, b', and k are the $m \times L$ coefficient matrix, the m-dimensional solution vector, and the L-dimensional vector over GF(2). It should be noted that k denotes unknowns and $K_R = (k^{(1)}, k^{(2)}, \ldots, k^{(s)}) \in GF(2)^s$.

We can get m equations from one D-*th* order differential because Eq. (6) is an m-bit equation. Therefore we need $(\lfloor L/m \rfloor)$ sets of the D-*th* order differential for the unique solution. Then we can construct

$$\mathbf{A}k = b, \tag{8}$$

where $\mathbf{A}$ and $\mathbf{b}$ are a $\lfloor L/m \rfloor \times L$ matrix and a $\lfloor L/m \rfloor$ vector.

Since one set of D-*th* order differential requires 2^D chosen plaintexts, the necessary number of plaintexts M is estimated as

$$M = 2^D \times \left\lfloor \frac{L}{m} \right\rfloor. \tag{9}$$

If we use the same technique shown in [7,8], Eq. (7) requires $2^D \times L$ times of $\tilde{E}(\cdot)$ calculation. Since we have to prepare $(\lfloor L/m \rfloor)$ sets of the D-th order differential, the computational cost is estimated as

$$T = 2^D \times L \times \left\lfloor \frac{L}{m} \right\rfloor. \tag{10}$$

3   Optimized Calculation for Algebraic Method

3.1   Basic Idea

In this section, we discuss an optimization of algebraic method. The number of chosen plaintexts and computational cost for algebraic method depends on the order D for the attack and the number of unknowns L in the attack equation. Let us consider a linear dependency for each unknown: for example $a_0 x_0 + a_0 x_1 = a_0(x_0 + x_1) = a_0 \tilde{x}_0$, where $\tilde{x}_0$ denotes a new redefined variable. Thus such a relationship can decrease the number of unknowns, so the complexity for algebraic method will decrease.

If we can analyze the number of independent unknowns $l(\leq L)$, in Eq. (8), it is enough for solving the attack equation using $\lfloor l/m \rfloor \times L$ matrix $\mathbf{A}$ and the $\lfloor l/m \rfloor$-dimensional vector $\mathbf{b}$. Therefore if $L > l$, we can decrease the computational cost and the necessary number of chosen plaintext to solve Eq. (8).

To know the number of independent unknowns l, we calculate the rank of the attack equation (6) using a following theorem.

Theorem 1: The expectation of d is less than $q + 2$, where d is defined as

$$\dim_{\mathrm{GF}(p)} \langle v_1, v_2, \cdots, v_d \rangle = q$$

for randomly chosen v_i in the q-dimensional vector space over $\mathrm{GF}(p)$.

Proof: Since a randomly chosen element in the q-dimensional vector space over $\mathrm{GF}(p)$ is contained in particular i-dimensional $(i \leq q)$ sub-space with probability $\dfrac{p^i}{p^q}$, we need to choose on average $\dfrac{1}{1 - \frac{p^i}{p^q}}$ elements in order to find one that is not in the sub-space. Thus the expectation of d can be evaluated as follows.

$$\sum_{i=0}^{q-1} \frac{1}{1 - \left(\frac{1}{p}\right)^{q-i}} = \sum_{i=0}^{q-1} \frac{p^{q-i}}{p^{q-i} - 1} = \sum_{i=0}^{q-1} \left(1 + \frac{1}{p^{q-i} - 1}\right)$$

$$= q + \sum_{i=1}^{q} \frac{1}{p^i - 1} = q + \frac{1}{p - 1} + \sum_{i=2}^{q} \frac{1}{p^i - 1}$$

$$< q + \frac{1}{p - 1} + \sum_{i=1}^{q-1} \frac{1}{p^{i-1}} = q + \frac{1}{p - 1} + \frac{1 - \left(\frac{1}{p}\right)^{q-1}}{p - 1}$$

$$\leq q + 2$$

$$\square$$

We prepare $L + 2$ linear equations according the attack equation (6) and construct the coefficient matrix by algebraic method. Then we can know the number of independent unknowns by the rank of the coefficient matrix from the above theorem.

Next, let us consider the complexity to know the number of independent unknowns. When we calculate the rank according to Theorem 1, we need $(L+2)$ sets of D-th order differentials. It requires $2^D \times \lfloor \frac{L+2}{m} \rfloor$ times of $\tilde{E}$ function operations. However we can reduce the complexity for the analysis as follows.

i-th order differential can be expressed as

$$\Delta_{V^{(i)}}^{(i)} E(X; K) = \Delta_{V^{(i-1)}}^{(i-1)} E(X; K) + \Delta_{V^{(i-1)}}^{(i-1)} E(X + A_i; K), \qquad (11)$$

where $V^{(i)} = \langle A_0, \cdots, A_{i-1}, A_i \rangle$ and $V^{(i-1)} = \langle A_0, \cdots, A_{i-1} \rangle$. Equation (11) shows that i-th order differential can be expressed as the sum of the first order differentials (*differential*). Since the attack equation (6) is derived from a D-th order differential, it also can be expressed as the sum of differential form. Since the term composed only unknown keys is regarded as a constant term, the attack equation never has such terms by calculation of the differential. It is enough for us to analyze $\sum \tilde{E}(X)$ to know the number of independent unknowns. Therefore, we do not have to prepare the ciphertexts corresponding to the chosen plaintexts for the attack but random ciphertexts. From the above, we only have to calculate the differential form as follows.

$$
\begin{aligned}
\mathcal{DF}(X, \mathrm{K}_R) &= \Delta\tilde{E}(X; \mathrm{K}_R) \\
&= \tilde{E}(X; \mathrm{K}_R) + \tilde{E}(X + A; \mathrm{K}_R)
\end{aligned}
\tag{12}
$$

Let R_{AEQ} and R_{DIF} be the rank of the coefficient matrix from the attack equation and differential form. Generally, we conjecture that $R_{AEQ} = R_{DIF}$. We confirmed it by computer simulation. If $\lfloor \frac{L+2}{m} \rfloor \ll 2^D$, the necessary computational is negligible compared with the original one.

3.2   Application for Probabilistic Attack Equation

Let us consider that we derive an attack equation derived by guessing a u-bit key. In such a case, the attack equation holds probabilistically and we must distinguish the true key from false keys.

Let's consider that we prepare additional α linearized equations and we construct a $\lfloor (l+\alpha)/m \rfloor \times L$ matrix $\mathbf{A}$ and a $(l+\alpha)$-dimensional vector $\mathbf{b}$, instead of a $\lfloor l/m \rfloor \times L$ matrix $\mathbf{A}$ and a l-dimensional vector $\mathbf{b}$ in equation (8).

We have already known that the rank of the matrix $\mathbf{A}$ equals to l. If the assumed key is true, the linearized attack equation always holds and the vector $\mathbf{b}$ is the element in the l-dimensional space. Otherwise, if the assumed key is false, we can regard the matrix $\mathbf{A}$ and the vector $\mathbf{b}$ as randomly chosen. Then, if the vector $\mathbf{b}$ is not the element in the l-dimensional space, such linearized attack equations never have a solution. Therefore we can remove false keys from the rank of the enlarged coefficient matrix of $l + \alpha$ linearized attack equations. The probability P, with which we can remove one false key, equals to a probability with which we pick up an element in l-dimensional space in $(l+\alpha)$-dimensional space. Therefore,

$$
P = \frac{2^l}{2^{l+\alpha}} = 2^{-\alpha}.
\tag{13}
$$

Since we assume a u-bit key, we can remove all false keys by preparing additional α equations which satisfies $2^{-\alpha} \times 2^u \ll 1$.

4 Higher Order Differential Attack of MISTY1 with Algebraic Method

4.1 MISTY1

MISTY1[6] is a 64-bit eight-round Feistel type block cipher with a 128-bit secret key and one of the candidates of the NESSIE project. Figure 1 shows its main structure and components. Note that it shows the equivalent FO and FI function and we call it MISTY in the followings. In this section, we use variables as shown in Fig. 2 and $\Lambda^{(L7)}$ denotes the left seven-bit variable of Λ.

4.2 Attack Equation for Five-Round MISTY1

In this section, we demonstrate an attack of MISTY1 using the optimized method. Due to the limited space, we omit the details in this paper.

We search for an effective chosen plaintext by S-Box oriented manner. As a result, we found that

$$\Delta_{V^{(14)}}^{(14)} Z_{R5}^{(L7)}(X) = 0, \tag{14}$$

where $X \in \mathrm{GF}(2)^{64}$ and $Z_{R5}^{(L7)} \in \mathrm{GF}(2)^7$ is the plaintext and the left seven-bit variable of the right side input Z_{R5} in the five-round (see Fig. 2). And

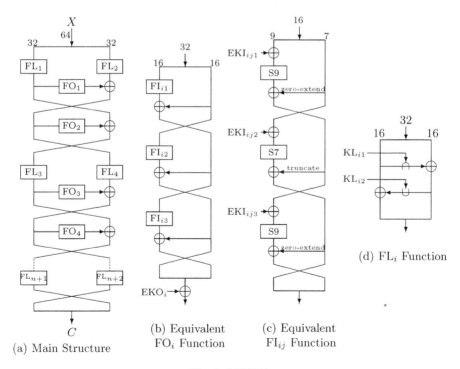

(a) Main Structure

(b) Equivalent FO$_i$ Function

(c) Equivalent FI$_{ij}$ Function

(d) FL$_i$ Function

Fig. 1. MISTY

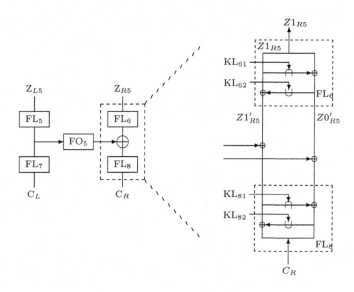

Fig. 2. Last round in our attack model(5-round MISTY)

$V^{(14)} = \langle A_0, \cdots, A_6, A_{16}, \cdots, A_{22} \rangle$, where $A_i \in \mathrm{GF}(2)^{64}$ denotes the all zero vector except the i-th bit.

Using above, we can derive an attack equation for five-round MISTY1. The attack equation can be expressed as follows (see Fig. 2).

$$0 = \sum Z1'^{(L7)}_{R5} + \sum \left[\left(Z0'^{(L7)}_{R5} \cup \mathrm{KL}^{(L7)}_{62} \right) \right], \qquad (15)$$

where $Z1'_{R5}$ and $Z0'_{R5}$ denote intermediate variables corresponding to the plaintext $X + A$ and $\sum$ denotes $\sum_{A \in V^{(14)}}$.

4.3   Analysis of Attack Equation

The second term of the left-hand side of Eq. (15) causes an increase in the complexity for the attack because we must consider additional 48-bit keys in the fifth round. So we ignore them by assuming $\mathrm{KL}^{(L7)}_{62} = 0x7\mathrm{f}$. Although $\mathrm{KL}^{(L7)}_{62} = 0x7\mathrm{f}$ holds with probability only 2^{-7}, we can increase the success probability of the attack by dividing the attack equation into 7 kinds of equations. This equation holds with probability 2^{-1}. Using those equations, we can determine a part of keys in the fifth round if $\mathrm{KL}^{(L7)}_{62} \neq 0x00$, which holds with probability $1 - 2^{-7}$. Otherwise we can determine $\mathrm{KL}^{(L7)}_{62} = 0x00$.

To apply algebraic method, we have to analyze the attack equation as an equation over $\mathrm{GF}(2)$. We analyzed it by the computer algebraic software RE-DUCE and the number of independent unknowns according to Sect. 3.1. The result is shown in Table 1.

Table 1. The analysis of the number of unknowns

Location	All Unknowns (L_i)	Independent unknowns (l_i)
0-*th bit*	545	173
1-*st bit*	521	169
2-*nd bit*	604	189
3-*rd bit*	521	169
4-*th bit*	545	173
5-*th bit*	554	177
6-*th bit*	569	181
all	1702	−

(*all* denotes the number of all unknowns in 7 kinds of linearized equation)

4.4   Complexity

In this section, we estimate the necessary number of plaintexts and the computational cost. In our attack, we use the 14-th order differential. Thus $D = 14$.

At first, we compare the original algebraic method with the optimized one. According to Eq. (9), the necessary number of plaintexts and computational cost are estimated as follows. Note that we have $m = 7$.

$$M = 2^{14} \times \left\lfloor \frac{1702}{7} \right\rfloor \simeq 2^{22.0} \tag{16}$$

$$T = 2^{14} \times 1702 \times \left\lfloor \frac{1702}{7} \right\rfloor \simeq 2^{32.7}. \tag{17}$$

Next we use an optimized method described in Sect. 3.1. We derive 7 kinds of equations using 7-bit sub-block as described in the previous section. Thus $m = 1$. Using the optimization and the equation including the minimum number of l_i, we can reduce those as follows.

$$M_{opt} = 2^{14} \times \left\lfloor \frac{169}{1} \right\rfloor \simeq 2^{21.4} \tag{18}$$

$$T_{opt} = 2^{14} \times 521 \times \left\lfloor \frac{169}{1} \right\rfloor \simeq 2^{30.5}. \tag{19}$$

As a result, our method requires much smaller complexity than the original one. And we confirmed that our optimization process requires little complexity.

Now let us consider the removal of the false key $KL_{62}^{(L7)}$. Our attack equations hold with probability 2^{-1} because we derive the attack equation by assuming a one-bit key in KL_{62}. So we need additionally α equations to determine the true key, which satisfies $2^{-\alpha+1} \ll 1$ from Sect. 3.2.

In this paper, we use all kinds of equations for the high success probability of the attack. Since one attack equation generates the 7-bit lineared equations, we can compute the coefficient matrix in those equations in parallel. Therefore,

Table 2. Summary of attacks on MISTY1

round	Complexity		Comment
	Data	Time	
4	$2^{22.3}$	2^{45}	Slicing Attack[4]
5	$2^{38.4}$	2^{116}	Higher Order Differential Attack[9]
5	$2^{21.7}$	$2^{32.4}$	This paper

we can get a unique solution for each equation by preparing η linear equations, where η denotes the maximum number of unknowns in those equations. The necessary number of plaintexts and computational cost are estimated as follows. Note that we adopt $\alpha = 10$ $(2^{-10+1} \ll 1)$.

$$M = 2^{14} \times (\eta + \alpha) \simeq 2^{21.7} \tag{20}$$
$$T = 2^{14} \times 1702 \times (\eta + \alpha) \simeq 2^{32.4}, \tag{21}$$

where $\eta = 189$. Our computer simulation according to our attack algorithm took about 6 minutes on Pentium III 800MHz.

5   Conclusion

We discuss the optimization for algebraic method. Using our method, we can reduce the complexity for the attack although that takes little complexity for the preparation.

Furthermore we demonstrate the method by an attack of MISTY1 and show that five-round one can be attacked. Table 2 shows results of attacks on MISTY1. As shown in that, our attack described in this paper works the best in known attacks.

References

1. K. Aoki, "Practical Evaluation of Security against Generalized Interpolation Attack," IEICE Trans. Fundamentals, Vol.E83-A, No. 1, pp. 33–38, January, 2000.
2. L.R. Knudsen and D. Wagner, "Integral Cryptanalysis," Fast Software Encryption 2002, FSE2002, Lenven, Belgium, February, 2002.
3. T. Jakobsen and L.R. Knudsen, "The Interpolation Attack on Block Cipher," Fast Software Encryption 4th International Workshop, LNCS.1008, Springer-Verlag, Berlin, 1996.
4. U. Kühn, "Improved Cryptanalysis of MISTY1," Fast Software Encryption 2002, FSE2002, Lenven, Belgium, February, 2002.
5. X. Lai, "Higher Order Derivatives and Differential Cryptanalysis," Communications and Cryptography, pp. 227–233, Kluwer Academic Publishers, 1994.
6. M. Matsui, "New Block Encryption Algorithm MISTY," FSE97, LNCS1267, Springer-Verlag, pp. 54–68, 1997.
7. S .Moriai, T. Shimoyama, and T. Kaneko, "Higher Order Differential Attack of a CAST Cipher," Fast Software Encryption Workshop '98, FSE '98, Paris, March, 1998.

8. T. Shimoyama, S. Moriai, T. Kaneko, and S. Tsujii, "Improving Higher Order Differential Attack and Its Application to Nyberg-Knudsen's Designed Block Cipher," IEICE Trans. Fundamentals, Vol.E82-A, No. 9, pp. 1971–1980, September, 1999.
9. H.Tanaka, C.Ishii, and T.Kaneko,"On the strength of Block Cipher KASUMI and MISTY," Symposium on Cryptography and Information Security, SCIS2001, Oiso, Japan, January, 2001. (In Japanese)

Fighting Two Pirates

Hans Georg Schaathun

Departement of Informatics, University of Bergen, N–5020 Bergen, Norway
georg@ii.uib.no
http://www.ii.uib.no/~georg/

Abstract. A pirate is a person who buys a legal copy of a copyrighted work and who reproduces it to sell illegal copies. Artists and authors are worried as they do not get the income which is legally theirs. It has been suggested to mark every copy sold with a unique fingerprint, so that any unauthorised copy may be traced back to the source and the pirate who bought it. The fingerprint must be embedded in such a way that it cannot be destroyed. Two pirates who cooperate, can compare their copies and they will find some bits which differ. These bits must be part of the fingerprint, and when the pirates can see and change these bits, they get an illegal copy with neither of their fingerprints. Collusion-secure fingerprinting schemes are designed to trace at least one of the pirates in such a collusion.

In this paper we prove that socalled $(2, 2)$-separating codes often are collusion-secure against two pirates. In particular, we consider the best known explicit asymptotic construction of such codes, and prove that it is collusion-secure with better rate than any previously known schemes.

1   Fingerprinting

Once upon a time, when computing logarithms was time consuming and tables of logarithms expensive, publishers found they had to protect the tables against illegal copying. They introduced tiny errors in the least significant digits to make every copy of the tables unique. In this way an illegal copy could be traced back to one legal original, and the customer who had bought this copy could be prosecuted.

This technique, known as fingerprinting, has been suggested for digital data. Research is going on within several fields to solve the various challenges involved. One problem is the embedding. How can the copy be marked without distorting the data, and without the users being able to change the fingerprint? We are not going to address this problem any further, just state some assumptions about its solution. The problem we are going to address is how to make the system resistant against a coalition of pirates.

A vendor holds the copyright to some work, it may be a sound recording, a digital image, a literary text, or something else. A copy is a digital file which resembles the work and has the same practical (and artistic) value. A user is a legal owner of a copy, presumably bought from the vendor. A pirate is a user who makes and distributes illegal copies of the work.

M. Fossorier, T. Hoeholdt, and A. Poli (Eds.): AAECC 2003, LNCS 2643, pp. 71–78, 2003.

A fingerprint is a word (tuple) of symbols uniquely identifying a user. The set of fingerprints is called an (n, M) code C, where n is the length of each word and M is the number of users (or fingerprints). We let d and m denote respectively the minimum and maximum Hamming distance between two codewords, and $\delta = d/n$ is the normalised minimum distance.

The fingerprint is supposed to be embedded into the copy in order to identify its owner. The embedding of one symbol of the fingerprint is called a mark. We assume that a user investigating a single copy is unable to locate or identify any mark, and therefore cannot change any mark. A coalition of users however, can compare their copies, and any difference between their copies must be a mark. The pirates can produce copies with a false fingerprint, but every mark has to match at least one of the legal copies held by the pirates. This is known as the marking assumption.

Let P be a coalition of pirates. Since each pirate is associated to a fingerprint, we write $P \subset C$. A position i is undetectable for P if all the elements of P match in position i. The feasible set $F(P)$ is the set of false fingerprint which may be produced by P, in other words

$$F(P) = \{(c_1, \dots, c_n) : \forall i = 1, \dots, n, \exists (x_1, \dots, x_n) \in P, \text{st. } c_i = x_i\}.$$

Note that $P \subset F(P)$. The elements of $F(P)$ are often called descendants of P.

The fingerprinting code C is assumed to be publicly known, however, the vendor uses a secret code C' chosen uniformly at random from the ensemble of codes equivalent to C. The codeword embedded in the copy is the codeword from C' corresponding to the codeword from C associated with the user. In this way, it is impossible for the pirates two know which coordinate position is corresponding to a given detectable mark, and which code symbol corresponds to a given value of the mark.

2   The Identifiable Parent Property

The goal in collusion-secure fingerprinting is to identify at least one pirate when discovering a false fingerprint produced by a coalition of at most t pirates. If this is possible, we say that the code has the t-identifiable parent property (t-IPP).

Let $\mathcal{P}_t(C)$ be the family of sets $P \subset C$ of cardinality at most t. Let $P_t(\mathbf{x}) \subset \mathcal{P}_t(C)$ be the family of coalitions which could have produced $\mathbf{x}$, i.e.

$$P_t(\mathbf{x}) = \{P \in \mathcal{P}_t(C) : \mathbf{x} \in F(P)\}.$$

If the elements of $P_t(\mathbf{x})$ has a non-empty intersection for any $\mathbf{x}$, then C is t-IPP. The following definition is equivalent, and standard.

Definition 1 (Identifiable Parent Property). *A code C is said to have the t-identifiable parent property (t-IPP) if there is an algorithm A such that for every $P \in \mathcal{P}_t(C)$ and every vector $\mathbf{x} \in F(P)$, $A(\mathbf{x})$ returns a member of P.*

The algorithmic issues are beyond the scope of this paper. As far as we are conserned, the algorithm A may be an exhaustive search through $\mathcal{P}_t(C)$.

Observe that t-IPP implies $(t-1)$-IPP. It is well-known that binary codes cannot be even 2-IPP [2]. More generally the following proposition is well known in the fingerprinting literature.

Proposition 1. *Let t and q be integers. If $t < q$, then there exist asymptotic families of q-ary codes with t-IPP. If $t \geq q$, there exists no q-ary code with t-IPP and more than t codewords.*

In order to get collusion-secure codes with more pirates, we use probabilistic fingerprinting schemes. We allow A to have a certain error probability ϵ. There are two types of error: we call it a *failure* if A returns void and a *mistake* if A returns a word which is not a member of P. Mistakes is a threat to justice, as it causes innocent users to be accused. If there is no probability of mistakes, the output of A is always reliable, but occasionally there is no output at all.

A t-IPP code with ϵ-error (or (t, ϵ)-IPP) is defined as a code where the probability of error is ϵ. Failures may be turned into mistakes by picking a random codeword whenever a failure should occur, and thus past literature rarely distinguish between failure and mistake.

We define (t, ϵ)-UPP (undisputable parent property) to be a code where the algorithm A has no risk of mistake and a probability ϵ of failure. Obviously, (t, ϵ)-UPP is stronger than (t, ϵ)-IPP.

We say that a word $\mathbf{x}$ is t-identifiable if it can be traced back to one undisputable parent, that is if

$$\bigcap_{P \in P_t(\mathbf{x})} P \neq \emptyset.$$

The set of t-identifiable words is denoted $\mathcal{I}_t(C)$.

When the pirates compare their copies, they find d' detectable bits. This number d' is the Hamming distance between their two fingerprints. As long as the embedding is kept secret by the vendor, it is impossible for the pirates to tell which detected mark corresponds to which coordinate position in the code.

When they construct a false fingerprint $\mathbf{x}$, they can only choose the distance $s = d(\mathbf{a}, \mathbf{x})$. Clearly $d(\mathbf{b}, \mathbf{x}) = d' - d(\mathbf{a}, \mathbf{x})$. The two pirates cannot be distinguished, so we can assume that $s \leq d'/2$. We call s the pirate strategy, and define $\sigma := s/n$ to be the normalised strategy, where n is the length of a fingerprint.

Once the strategy is chosen, a fingerprint is produced uniformly at random from a set of $2\binom{s}{d'}$ feasible words. (If d' is even and $s = d'/2$, there are $\binom{s}{d}$ feasible words.) There is a certain probability $p_s(P)$ that the produced false fingerprint is identifiable. The pirates will obviously choose s to minimise $p_s(P)$. The probability that the pirates gets away with their forgery is $1 - p(P)$, where $p(P) = \min_s p_s(P)$.

For simplicity, we assume that the pirates know which two codewords they posess. This allows them to make a perfect minimisation of $p_s(P)$, which might not be possible in reality. Hence the $p(P)$ defined here is a lower bound on the true probability.

Theorem 1. *A q-ary code cannot have (t, ϵ)-UPP for any $t > q$ and $\epsilon < 1$.*

Proof. Consider a code and a coalition of $q + 1$ pirates. For each coordinate position, there is at least one symbol which appears in at least two of the pirate codewords. Thus the pirates has a feasible word which matches at least two pirates in each coordinate position. Since this false fingerprint is feasible for any subset of q pirates, none of the pirates are undisputable parents.

3   Separating Codes

Much of the fingerprinting literature has focused on properties which are related to, but weaker than, t-IPP. The most important one of these properties is (t, t')-separation. Resently it was proved that (t, t)-separating codes can be used for constructing (t, ϵ)-IPP codes [1]. We shall see that some good $(2, 2)$-separating codes are actually good $(2, \epsilon)$-IPP codes in themselves with better rates than the codes from [1].

Definition 2 (Separating Code). *Let $\mathbf{t} = (t_1, \dots, t_z)$ be a tuple of natural numbers. A sequence $(T_1, \dots, T_z)$ of pairwise disjoint vector sets is called a $\mathbf{t}$-configuration if $\#T_j = t_j$ for all j. Such a configuration is separated if there is a position i, such that for all $l \neq l'$ every vector of T_l is different from every vector of $T_{l'}$ on position i.*

A code is $\mathbf{t}$-separating (a $\mathbf{t}$-SS) if every $\mathbf{t}$-configuration is separated.

If $t_i = 1$ for all i, then $\mathbf{t}$-separation is equivalent to z-hashing. For $z = 2$ there is a vast literature, in particular on $(2, 1)$- and $(2, 2)$-SS, it dates back at least to '69 [6]. See [9] for a survey.

If a code C is not $(t, 1)$-separating, there is a pirate coalition T_1 of t users who are able to forge a fingerprint $\mathbf{x}$ which belongs to a user not member of T_1. To see this, just let $(T_1, \{\mathbf{x}\})$ be a $(t, 1)$-configuration which is not separated. We say that $\mathbf{x}$ is framed by T_1, and $(t, 1)$-SS are often called t-frameproof codes in the fingerprinting literature.

If a code is (t, t)-separating, it means in fingerprinting terms, that two disjoint coalitions $T_1, T_2 \in \mathcal{P}_t(C)$ cannot produce the same false fingerprint, i.e. $F(T_1) \cap F(T_2) = \emptyset$. These codes where called t-secure frameproof in some early fingerprinting literature.

Definition 3 (Separating Weights). *Let $(T_1, \dots, T_z)$ be a $\mathbf{t}$-configuration. The separating weight $\theta_{\mathbf{t}}(T_1; \dots; T_z)$ is the number of positions where the configuration is separated.*

If C is an (n, M) code, its minimum separating weight $\theta_{\mathbf{t}}$ is the least separating weight for any $\mathbf{t}$-configuration from C. The normalised separating weight is $\tau_{\mathbf{t}} := \theta_{\mathbf{t}}/n$.

Obviously, a code is $\mathbf{t}$-separating if and only if $\theta_{\mathbf{t}} > 0$.

Proposition 2. *For a binary code, we have* $\theta_{2,1} \geq d - m/2$.

This result was found by Sagalovich [8], but we include a proof for the reader's convenience.

Proof. Let $(\mathbf{c}', \mathbf{c}, \mathbf{a})$ be any three codewords. Since separating weights are invariant over the ensemble of equivalent codes, we can by translation assume that $\mathbf{c}' = \mathbf{0}$. We shall find a lower bound on the separating weights $\theta(\mathbf{0}, \mathbf{c}; \mathbf{a})$.

First we take $(2, 1)$-separation. Let $\mathbf{0}$, $\mathbf{c}$, and $\mathbf{a}$ be rows of a matrix. There are three types of columns; Type R is $(001)^{\mathrm{T}}$ which are the ones giving separation, Type 0 is $(000)^{\mathrm{T}}$, and Type I is $(010)^{\mathrm{T}}$ and $(011)^{\mathrm{T}}$. Let v_i be the number of columns of Type i. We have that $\theta(\mathbf{0}, \mathbf{c}; \mathbf{a}) = v_{\mathrm{R}}$ and $w(\mathbf{c}) = v_{\mathrm{I}}$.

Define

$$\Sigma := d(\mathbf{0}, \mathbf{a}) + d(\mathbf{c}, \mathbf{a}) = 2v_{\mathrm{R}} + v_{\mathrm{I}} = 2\theta(\mathbf{0}, \mathbf{c}; \mathbf{a}) + w(\mathbf{c}).$$

Since Σ is the sum of two distances, we have

$$2d \leq \Sigma \leq 2m,$$

so

$$2\theta(\mathbf{0}, \mathbf{c}; \mathbf{a}) = \Sigma - w(\mathbf{c}) \geq 2d - m.$$

It has also be shown that $\theta_{2,2} \geq 2d - 3m/2$ in a similar way. A corollary is that if $\delta > \frac{3}{4}$, then $\theta_{2,2}$ is non-zero, and the code is $(2, 2)$-separating.

Proposition 3. *Let* $\mathbf{t}$ *be a tuple of natural numbers. If* C_1 *is a* M'-*ary* $[n_1, M]$ *code with separating weight* $\theta'_{\mathbf{t}}$, *and* C_2 *is a* q-*ary* $[n_2, M']$ *code with separating weight* $\theta''_{\mathbf{t}}$, *then the concatenation* C *of the two codes is a* $[n, M]$ *code with* $n = n_1 n_2$ *and separating weight* $\theta_{\mathbf{t}} \geq \theta'_{\mathbf{t}} \theta''_{\mathbf{t}}$.

Proof. Let $(T_1, \ldots, T_z)$ be any $\mathbf{t}$-configuration from C. Then there is a corresponding $\mathbf{t}$-configuration $(T'_1, \ldots, T'_z)$ in C_1, which is separated in at least $\theta'_{\mathbf{t}}$ positions.

Now consider a single position i, where $(T'_1, \ldots, T'_z)$ is separated. Each symbol in this position corresponds to a word in C_2, so $(T'_1, \ldots, T'_z)$ corresponds to a collection of subsets $(T''_1, \ldots, T''_z)$ in C_2. Since $(T'_1, \ldots, T'_z)$ is a separated $\mathbf{t}$-configuration, $(T''_1, \ldots, T''_z)$ must also be a separated $\mathbf{t}$-configuration, and since C_2 has separating weight $\theta''_{\mathbf{t}}$, it follows that $(T''_1, \ldots, T''_z)$ is separated in at least $\theta''_{\mathbf{t}}$ positions.

We conclude that $(T_1, \ldots, T_z)$ is separated in at least $\theta'_{\mathbf{t}} \theta''_{\mathbf{t}}$ positions, and since this holds for any $\mathbf{t}$-configuration, the proposition follows.

Corollary 1. *The concatenation of two* $\mathbf{t}$-*SS is a* $\mathbf{t}$-*SS.*

The current best constructible rate for asymptotic $(2,2)$-SS is 0.026. This was constructed in [5] by concatenating an asymptotic code with $\delta > \frac{3}{4}$ with a small inner code which had been explicitly confirmed to be $(2,2)$-separating. However, Sagalovich [9] had already given a different construction of $(2,2)$-SS with this rate.

The outer code used in the construction is one due to Tsfasman. He showed in [10], that there is an asymptotic class of q-ary codes with rate R and minimum distance δ whenever

$$R + \delta < 1 - (\sqrt{q} - 1)^{-1}.$$

The inner code is the punctured dual C' of a two-error-correcting BCH code with parameters $[126, 14, 55]$. This code was proven to be 3-wise intersecting in [4], a property which is equivalent to $(2,2)$-separation [3]. To see that C' is $(2,2)$-separating, we recall that the dual of 2-BCH has only two weights, $2^{2t} - 2^t$ and $2^{2t} + 2^t$. Consequently C' has $d \geq 2^{2t} - 2^t - 1$ and $m \leq 2^{2t} + 2^t$, and

$$4d - 3m \geq 2^{2t} - 7 \cdot 2^t - 4,$$

which is greater than zero whenever $t \geq 3$. Our code C', has $m = 72$, so $\theta_{2,1} \geq 55 - 72/2 = 19$.

The specific outer code shall have $q = 2^{14}$, and since we require $\delta \approx 0.75$, we get $R \approx 1 - 127^{-1} - 0.75 \approx 0.2421$. The $(2,1)$-separating weight is $\tau_{2,1} \geq 0.75 - 0.5 = 0.25$. The concatenated code $\mathfrak{C}$ will have $R \approx 0.026$, $\delta \approx 0.3274$, and $\tau_{2,1} \geq 0.03770$. As mentioned, this construction is not new. The new result, which will be proved in the next section, is that $\mathcal{C}$ is $(2, \epsilon)$-UPP where ϵ tends to 0 with increasing n.

4 Binary Separating Codes for Fingerprinting

In the sequel, we assume a binary code. Let $m(\mathbf{a}, \mathbf{b}, \mathbf{c})$ be the word obtained by majority voting of the three vectors $\mathbf{a}$, $\mathbf{b}$, and $\mathbf{c}$. That is, in each position i in $m(\mathbf{a}, \mathbf{b}, \mathbf{c})$ contains the symbol which occurs in position i of at least two of the three vectors $\mathbf{a}$, $\mathbf{b}$, and $\mathbf{c}$.

Lemma 1. *If C is an (n, M) $(2,2)$-SS, then for any $P = \{\mathbf{a}, \mathbf{b}\} \subset C$, we have*

$$F(P) \backslash \mathcal{I}_2(C) = \{m(\mathbf{a}, \mathbf{b}, \mathbf{c}) : \mathbf{c} \in C \backslash P\},$$

and

$$\#(F(P) \backslash \mathcal{I}_2(C)) = M - 2.$$

This result was pointed out in [7].

Proof. Let $\mathbf{x}$ be a vector which is not identifiable. Because C is $(2,2)$-separating, the possible parent sets of $\mathbf{x}$ must form a triangle, i.e. $\{\mathbf{a}, \mathbf{b}\}$, $\{\mathbf{a}, \mathbf{c}\}$, and $\{\mathbf{b}, \mathbf{c}\}$. The only vector $\mathbf{x}$ which is feasible for any of the three sets is $m(\mathbf{a}, \mathbf{b}, \mathbf{c})$.

Given a pirate coalition $P = \{\mathbf{a}, \mathbf{b}\}$, there are $M - 2$ possible triangles, for $\mathbf{c} \in C \backslash P$. For each triangle there is one word which is not identifiable.

Lemma 2. *For any strategy $s < \theta_{2,1}$ we get $p_s(P) = 1$.*

Proof. Let $\mathbf{a}$ and $\mathbf{b}$ be a pirate. If the pirates manage to forge a fingerprint $\mathbf{x}$ forming a triangle with a third codeword $\mathbf{c}$, then $\mathbf{c}$ must match $\mathbf{a}$ in s out of the $d(\mathbf{a}, \mathbf{b})$ detectible marks and match $\mathbf{b}$ on the others. These positions are exactly the ones where $(P, \{\mathbf{c}\})$ is separated, so if $s < \theta_{2,1}$, then no such $\mathbf{c}$ can exist.

Theorem 2. *Let C be a $(2,2)$-SS, and let $P \subset C$ be any pirate coalition of size at most $t = 2$. For any strategy s, the probability that P escapes detection is*

$$1 - p_s(P) \le (M-2) \min\left\{ \frac{1}{2}\binom{n\delta}{n\tau_{2,1}}^{-1}, \binom{n\delta}{n\delta/2}^{-1} \right\}.$$

Proof. Let $\kappa(s)$ be the probability of chosing a particular fingerprint given a strategy s. By Lemma 2, we can assume $s \ge \theta_{2,1}$. We have

$$\kappa(d(\mathbf{a},\mathbf{b})/2) = \binom{d(\mathbf{a},\mathbf{b})}{d(\mathbf{a},\mathbf{b})/2}^{-1} \le \binom{n\delta}{n\delta/2}^{-1}.$$

If $s \ne d(\mathbf{a},\mathbf{b})/2$, we get

$$\kappa(s) = \frac{1}{2}\binom{d(\mathbf{a},\mathbf{b})}{s}^{-1} \le \frac{1}{2}\binom{n\delta}{n\tau_{2,1}}^{-1},$$

The number of non-identifiable words is $\mu = M - 2$, so there cannot be more than μ feasible false fingerprints allowing the pirates to escape. Multiplying μ with $\kappa(s)$ we get the theorem.

If we take assymptotic values for increasing n, we arrive at the following corollary, where H is the natural entropy function.

Corollary 2. *Any $(2,2)$-SS is a $(2,\epsilon)$-UPP with*

$$\epsilon \le e^{\lambda n},$$

where

$$\lambda = R\ln 2 - H(\tau_{2,1}/\delta)\delta.$$

Considering our $(2,2)$-SS $\mathfrak{C}$ with $R \approx 0.026$ and $\tau_{2,1} \ge 0.03770$, we get the following values:

$$\lambda \approx -0.09891,$$
$$\epsilon \le 0.9058^n,$$

which leads to the following theorem.

Theorem 3. *There is a constructible asymptotic binary code with $(2,\epsilon)$-UPP with rate $R \approx 0.026$ and failure rate $\epsilon \le 0.9058^n$.*

The code $\mathfrak{C}$ has better rate than any $(2,\epsilon)$-IPP known from past literature. Though the code has been known, it is a new result that it has UPP, or even IPP. Unfortunately, the results does not extend very well, since Theorem 1 rules out any q-ary (t,ϵ)-UPP codes for $t > q$.

5   Discussion

We have introduced a probabilistic 2-IPP code with a rate better than anything we have managed to locate in the literature. Furthermore, with this code, there is no risk of accusing an innocent user. It still remains to construct an efficient tracing algorithm usable with the present code.

We have seen that codes with (t, ϵ)-UPP cannot exist for $t > q$, but it is an open question whether the present techniques can be modified to construct (t, ϵ)-IPP codes with $t > q$. It would also be interesting to construct q-ary codes with (q, ϵ)-UPP for arbitrary q.

References

1. A. Barg, G.R. Blakley, and G. Kabatiansky. Good digital fingerprinting Codes. Technical report, DIMACS, 2001.
2. Dan Boneh and James Shaw. Collusion-secure fingerprinting for digital data. *IEEE Trans. Inform. Theory*, 44(5):1897–1905, 1998. Presented in part 1995, see Springer LNCS.
3. Bella Bose and T.R.N. Rao. Separating and completely separating systems and linear Codes. *IEEE Trans. Comput.*, 29(7):665–668, 1980.
4. Gérard Cohen and Gilles Zemor. Intersecting Codes and independent families. *IEEE Trans. Inform. Theory*, 40:1872–1881, 1994.
5. Gérard D. Cohen, Sylvia B. Encheva, Simon Litsyn, and Hans Georg Schaathun. Intersecting Codes and separating Codes. *Discrete Applied Mathematics*, 2001. To appear.
6. A.D. Friedman, R.L. Graham, and J.D. Ullman. Universal single transition time asynchronous state assignments. *IEEE Trans. Comput.*, 18:541–547, 1969.
7. Jacob Löfvenberg. *Codes for Digital Fingerprinting*. PhD thesis, Linköpings Universitet, 2001. http://www.it.isy.liu.se/publikationer/index.html.
8. Yu.L. Sagalovich. A method for increasing the reliability of finite automata. *Problems of Information Transmission*, 1(2):27–35, 1965.
9. Yu.L. Sagalovich. Separating systems. *Problems of Information Transmission*, 30(2):105–123, 1994.
10. Michael A. Tsfasman. Algebraic-geometric Codes and asymptotic Problems. *Discrete Appl. Math.*, 33(1-3):241–256, 1991. Applied algebra, algebraic algorithms, and error-correcting codes (Toulouse, 1989).

Copyright Control and Separating Systems

Sylvia Encheva[1] and Gérard Cohen[2]

[1] HSH, Bjørnsonsg. 45, 5528 Haugesund, Norway
[2] ENST, 46 rue Barrault, 75634 Paris Cedex 13, France

Abstract. Separating systems have earlier been shown to be useful in designing asynchronous sequential circuits, finite automata and fingerprinting. In this paper we study the problem of constructing (s,1)-separating systems from codes and designs.

Keywords: separating systems, frameproof codes, designs, copyright protection.

1 Introduction

Monitoring and owner identification applications place the same watermark in all copies of the same content. However, electronic distribution of content allows each copy distributed to be customized for each recipient. This capability allows a unique watermark to be embedded in each individual copy. Transactional watermarks, also called fingerprints, allow a content owner or content distributor to identify the source of an illegal copy. This is potentially valuable both as a deterrent to illegal use and as technological aid to investigation.

In the case of collusion attacks, a hacker uses several copies of one piece of media, each with a different watermark, to construct a copy with another watermark. Resistance to collusion attacks can be critical in a fingerprinting application, which entails putting a different mark in each copy of a piece of media. For example, in the DiVX application, a hacker can buy a number of DiVX players, and play one movie on all of them to obtain differently watermarked copies. If hackers can embed valid authentication marks, they can cause the watermark detector to accept bogus or modified media. "Frameproof" codes were introduced in [2] (see also [10]) as a method of "digital fingerprinting" which prevents a coalition of a given size from forging a copy with no member of the coalition being caught, or from framing an innocent user.

The outline of the paper is as follows. Definitions and basic results are presented in Sect. 2. In Sect. 3, some equidistant codes are shown to provide protection against coalitions of size 3. In Sect. 4 designs are used to deal with the general case. Section 5 presents asymptotic constructions *via* concatenation, for which seeds described in Sect. 3 can be used.

2 Definitions and Basic Results

We use the notation of [10] for fingerprinting issues and of [7] for codes, designs and Hadamard matrices.

M. Fossorier, T. Hoeholdt, and A. Poli (Eds.): AAECC 2003, LNCS 2643, pp. 79–86, 2003.

We identify a vector with its *support*, set of its non-zero positions; for a pair of vectors (u, v), we denote by $u \cap v$ the vector having for support the intersection of the supports of u and v. For any positive real number x we shall denote by $\lceil x \rceil$ the smallest integer at least equal to x.

A set $\Gamma \subseteq GF(q)^n$ is called an (n, M, d)-*code* if $|\Gamma| = M$ and the minimum Hamming distance between two of its elements (codewords) is d.

Suppose $\mathcal{C} \subseteq \Gamma$. For any position i define the *projection* $P_i(\mathcal{C}) = \bigcup_{a \in \mathcal{C}} a_i$.

Define the *feasible set* of $\mathcal{C}$ by

$$F(\mathcal{C}) = \{x \in GF(q)^n : \forall i, x_i \in P_i(\mathcal{C})\}.$$

The feasible set $F(\mathcal{C})$ represents the set of all possible n-tuples (descendants) that could be produced by the coalition $\mathcal{C}$ by comparing the codewords they jointly hold. Observe that $\mathcal{C} \subseteq F(\mathcal{C})$ for all $\mathcal{C}$, and $F(\mathcal{C}) = \mathcal{C}$ if $|\mathcal{C}| = 1$.

Now, if there is a codeword $a \in F(\mathcal{C}) \setminus \mathcal{C}$, then the user who owns codeword a can be "framed" because the coalition $\mathcal{C}$ can actually construct a; the following definition from [2] to forbid this situation.

Definition 1. *A code Γ is called $(s, 1)$-separating if, for every $\mathcal{C} \subseteq \Gamma$ such that $|\mathcal{C}| \leq s$, holds $F(\mathcal{C}) \cap \Gamma = \mathcal{C}$.*

3   Separation via Equidistant Codes

Positions where at least three codewords coincide are denoted $\overbrace{*...*}$, positions where two codewords coincide are denoted $\overbrace{*...*}$ or $\overbrace{...}$ and positions where they have different coordinates are left empty.

Proposition 1. *Any equidistant binary code is $(2, 1)$-separating.*

Proof. Suppose c_1, c_2 are codewords in an equidistant binary code C with distance d. Any third codeword c_3 that would be framed by c_1, c_2 should have nonzero support in the same positions where c_1, c_2 intersect. Since c_1 and c_3 have different values at d positions and c_2 and c_3 have different values at another d positions, then c_1 and c_2 would have different values at $2d$ positions. It is a contradiction.   □

Proposition 2. *An equidistant q-ary code with odd distance is $(3, 1)$-separating.*

Proof. Suppose C is not $(3,1)$-separating. Then w.l.o.g. we may assume that there are four codewords $c_1, c_2, c_3, c_4 \in C$ such that c_4 is a descendant of the coalition c_1, c_2, c_3. Let

c_1, c_2 coincide on positions $1, 2, ..., \alpha + \beta = n - d$,

c_1, c_3 coincide on positions $1, 2, ..., \alpha; \alpha + \beta + 1, ..., \alpha + 2\beta$,

c_2, c_3 coincide on positions $1, 2, ..., \alpha; \alpha + 2\beta + 1, ..., \alpha + 3\beta$,

c_1, c_4 coincide on positions $1, 2, ..., \alpha + x; \alpha + \beta + 1, ..., \alpha + \beta + y$,

$\alpha + 2\beta + z + 1, ..., \alpha + 3\beta + r_1$,

c_2, c_4 coincide on positions $1, 2, ..., \alpha + x$; $\alpha + 2\beta - y + 1, ..., \alpha + 2\beta + z$; $\alpha + 3\beta + r_1 + 1, ..., \alpha + 3\beta + r_1 + r_2$;

c_3, c_4 coincide on positions $1, 2, ..., \alpha$; $\alpha + \beta - x + 1, ..., \alpha + \beta + y$; $\alpha + 2\beta + 1, ..., \alpha + 2\beta + z$; $\alpha + 3\beta + r_1 + r_2 + 1, ..., \alpha + 3\beta + r_1 + r_2 + r_3 = n$.

Thus the 4 times n array, obtained from c_1, c_2, c_3, c_4, may be described as follows

Combining $\alpha + \beta = n - d$ (c_1, c_2 coincide on $n - d$ positions), with
$\alpha + x + y + \beta - z + r_1 = n - d$ (c_1, c_4 coincide on $n - d$ positions),
$\alpha + x + \beta - y + z + r_2 = n - d$ (c_2, c_4 coincide on $n - d$ positions) and
$\alpha + \beta - x + y + z + r_3 = n - d$ (c_3, c_4 coincide on $n - d$ positions)
we obtain that
$$z = x + y + r_1$$
$$y = x + z + r_2$$
$$x = y + z + r_3,$$
i.e. $x = y = z = r_1 = r_2 = r_3 = 0$.

Thus, counting the coordinates of c_1, we obtain $\alpha + 3\beta = n$. This last equality together with $\alpha + \beta = n - d$ gives $d = 2\beta$. Therefore, if d is odd, C is (3,1)-separating. $\quad\square$

Example 1. The equidistant $(4, 9, 3)_3$ tetracode is (3,1)-separating.

Corollary 1. *Let C be an equidistant q-ary code with distance d. Provided that no four codewords coincide on $n - 3\frac{d}{2}$ coordinates, then C is (3,1)-separating.*

Proof. The proof of Proposition 2 shows that the following two conditions $\alpha + \beta = n - d, \alpha + 3\beta = n$ should be satisfied for any code that is not (3,1)-separating, α being the number of positions on which four codewords coincide. Thus, if $\alpha \neq n - 3\frac{d}{2}$, the code C is (3,1)-separating. $\quad\square$

The case d even should be considered carefully. Below we deal with two such equidistant codes and show that one of them is (3,1)-separating and the other one is not.

Example 2. Let $H_{4(2r-1)}$ be the matrix obtained from a Hadamard matrix of Paley type [7] of order $4(2r - 1)$ when +1's are replaced by 0's and -1's by 1's. Then, the rows of $H_{4(2r-1)}$ constitute a binary $(3, 1)$-separating code. A proof is given in Appendix.

Let H_n be a Sylvester type matrix of order $n = 2^i$ [7]. If $+1$'s are replaced by 0's and -1's by 1's, H_n is changed into the binary matrix A_n [7]; then A_n is not $(3,1)$-separating (in fact it is equivalent to a linear code, and as such cannot be by Proposition 7).

4   Designs and Separation

Proposition 3. *[8] An equidistant q-ary code with parameters*

$$n = \frac{q^2\mu - 1}{q - 1}, M = q^2\mu, d = q\mu$$

exists if and only if there exists an affine design with parameters

$$v = q^2\mu, k = q\mu, \lambda = \frac{q\mu - 1}{q - 1}, r = \frac{q^2\mu - 1}{q - 1}, b = \frac{q^3\mu - q}{q - 1}.$$

Proposition 4. *Equidistant q-ary codes with parameters*

$$n = \frac{q^2\mu - 1}{q - 1}, M = q^2\mu, d = q\mu$$

are $(s,1)$-separating, provided $q \geq s \geq 2$.

Proof. Follows immediately by Proposition 8 and Proposition 3 for all $q \geq s \geq 2$. □

Example 3. The enumeration of all affine 2-(27,9,4) designs in [6] shows that there are exactly 68 inequivalent optimal equidistant codes with parameters

$$n = 13, M = 27, d = 9, q = 3.$$

By Proposition 8 they are $(3,1)$-separating.

Proposition 5. *Let D be a block design, such that any two blocks intersect on $0 \leq i_1 < i_2 < ... < i_l$ positions. D is $(s,1)$-separating if*

$$si_l < k.$$

Proof. Take any $s + 1$ blocks in D. Since any two blocks intersect on at most i_l positions and $si_l < k$, the $s + 1$-st block will contain at least $k - si_l$ points that do not belong to any of the other s blocks. Therefore, there are at least $k - si_l$ columns of type $\overbrace{0...0}^{s}1$, i.e. D is $(s,1)$-separating. □

Proposition 6. *Let D be a block design, such that any two blocks intersect on $0 \leq i_1 < i_2 < ... < i_l$ positions. D is $(2,1)$-separating if*

$$i''' + i'' - i' \neq k$$

for any $i', i'', i''' \in \{i_1 < i_2 < ... < i_l\}$ with $i' \leq min\{i'', i'''\}$.

Proof. If D is not $(2,1)$-separating, then there are no columns of either type $(001)^t$ or $(110)^t$ in a particular array c_1, c_2, c_3, where c_1, c_2, c_3 are blocks in D. Suppose $|c_1 \cap c_2| = i', |c_1 \cap c_3| = i'', |c_2 \cap c_3| = i'''$. To avoid columns $(110)^t$ we need $i' \leq i''$ and $i' \leq i'''$. A column $(001)^t$ will occur if

$$i' + i'' - i' + i''' - i' = i''' + i'' - i' < k.$$

Since the number of points in a block is exactly k we conclude that D is $(2,1)$-separating if $i''' + i'' - i' \neq k$, with $i', i'', i''' \in \{i_1 < i_2 < ... < i_l\}$ and $i' \leq min\{i'', i'''\}$. $\square$

Example 4. The set of codewords of weight 11 in the [47,24,11] binary code is a design in which blocks intersect on 1, 3 or 5 positions. It thus yields a (47, 4324) code which is $(2,1)$-separating.

Corollary 2. *A quasi-symmetric design is a $(2,1)$-separating if $2i'' - i' \neq k$.*

Proof. The condition $i''' + i'' - i' \neq k$ in Proposition 6 is satisfied for any quasi-symmetric design with $2i'' - i' \neq k$ since such designs have exactly two intersection numbers. $\square$

Example 5. A quasi-symmetric 2-(31,7,7) design with $i' = 1, i'' = 3$ [12] is a (31, 155) code and $(2,1)$-separating.

5   Infinite Constructions via Linear Codes

We denote by $C[n, k, d]_q$ (or simply $C[n, k]_q$ when d is irrelevant) a *linear* code of length n, dimension k over $GF(q)$ and minimum distance d. The *rate* of C, linear or not, is defined as $n^{-1} \log_q |C|$.

Proposition 7. *If C is a q-ary linear $(s,1)$-separating code, then $s \leq q$.*

Proof. Let c_1, c_2 be two linearly independent codewords and $T = \{c_1 \cup c_2 + \gamma c_1, \gamma \in GF(q)\}$. We now prove that in every position i, at least one codeword in T has a 0. If $c_2^i = 0$, this holds, so assume $c_2^i \neq 0$. Then $c_2 + (-c_1^i)^{-1} c_2^i c_1$ has 0 in position i, as required. Hence $\mathbf{0} \in F(T)$, with $|T| = q + 1$. $\square$

Proposition 8. *[11] Let C be a code of length n and minimum distance d. If $sd \geq (s-1)n + 1$, then C is $(s,1)$-separating.*

The following example shows a family of codes meeting with equality the bounds given in the previous two propositions.

Example 6. The Maximum-distance separable codes (satisfying the Singleton bound $d \leq n - k + 1$ with equality; see, e.g. [7]) $[q+1, 2, q]_q$ are $(q, 1)$-separating.

We now construct infinite families of linear $(q, 1)$-separating codes.

The ternary construction will make use of three ingredient codes, and apply twice the concatenation method.

Proposition 9. *[5] If Γ_1 is a $(s, 1)$-separating, M'-ary (n_1, M) code and Γ_2 a $(s, 1)$-separating, q-ary, (n_2, M') code, then the concatenated code $\Gamma := \Gamma_2 \circ \Gamma_1$ is a $(s, 1)$-separating $(n_1 n_2, M)_q$ code.*

The first seed is the remarkable $[4, 2, 3]_3$ tetracode T, $(3, 1)$-separating by Proposition 2. Let R_1 be the $[10, 4, 7]_{3^2}$ Reed-Solomon code, which is $(3, 1)$-separating by Proposition 8.

The concatenated code $T \circ R_1$ has parameters $[40, 8]_3$ and, by Proposition 8, is $(3, 1)$-separating. In order to produce infinite families of separating codes, we need the following constructive result from Tsfasmann [13].

Lemma 1. *For any $\alpha > 0$ there is an infinite family of codes $\mathcal{N}$ with parameters $[N, NR, N\delta]_q$ for $N \geq N_0(\alpha)$ and*

$$R + \delta \geq 1 - (\sqrt{q} - 1)^{-1} - \alpha.$$

Then we take the infinite family $\mathcal{N}$ of codes with parameters $[N, K, D = \lceil 2N/3 \rceil + 1]_{3^8}$ of rate $1/3 - (3^4 - 1)^{-1}$, and the concatenated code $T \circ R_2 \circ \mathcal{N}$ gives an infinite family of linear ternary $(3, 1)$-separating codes with rate approximately $77/1200 \approx 0.0642$.

We sketch the constructions for the cases $q = 4$ and $q = 5$. As in the previous example, we concatenate three codes to build the infinite family. Each one has $d/n > (q - 1)/q$ and thus is $(q, 1)$-separating by Proposition 8.

For $q = 4$, the first two are doubly extended Reed-Solomon codes. We take successively: $C_1[5, 2, 4]_4$ of Example 6 with $q = 4$; then $C_2[17, 5, 13]_{4^2}$, getting $C_1 \circ C_2[85, 10]_4$; finally, $C(N)[N, K, D = \lceil 3N/4 \rceil + 1]_{4^{10}}$ with rate $\approx 1/4 - (4^5 - 1)^{-1} \approx 1/4$. The outcome is an infinite constructive family of linear quaternary $(4, 1)$-separating codes with rate approximately $1/34 \approx 0.0294$.

For $q = 5$, we consider $C_1[6, 2, 5]_5$ of Example 6 with $q = 5$.

Then $C_2[26, 6, 21]_{5^2}$, getting $C_1 \circ C_2[156, 12]_5$ of rate $1/13$; finally the family $C(N)$ of codes of rate $\approx 1/5$ over $q = 5^{12}$, getting linear quinternary $(5, 1)$-separating codes with rate $1/65 \approx 0.0153$.

6   Non-constructive Lower Bounds

There are classical probabilistic methods with expurgation producing lower bounds. A specialization of a result from [1] gives:

Table 1. Asymptotic lower bounds on rates for $(s,1)$-separation

		Lower bounds			
		Linear		Non-Linear	
s	q	Const.	Non-const.	Const.	Non-const.
2	2	0.156[3]	0.205	—	0.205
3	2	∃	∃	0.043 [4]	0.063
4	2	∃	∃	0.018 [4]	0.022
3	3	0.064	—	—	0.106
4	4	0.029	—	—	0.099
5	5	0.015	—	—	0.051

Proposition 10. *There exist infinite families of $(s,1)$-separating codes over $GF(q)$ with rate $R(s,q) \geq 1 - s^{-1}\log_q(q^s - (q-1)^s)$.*

Appendix
We have the following result from [7].

Theorem 1. *If H_n is an n times n Hadamard matrix with $n > 1$, then n is even and for any two distinct rows of H there are precisely $\frac{n}{2}$ columns in which the entries in the two rows agree. Further, if $n > 2$ then n is divisible by 4, and for any three distinct rows of H there are precisely $\frac{n}{4}$ columns in which the entries in all three rows agree. The same statements hold for columns of H.*

Let $H_{4(2r+1)}$ be the matrix obtained from a Hadamard matrix [7] of order $4(2r+1)$ when $+1$'s are replaced by 0's and -1's by 1's.

Proposition 11. *$H_{4(2r+1)}$ is a binary (3,1)-separating code.*

Proof. Suppose $H_{4(2r+1)}$ is not (3,1)-separating, i.e. there are four rows $c_1, c_2, c_3, c_4 \in H_{4(2r+1)}$ yielding no column of either type $(0001)^t$ or $(1110)^t$. Two cases may occur. If no c_i is the all 1 vector, then, w.l.o.g. we may assume that these four rows form the following array:

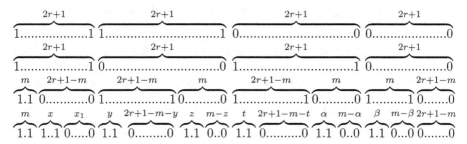

where $2r+1-m-x = x_1$. By Theorem 1 any three rows in $H_{4(2r+1)}$ agree in $2r+1$ positions. Applying Theorem 1 to the rows c_2, c_3, c_4 we obtain $m+t = z$. Since $z \leq m$ we conclude that $t = 0$ and $m = z$.

In a similar way, considering c_1, c_2, c_4 gives $x = 0$ and $m = \beta$, and c_1, c_3, c_4 gives $m = \alpha$. Any two rows of weight $2(2r+1)$ in $H_{4(2r+1)}$ have $2r+1$ columns of type $(11)^t$. Therefore c_4 should have $2r+1$ 1's in positions $\{2(2r+1)+1, ..., 4(2r+1)\}$, which leads to $t + \alpha + \beta = 2r + 1$. This contradicts our assumption that $\alpha = \beta = m$ and $t = 0$. Therefore, there is a column of either type $(0001)^t$ or $(1110)^t$ in $H_{4(2r+1)}$.

The second case, where one of c_1, c_2, c_3, c_4 is the all 1 vector, leads to a contradiction in a similar way. This concludes the proof that $H_{4(2r+1)}$ is (3,1)-separating.   □

Acknowledgments. The authors wish to thank the referees for their useful comments.

References

1. A. Barg, G. Cohen, S. Encheva, G. Kabatiansky and G. Zémor, "A hypergraph approach to the identifying parent property", *SIAM J. Disc. Math.*, vol. 14, no. 3, pp. 423–431 (2001).
2. D. Boneh and J. Shaw, "Collusion-secure fingerprinting for digital data", *Springer-Verlag LNCS* 963 pp. 452–465 (1995).
3. G. Cohen and G. Zémor,"Intersecting codes and independent families", *IEEE Trans. Inform. Theory*, 40, pp. 1872–1881, (1994).
4. G. Cohen and S. Encheva, "Efficient constructions of frameproof codes", *Electronics Letters* 36 (22) pp. 1840–1842 (2000).
5. G. Cohen, S. Encheva and H.G. Schaathun, "More on (2,2)-separating systems", *IEEE Trans. Inform. Theory*, 48, pp. 2606–2609, (2002).
6. C. Lam and V.D.Tonchev, "Classification of affine resolvable 2-(27,9,4) designs", *J. Statist. Plann. Inference* 56 (2), pp. 187–202 (1996).
7. F.J. MacWilliams and N.J.A. Sloane, *The Theory of Error-Correcting Codes*, North-Holland, Amsterdam (1977).
8. V.S. Pless, W.C. Huffman, Editors, *Handbook of Coding Theory*, Elsevier, Amsterdam (1998).
9. Yu.L. Sagalovitch, "Separating systems", *Problems of Information Transmission* 30 (2) pp. 105–123 (1994).
10. D.R. Stinson, Tran Van Trung and R. Wei, "Secure Frameproof Codes, Key Distribution Patterns, Group Testing Algorithms and Related Structures", *J. Stat. Planning and Inference* 86 (2) pp. 595–617 (2000).
11. D.R. Stinson and R. Wei, "Combinatorial properties and constructions of traceability schemes and frameproof codes", *SIAM J. Discrete Math* 11 pp. 41–53 (1998).
12. V. Tonchev, "Quasi-symmetric 2-(31,7,7) designs and a revision of Hamada's conjecture", *J. Comb. Theory* Ser. A 42 pp. 104–110 (1986).
13. M.A. Tsfasmann, "Algebraic-geometric codes and asymptotic problems", *Discrete Appl. Math.* 33 241–256 (1991).

Unconditionally Secure Homomorphic Pre-distributed Commitments

Anderson C.A. Nascimento[1], Akira Otsuka[1], Hideki Imai[1], and
Joern Mueller-Quade[2]

[1] Institute of Industrial Science, The University of Tokyo
4-6-1, Komaba, Meguro-ku, Tokyo, 153-8505 Japan
{anderson,otsuka,imai}@iis.u-tokyo.ac.jp
[2] Universitaet Karlsruhe, Institut fuer Algorithmen und Kognitive Systeme
Am Fasanengarten 5, 76128 Karlsruhe, Germany
{muellerq}@ira.uka.de

Abstract. In this paper we deal with unconditionally secure commitment schemes based on pre-distributed data. We provide bounds for the amount of data which has to be pre-distributed to the participants of the commitment, thus solving an open problem stated in the literature. We also introduce the issue of homomorphism in unconditionally secure commitment schemes. We provide a definition and a construction based on modules of mappings over finite rings. As an application of our constructions, we provide a new pre-distributed primitive which yields non-interactive unconditionally secure zero knowledge proofs of any polynomial relation among commitments.

1 Introduction

Pre-distribution of secret correlated data is a major resource for the implementation of secure cryptographic protocols. Examples are: pre-distribution of cryptographic keys [1], digital signatures [5], broadcast channels [2], among others. A significant advantage of protocols based on pre-distributed data is the possibility of achieving information theoretical security. This is an interesting feature when highly important transactions are in question and unproven number theoretical assumptions are not desired.

Recently, Rivest [3] proved that pre-distributed data is sufficient for the implementation of unconditionally secure bit commitment schemes. He proposed a very simple protocol where two parties, here on called Alice and Bob, receive secret data from a trusted initializer in a setup phase and later on, are able to implement an information theoretically secure bit commitment scheme.

Later, Blundo et al. [4] proposed a model for all bit commitments based on pre-distributed data and proved a lower bound on the cheating probability for the sender of the commitment. In [4], it was stated as an open problem to characterize the communication complexity between the trusted initializer and the players involved in the commitment.

M. Fossorier, T. Hoeholdt, and A. Poli (Eds.): AAECC 2003, LNCS 2643, pp. 87–97, 2003.

In this contribution we solve this problem. By using information theoretical techniques, we bound the minimum amount of data that has to be transmitted to Alice and Bob during the setting phase. Also, we prove new bounds on the cheating probability for the sender of the commitment. It is noteworthy that to prove our bounds, we show a very interesting connection between pre-distributed commitment schemes and hypothesis testing.

We then introduce the problem of implementing unconditionally secure homomorphic commitment schemes. Homomorphic commitments are very useful tools in two and multiparty computation protocols and zero knowledge proofs of knowledge. We propose a new construction of unconditionally secure non-interactive commitment schemes based on modules of mappings and show that this construction yields homomorphic commitment schemes.

Finally, we demonstrate an application of our construction by showing concrete protocols implementing non-interactive information theoretically secure zero knowledge proofs of any polynomial relation among commitments.

This paper is organized as follows: in Sect. 2 we introduce our notation and preliminaries, in Sect. 3, we introduce our model for pre-distributed commitments. In Sect. 4, we prove our new bounds. In Sect. 5, we show a construction for homomorphic commitments. Finally, in Sect. 6, we show protocols implementing zero knowledge proofs of polynomial relations among commitments.

2   Preliminaries

In this section, we briefly review Rivest's scheme, some basic notions of hypothesis testing and some results from [4].

2.1   Commitment Schemes

A bit commitment scheme consists of two phases, a commit and an unveiling phase. In the commit phase, a sender, usually called Alice, commits to a bit b and sends a proof of this commitment to a receiver, usually called Bob. During the unveiling phase, Alice opens the commit and reveals the value of the bit b to Bob. A secure bit commitment protocol is secure if it is concealing, binding and correct. A bit commitment is binding if Alice is not able to unveil both b and $\bar{b}$ with high probability. It is concealing if after receiving Alice's commitment, Bob is unable to learn anything on the committed bit before the unveiling phase. And it is correct if its probability of failure for honest Alice and honest Bob is zero.

In our paper we deal with generalized commitments where Alice wants to commit to a number belonging to Z_p, for some prime p.

The security of commitment schemes can be based on some unproven computational assumptions. In this case, the players have no probability of successfully cheating as far as some computational tasks are hard to be performed (like factoring or computing discrete logarithms).

Rivest showed that if some pre-distributed data is available to Alice and Bob (he called it the trusted initializer model) unconditionally secure commitments can be implemented. We briefly review Rivest's scheme. It allows Alice to commit to an element of Z_p, for some prime p.

Initialization Phase

1. The Trusted Initializer (TI) selects at random $a, b, d \in Z_p$ and forms the function: $f(x) = ax + b \bmod p$.
2. The TI sends $f(x)$ to Alice and $t = f(d) = ad + b \bmod p$ and d to Bob over secure channels. We call $f(x)$ a commit key and the pair (t, d) a verification key.

Commit Phase

1. If Alice wants to commit to a number $g \in Z_p$, she sends $h = f(g) \bmod p$ to Bob.

Opening Phase

1. Alice sends the function $f(x)$ and g to Bob. Bob accepts the commitment iff $f(g) = h \bmod p$ and $t = f(d) \bmod p$

Is is easy to see that a cheating Alice is detected at the opening phase with probability $1 - 1/p$ (she only succeeds if she can correctly guess the value d which was given to Bob, so for large p the protocol is binding. However, note that the protocol proposed by Rivest is not perfectly concealing, since Bob learns if Alice's commitment is equal to d or not.

2.2   Hypothesis Testing

It is important to note that at the end of the opening phase, Bob's acceptance of the commitment depends upon the result of a test. The problem of deciding which one of two hypothesis (H_0 and H_1) is true given some data represented by a random variable U (which may be the outcome of a measurement) is a well known problem in information theory and is called hypothesis testing. In the case of a bit commitment scheme, Bob has to decide between two hypothesis (1) Alice has not cheated during the bit commitment protocol and (2) Alice has changed the value of the commitment between the commit and the open phase. We now review some basic results from hypothesis testing.

Hypothesis testing is the task of deciding between two hypothesis (usually called H_0 and H_1) given some data represented by a random variable U. The probability distribution for U when H_0 (H_1) is the right hypothesis is denoted by $Q(Z)$. Let G be a decision rule such that if $G(U) = 0$ ($G(U) = 1$) it means that H_0 (H_1) is accepted as being the correct hypothesis. We define two probabilities of error: α and β. α stands for the probability that $G(U) = 0$ given that H_1 is the correct hypothesis whereas β denotes the probability that $G(U) = 1$ when H_0 is the correct hypothesis.

It follows from the Neyman-Pearson theorem that the optimal test is to assume H_0 as being the right hypothesis iff $\log \frac{Q(u)}{Z(u)} \geq T$, where T is a threshold depending on α and u is the outcome of the measurement.

One of the major results in hypothesis testing establishes a trade-off between α and β. [8]

Theorem 1. *In any hypothesis testing, the error probabilities α and β satisfy $d(\alpha, \beta) \leq L(Q; Z)$, where $d(\alpha, \beta) = \alpha \log \frac{\alpha}{1-\beta} + (1 - \alpha) \log \frac{1-\alpha}{\beta}$ and $L(Q; Z) = \sum_u Q(u) \log \frac{Q(u)}{Z(u)}$.*

The last theorem can be generalized to the case that some side information C is available. [8]

Theorem 2. *The average error probabilities $\bar{\alpha} = \sum_v P_v(v)\alpha(v)$ and $\bar{\beta} = \sum_v P_v(v)\beta(v)$ satisfy $d(\bar{\alpha}, \bar{\beta}) \leq \sum_v P_v(v)L(Q; Z|V = v)$, where $L(Q; Z|V = v) = \sum_u Q(u|v) \log \frac{Q(u|v)}{Z(u|v)}$.*

In [4] a model was proposed for pre-distributed bit commitments. It was proved that in their model, no bit commitment protocol can be perfectly binding. In the next section we propose a new model based on information theory where we generalize the results of [4] and provide new bounds.

3   A Model

In our model there are three parties, a trusted initializer, Ted, a sender of the commitment, Alice, and a receiver, Bob. At the beginning of the protocol, Alice and Bob receive some secret data from Ted. We designate Alice's and Bob's secret data and theirs respective random variables by U_a and U_b, respectively. We assume that Ted chooses U_a and U_b from the same domain $\mathcal{U}$. The domain where the committed values are chosen is denoted by $\mathcal{B}$. The value that Alice wants to commit to as well as its random variable is represented by S.

The commit algorithm consists of Alice using a publicly known algorithm $Commit(S, U_a) = C$ which generates an output value which is sent to Bob. The domain where the commitments are taken from is denoted by $\mathcal{C}$. We denote the random variable associated to the output of the algorithm $Commit$ by C.

The opening phase consists of Alice announcing the secret data received from Ted, U_a, and the value of the commitment, s to Bob. Bob performs a test by using a publicly known algorithm $Test(U_a, U_b, S, C)$ and depending on the result accepts Alice's commitment or not.

The protocol is concealing if the information sent in the committing phase, C, reveals nothing about the committed value s. Mathematically, $I(C : S|U_b) = 0$, where $I(\cdot : \cdot|\cdot)$ is the Shannon conditional mutual information.

The protocol is β-binding if the probability that a dishonest Alice is not caught is at most β, that is: $P(Test(U_a, U_b, S, C) = $accept $\wedge$ $Test(U_a, U_b, S^*, C) = $ accept; $S^* \neq S) < \beta$.

Finally, the protocol is correct if an honest Alice cannot have her commitment rejected, $P(Test(U_a, U_b, S, C) =$reject$|$ Alice is honest$) = 0$.

We denote the average probability of successfully cheating for Alice by $\bar{\beta}$, where the average is taken over all the possible values U_a, U_b and s may take.

Definition 1. *A pre-distributed bit commitment scheme is $\beta-secure$ if the following conditions hold:*

$$I(C : S|U_b) = 0$$
$$P(Test(U_a, U_b, S, C) = accept \wedge Test(U_a, U_b, S^*, C) = accept; S^* \neq S) < \beta$$
$$P(Test(U_a, U_b, S, C) = reject| \text{ Alice is honest}) = 0$$

where U_a and U_b are the data pre-distributed to Alice and Bob, S is the value Alice is committing to, C is the output of a publicly known Commit algorithm and Test is another publicly known algorithm. A $\bar{\beta}-secure$ protocol is defined in the same way, but it bounds the average probability of successfully cheating for Alice instead of the maximum probability of successfully cheating.

4   Bounds

In this section we prove bounds on the size of the data which has to be distributed to Alice and Bob to implement a $\bar{\beta}-$ secure protocol.

First we review a bound on Alice's cheating probability which was proved in [4].

Proposition 1. *Alice's average cheating probability $\bar{\beta}$ is lowerbounded by $2^{-H(U_b|U_a)}$.*

Proof. It is clear that if Alice can correctly guess Bob's secret data U_b she should be able to cheat since the security of the protocol comes from the secrecy of the pre-distributed data U_a, U_b. Alice's average probability of error when guessing Bob's secret data is lowerbounded by $\bar{\beta} \geq 2^{-H(U_b|U_a)}$. ∎

Following a similar approach used to study authentication codes [6], we now use the theory of hypothesis testing to prove a new lower bound on Alice's cheating probability. The intuition behind the result is that a possible strategy (maybe not optimal) for Alice is to completely ignore the data she received from Ted and guess a new value for the data she is going to use to commit to Bob as well as to guess Bob's secret information. Although it seems to be not a good strategy for Alice, we use this result to prove other lower bounds on the dimensions of the data that has to be pre-distributed.

As explained in Sect. 2, the test which is performed by Bob at the end of the opening phase can be understood as a hypothesis testing problem. The receiver of the commitment wants to distinguish between two situations: (I) The sender behaved honestly (hypothesis H_0) and (II) The sender has cheated (hypothesis H_1). Under hypothesis H_0 the probability distribution that generates the commit and open information $[U_a, U_b]$ is the true probability distribution generated by Ted's choices $P(U_a, U_b)$. Under hypothesis H_1, $[U_a, U_b]$ is generated by another probability distribution $Z(U_a, U_b)$, where $Z(\cdot)$ represents a cheating strategy for Alice.

Proposition 2. *In any pre-distributed bit commitment scheme, the sender can always cheat with average probability larger or equal than $\bar{\beta} \geq 2^{-I(U_a : U_b)}$*

Proof. We first remember that α equals the probability that Bob rejects a legal commitment from Alice and β equals the probability that Bob accepts a commitment from a dishonest Alice. ¿From the correctness condition of a bit commitment scheme we have that $\alpha = 0$. By making $Z(u) = P(U_a).P(U_b)$ and $Q(u) = P(U_a, U_b)$ in Theorem 2 we have that:

$$d(\bar{\alpha}, \bar{\beta}) \leq \sum_{U_a, U_b} P(U_a, U_b) \log \frac{P(U_a, U_b)}{P(U_a).P(U_b)} = I(U_a : U_b)$$

By using the definition of $d(\bar{\alpha}, \bar{\beta})$ we prove our result. ∎

By using Propositions 1 and 2, one can prove lower bounds on the amount of data that has to be pre-distributed to Alice and Bob.

Proposition 3. *In any β−secure bit commitment scheme based on pre-distributed data the following inequalities hold: $|U_b| \geq (\frac{1}{\beta})^2, |U_a| \geq (\frac{1}{\beta})(|\mathcal{B}|)$*

Corollary 1. *In any pre-distributed bit commitment scheme, the following bounds hold:$\beta \geq 2^{-I(U_a : U_b)}, |U_b| \geq (\frac{1}{\beta})^2, |U_a| \geq (\frac{1}{\beta})(|\mathcal{B}|)$ where β is the maximum cheating probability for Alice.*

Proof. To prove the corollary, we just need to remember that $\beta \geq \bar{\beta}$. ∎

We remark that this result proves the optimality of Rivest's scheme in terms of its required memory size. In Rivest's protocol $|\mathcal{B}| = p$, $\beta = 1/p$, $|U_b| = p^2$ and $|U_a| = p^2$.

5   Homomorphic Commitment Schemes

Consider the following problem, Alice commits to two values b_0, b_1 by sending $C(U_a, b_0)$ and $C(U_a', b_1)$ to Bob (to make the notation simple, we denote $C(U_a, b_0)$ and $C(U_a', b_1)$ by $[b_0]$ and $[b_1]$ respectively). She wants to prove that $b_0 = b_1$ without revealing any further information about b_0 and b_1.

If the commitment used by Alice has the property that $[b_0] - [b_1] = [b_0 - b_1]$ (note that Rivest scheme does not have this property), there is a very simple solution for that problem. Alice just asks Bob to compute $[b_0 - b_1]$ and unveils it to be zero.

Schemes where operations performed on commitments become commitments to the operation performed on committed values are called homomorphic (a formal definition is given below) and are very useful tools in modern cryptography.

We formalize the definition of a homomorphic commitment scheme below.

Definition 1. *Let $\oplus_1, \oplus_2$ and $\oplus_3$ be binary operations on elements from the committed values domain $\mathcal{B}$, the commitments domain $\mathcal{C}$ and the pre-distributed data domain $\mathcal{U}$, respectively. We say that a pre-distributed commitment scheme is $(\oplus_1, \oplus_2, \oplus_3)-$homomorphic if whenever $C(U_a, b_0) = c_0$ and $C(U_a', b_1) = c_1$ then $c_0 \oplus_2 c_1 = C(U_a \oplus_3 U_a', b_0 \oplus_1 b_1)$. When $\oplus_1 = \oplus_2 = \oplus_3 = +$ we say we have a $(+)-$homomorphic commitment.*

We now show a construction of homomorphic commitments based on modules of mappings.

Let $\mathcal{F} = \{f_i\}$ be a set of functions, $f_i : R \to R$, where R is a finite commutative ring with unity. We also require $\mathcal{F}$ to be a D-module, where D is a subring of R . For any $s, a, k, h \in R, h \neq a, f_i \neq f_j \in \mathcal{F}$ define $\gamma_{s,a}$ and μ as

$$\gamma_{s,a} = \{f \in \mathcal{F} : f(s) = a\} \text{ and } \mu_{s,a} = |\{f \in \gamma_{s,a} : f(k) = h\}|$$

Define β as:

$$\beta = \max_{c,d,w} \frac{|\{f_j, f_i \in \mathcal{F} : f_j(c) = f_i(c) \text{ and } f_j(d) = w \neq f_i(d)\}|}{|\{f \in \mathcal{F} : f(a) = s\}|} =$$
$$\max_{c,d,w} \frac{|\{f_j, f_i \in \mathcal{F} : f_j(c) = f_i(c) \text{ and } f_j(d) = w \neq f_i(d)\}|}{|R|}$$

where $c, d, w \in R$

Construction 1.

Setup Phase

1. Ted makes $c \in D$ public and then distributes f_i to Alice and $t, d = f_i(t)$ to Bob, where t and f_i are random elements belonging to R and $\mathcal{F}$, respectively.

Commit Phase

1. To commit to $m \in R$, an honest Alice sends $h = f_i(c) + m$ to Bob.

Opening Phase

1. Alice sends f_i and m to Bob. Bob accepts the commitment iff $h = f_i(c) + m$ and $f_i(t) = d$

Before proving the security of the protocol, we note that as Alice does not learn anything about t, in repeated executions of the protocol, Bob has to receive two elements from R just in the first execution of the commitment protocol. From the second execution on, Bob can receive only one value, since t can be repeatedly used without security problems.

Theorem 3. *If for any $s, a, k, h \in R$, $|\gamma_{s,a}| = \delta \geq |R|$ and $\mu = \nu \geq 1$, the commitment scheme described above is β-secure, $\beta = \nu/\delta$. Moreover, it is $(+)$-homomorphic.*

Proof.

Bindingness. Suppose that in the commit phase, Alice commits to a number $m \in R$. She wants to successfully open a commitment to a value $y \neq m$. First, we note that in order to successfully cheat, Alice has to find a function $f_j \neq f_i$ such that $f_j(t) = d$ and $f_j(c) = y - f_i(c) - m$. As Alice does not know t the only way she can do that is by guessing the function f_j. She succeeds with probability at most ν/δ.

Concealingness. Since f_i is chosen at random from $\mathcal{F}$, and for any $s, a, k, h \in R, |\gamma_{s,a}| = \delta \geq |R|$ and $|\mu| = \nu \geq 1$, we have that $f_i(c)$ is uniformly distributed over R.

Correctness. Evident from the protocol.

Homomorphism. Suppose Alice is committed to m and m', so Alice has received f_i, f_j from the Ted, while Bob received $d = f_i(t), d' = f_j(t)$ and t from the Ted and $h = f_i(c) + m, h' = f_j(c) + m'$ from Alice. Let λ be a public known constant belonging to D.
    To construct a commitment to $m + \lambda m'$, Bob computes

$$h'' = h + \lambda h' = m + \lambda m' + (f_i(c) + \lambda f_j(c)) = m + \lambda m' + (f_i + \lambda f_j)(c)$$

h'' is a commitment to $m + \lambda m'$. The opening information is $f_i + \lambda f_j$ and $m + \lambda m'$. Bob accepts the commitment if and only if $(f_i + \lambda f_j)(t) = d + \lambda d'$, if Bob is honest, this is the case, due to the fact that $\mathcal{F}$ is a D-module. ∎

In our commitment protocol, Alice can prove to Bob that two commitments are the same by asking Bob to subtract the commitments and unveil the commitments difference to be equal to zero.

As a particular case of our general construction, we show a protocol which was proposed in [4] (in [4] it is not stated that their protocol is homomorphic). Let $R = D = Z_p$, $f_i(x) = a_i x + b_i, a_i, b_i \in Z_p$ and $c = 0$ in our general construction. The resulting protocol is a linear commitment over Z_p. By linear we mean that the equality $[b] + \lambda[c] = [b + \lambda c]$ holds for any publicly known $\lambda \in Z_p$.

Protocol 1.

Setup Phase

1. Ted distributes r and s to Alice and $g = rt + s \bmod p$ and t to Bob, where r, s and t are random elements belonging to Z_p, for some prime p. All the computations are over Z_p.

Commit Phase

1. To commit to m, an honest Alice sends $h = m + s \bmod p$ to Bob.

Opening Phase

1. Alice sends r, m and s to Bob. Bob accepts the commitment iff $g = rt + s \bmod p$ and $h = m + s \bmod p$

6 Zero Knowledge Proofs of Polynomial Relations among Commitments

In this section we show how the proposed homomorphic commitments can be used to provide non-interactive zero knowledge proofs of any polynomial relation among commitments.

First, we note that the Ted can commit Alice to a random number m by doing the following:

Protocol 2.

Setup Phase

1. Ted distributes r, s and m to Alice and $g = rt - s \bmod p$, t and $h = m + r$ to Bob, where r, s, t and m are random elements belonging to Z_p, for some prime p. All the computations are over Z_p.

The opening phase is as in Protocol 1.

Although a specific protocol was used in the example, it is easy to see that in any pre-distributed commitment scheme, Ted is able to commit Alice and to specific values.

We now introduce a new pre-distributed cryptographic primitive called a *one time multiplication proof* (OTMP). We use OTMPs to implement our protocols for zero knowledge proofs of polynomial relations among commitments.

OTMP

1. By using Protocol 2, Ted commits Alice to three random numbers l, l' and l'', such that $l'' = ll''$. All the computations are over Z_p.

Now we present our protocol which provides Alice with zero knowledge proofs of multiplication among commitments. We heavily use the linearity of our commitment scheme. In the following we denote a linear commitments to a number $b \in Z_p$ by $[b]$. We recall that in linear commitment schemes, Alice can prove to Bob that two commitments are the same by asking Bob to subtract the commitments and unveil the commitments difference to be equal to zero.

We now show a protocol which enables Alice to prove multiplicative relations among commitments.

Protocol 3.

Setup Phase

1. Alice and Bob receive a number I of pre-distributed linear commitments and a number Z of OTMPs.

Commit Phase

2. Alice commits to all the inputs $x_1, ..., x_I$ of the polynomial $P(x_1, ..., x_I)$.

Proving Linear Relations among Commitments

3. Given a known constant $\lambda \in Z_p$ and two commitments $[x_i]$ and $[x_j]$, to obtain a commitment to $[x_i + \lambda x_j]$, Bob computes $[x_i] + \lambda [x_j]$. The result should be equal to $[x_i + \lambda x_j]$ due to the linearity of the commitment.

Proving Multiplicative Relations among Commitments

4. Assuming that Alice and Bob have an OTMP, Bob should posses commitments to three random values l, l' and l'', such that $l'' = ll''$. Given two commitments $[x_i]$ and $[x_j]$, to obtain a commitment to $[x_i x_j]$, Bob asks Alice to send him $y = x_i - l$ and $y' = x_j - l'$ and to prove the correctness of these equalities. Bob calculates $[ll'] + y[l'] + y'[l] + yy'$, this value can be shown to be equal to $[x_i x_j]$ (proof omitted). Alice's unveil information is computed according to $[ll''] + y[l'] + y'[l] + yy'$.

By applying addition of commitments and multiplication of commitments we can obtain commitments for arbitrary polynomial relations among commitments. E.g. to prove that $mm' = m''$ Alice lets Bob compute $[mm'] - [m''] = [mm' - m'']$ and unveils this to be zero.

References

1. T. Matsumoto and H. Imai, On the Key Predistribution Systems: A Practical Solution to the Key Distribution Problem, Advances in Cryptology – CRYPTO '87, Santa Barbara, California, USA, 1987, LNCS 293 Springer 1988,
2. B. Pfitzmann and M. Waidner. Information theoretic pseudosignatures and byzantine agreement for t $>= n/3$, Reserach Report IBM Reserach, Nov. 1996

3. R.L. Rivest, Unconditionally secure commitment and oblivious transfer schemes using concealing channels and a trusted initializer, pre-print available from `http://theory.lcs.mit.edu/~{}rivest/Rivest-commitment.pdf`.
4. C. Blundo, B. Masucci, D. R. Stinson, and R. Wei , Constructions and Bounds for Unconditionally Secure Non-Interactive Commitment Schemes, Designs, Codes, and Cryptography, (special issue in honour of Ron Mullin), Vol. 26, pp. 97–110, 2002.
5. G. Hanaoka, J. Shikata, Y. Zheng, H. Imai: Unconditionally Secure Digital Signature Schemes Admitting Transferability. Advances in Cryptology – ASIACRYPT 2000, Kyoto, Japan, 2000, LNCS 1976, Springer 2000
6. U. Maurer, Authentication Theory and Hypothesis Testing, IEEE Transaction on Information Theory, vol. 46, no. 4, pp. 1350–1356, 2000.
7. J. Kilian, A Note on Efficient zero-knowledge Proofs and Arguments, 24th ACM Symposium on Theory of Computation, 1992, pp. 723–732
8. R. Blahut. Principles and Practice of Information Theory, Addison-Wesley Publishing Company, 1987.

A Class of Low-Density Parity-Check Codes Constructed Based on Reed–Solomon Codes with Two Information Symbols⋆

Ivana Djurdjevic, Jun Xu, Khaled Abdel-Ghaffar, and Shu Lin

Department of Electrical and Computer Engineering
University of California, Davis
Davis, CA 95616, USA
{idjurdje,junxu,ghaffar,shulin}@ece.ucdavis.edu

Abstract. This paper presents an algebraic method for constructing regular low-density parity-check (LDPC) codes based on Reed–Solomon codes with two information symbols. The construction method results in a class of LDPC codes in Gallager's original form. Codes in this class are free of cycles of length 4 in their Tanner graphs and have good minimum distances. They perform well with iterative decoding.

1 Introduction

LDPC codes were discovered by *Gallager* in early 1960's [1]. After being over-looked for almost 35 years, this class of codes has been recently rediscovered and shown to form a class of *Shannon limit* approaching codes [2,3,4,5,6,7,8]. This class of codes decoded with iterative decoding, such as the *sum-product algorithm* (SPA) [1,4,5,6], performs amazingly well. Since their rediscovery, LDPC codes have become a focal point of research.

In this paper, an algebraic method for constructing regular LDPC codes is presented. This construction method is based on the simple structure of *Reed–Solomon* (RS) codes with two information symbols. It guarantees that the Tanner graphs [9] of constructed LDPC codes are free of cycles of length 4 and hence have girth at least 6. The construction results in a class of LDPC codes in Gallager's original form [1]. These codes are simple in structure and have good minimum distances. They perform well with iterative decoding.

The paper is organized as follows. Section 2 gives a very brief review of RS codes and discusses the structure of RS codes with two information symbols. Section 3 presents the proposed algebraic method for constructing LDPC codes. Some example codes and their error performances are given in Section 4. Section 5 concludes the paper.

⋆ This research was supported by NSF under Grants CCR-0096191, CCR-0117891, ECS-0121469, NASA under Grant NAG 5-10480, and fellowship provided by Accel Partners.

M. Fossorier, T. Hoeholdt, and A. Poli (Eds.): AAECC 2003, LNCS 2643, pp. 98–107, 2003.

2   RS Codes with Two Information Symbols

Consider the Galois field $GF(p^s)$ where p is a prime and s is a positive integer. Let α be a primitive element of $GF(p^s)$. Let $q = p^s$. Then $0 = \alpha^\infty$, $1 = \alpha^0, \alpha^1, \alpha^2$, $\cdots$, α^{q-2} form all the elements of $GF(p^s)$. Let ρ be a positive integer such that $2 \leq \rho < q$. Then the generator polynomial [10] of the cyclic $(q-1, q-\rho+1, \rho-1)$ RS code $\mathcal{C}$ of length $q-1$, dimension $q-\rho+1$, and minimum distance $\rho-1$ is:

$$g(X) = (X - \alpha)(X - \alpha^2) \cdots (X - \alpha^{\rho-2})$$
$$= g_0 + g_1 X + g_2 X^2 + \cdots + X^{\rho-2},$$

where $g_i \in GF(p^s)$. The generator polynomial $g(X)$ is a minimum weight code polynomial in $\mathcal{C}$ and hence all its $\rho - 1$ coefficients are nonzero. The generator matrix G of $\mathcal{C}$ is a $(q-\rho+1) \times (q-1)$ matrix obtained by cyclically shifting the $(q-1)$-tuple generator vector $g = (g_0, g_1, \cdots, 1, 0, \cdots, 0)$ to the right $q - \rho + 1$ times,

$$G = \begin{bmatrix} g_0 & g_1 & g_2 & \cdots & 1 & 0 & \cdots & 0 \\ 0 & g_0 & g_1 & g_2 & \cdots & 1 & 0 & \cdots \\ \vdots & \vdots & \vdots & \vdots & \vdots & \vdots & \vdots & \vdots \\ 0 & \cdots & 0 & g_0 & g_1 & g_2 & \cdots & 1 \end{bmatrix}.$$

Suppose we shorten $\mathcal{C}$ by deleting the first $q - \rho - 1$ information symbols from each codeword of $\mathcal{C}$ [10]. Then we obtain a $(\rho, 2, \rho - 1)$ shortened RS code $\mathcal{C}_b$ with only two information symbols whose generator matrix is simply the first two rows of G with the last $q - \rho - 1$ columns deleted,

$$G_b = \begin{bmatrix} g_0 & g_1 & g_2 & \cdots & 1 & 0 \\ 0 & g_0 & g_1 & g_2 & \cdots & 1 \end{bmatrix}.$$

All the linear combinations of the two rows of G_b over $GF(p^s)$ give all the p^{2s} codewords of $\mathcal{C}_b$. The nonzero codewords of $\mathcal{C}_b$ have two different weights, $\rho - 1$ and ρ.

In the following, we develop a number of structural properties of $\mathcal{C}_b$ which are keys to the construction of a class of regular LDPC codes whose Tanner graphs are free of cycles of length 4 (i.e., their girths are at least 6). First, since the minimum distance of $\mathcal{C}_b$ is $\rho - 1$, two codewords in $\mathcal{C}_b$ have at most one location with the same code symbol, i.e., they agree at most at one location. Let c be a codeword in $\mathcal{C}_b$ with weight ρ. Then the set

$$\mathcal{C}_b^{(1)} = \{\beta c : \beta \in GF(p^s)\}$$

of p^s codewords in $\mathcal{C}_b$ forms a one-dimensional subcode of $\mathcal{C}_b$ and is a $(\rho, 1, \rho)$ maximum distance separable (MDS) code. Each nonzero codeword in $\mathcal{C}_b^{(1)}$ has weight ρ. Two codewords in $\mathcal{C}_b^{(1)}$ differ at every location. Partition $\mathcal{C}_b$ into p^s cosets based on the subcode $\mathcal{C}_b^{(1)}$. Let $\mathcal{C}_b^{(1)}, \mathcal{C}_b^{(2)}, \cdots, \mathcal{C}_b^{(p^s)}$ denote these p^s cosets of $C_b^{(1)}$. Then two codewords in any coset $C_b^{(i)}$ must differ in all the locations. If we arrange the p^s codewords of a coset $C_b^{(i)}$ as a $p^s \times \rho$ array, then all the p^s elements of any column of the array are different (they are p^s different elements of $GF(p^s)$).

3   RS-Based Gallager-LDPC Codes

Consider the p^s elements, $\alpha^\infty, \alpha^0, \alpha^1, \cdots, \alpha^{p^s-2}$, of $GF(p^s)$. Let

$$z = (z_\infty, z_0, z_1, \cdots, z_{p^s-2})$$

be a p^s-tuple over $GF(2)$ whose components correspond to the p^s elements of $GF(p^s)$, i.e, z_i corresponds to the field element α^i. We call α^i the *location number* of z_i. For $i = \infty, 0, 1, \cdots, p^s - 2$, we define the *location vector* of α^i as a p^s-tuple over $GF(2)$:

$$z(\alpha^i) = (0, 0, \cdots, 0, 1, 0, 0, \cdots, 0),$$

for which the ith component z_i is equal to 1 and all the other components are equal to zero. It follows from the definition that the 1-components of the location vectors of two different elements in $GF(p^s)$ are at two different locations.

Let $b = (b_1, b_2, \cdots, b_\rho)$ be a codeword in C_b. For $1 \leq j \leq \rho$, replacing each component b_j of b by its location vector $z(b_j)$, we obtain a ρp^s-tuple over $GF(2)$,

$$z(b) = (z(b_1), z(b_2), \cdots, z(b_\rho)),$$

with weight ρ, which is called the *symbol location vector* of b. Since any two codewords in C_b have at most one location with the same code symbol, consequently their symbol location vectors have at most one 1-component in common. Let

$$\mathcal{Z}(C_b^{(i)}) = \{z(b) : b \in C_b^{(i)}\}$$

be the set of symbol location vectors of the p^s codewords in the ith coset $C_b^{(i)}$ of $C_b^{(1)}$. It follows from the structural properties of the cosets of $C_b^{(1)}$ developed in Section 2 that two symbol location vectors in $\mathcal{Z}(C_b^{(i)})$ do not have any 1-component in common.

For $1 \leq i \leq p^s$, form a $p^s \times \rho p^s$ matrix A_i over $GF(2)$ whose rows are the p^s symbol location vectors in $\mathcal{Z}(C_b^{(i)})$. Since the weight of each vector in $\mathcal{Z}(C_b^{(i)})$ is ρ, the total number of 1-entries in A_i is ρp^s. Since no two symbol location vectors in $\mathcal{Z}(C_b^{(i)})$ have any 1-component in common, the weight of each column of A_i is one. Therefore, A_i is a $(1, \rho)$-regular matrix with column and row weights 1 and ρ, respectively. In fact, it follows from the definition of the symbol location vector of a codeword in C_b and the structural properties of the codewords in each coset $C_b^{(i)}$ that A_i consists of ρ *column-permuted $p^s \times p^s$ identity matrices* (called *permutation matrices*) in a row. Matrix $\boldsymbol{A}_i$ is called the *symbol location matrix* of the coset $\mathcal{C}_b^{(i)}$. The class of symbol location matrices,

$$\mathcal{A} = \{\boldsymbol{A}_1, \boldsymbol{A}_2, \cdots, \boldsymbol{A}_{p^s}\},$$

has the following structural properties: (1) no two rows in the same matrix $\boldsymbol{A}_i$ have any 1-component in common; and (2) no two rows from two different member matrices, $\boldsymbol{A}_i$ and $\boldsymbol{A}_j$, have more than one 1-component in common.

Let γ be a positive integer such than $1 \leq \gamma \leq p^s$. Form the following $\gamma p^s \times \rho p^s$ matrix over GF(2):

$$\boldsymbol{H}_{GA}(\gamma) = \begin{bmatrix} \boldsymbol{A}_1 \\ \boldsymbol{A}_2 \\ \vdots \\ \boldsymbol{A}_\gamma \end{bmatrix}.$$

This matrix is a (γ, ρ)-regular matrix with column and row weights γ and ρ, respectively. It follows from the structural properties of the member matrices $\boldsymbol{A}_i$ in the class $\mathcal{A}$ that no two rows (or two columns) of $\boldsymbol{H}_{GA}(\gamma)$ have more than one 1-component in common. This implies that there are no 4 ones in $\boldsymbol{H}_{GA}(\gamma)$ at the 4 corners of a rectangle. This ensures that the associated Tanner graph of $\boldsymbol{H}_{GA}(\gamma)$ is free of cycles of length 4 and hence its girth is at least 6. We note that $\boldsymbol{H}_{GA}(\gamma)$ is exactly in Gallager's original form [1] for the parity check matrix of a (γ, ρ)-regular LDPC code. Therefore, the null space of this matrix gives a (γ, ρ)-regular *Gallager-LDPC code*, denoted $\mathcal{C}_{GA}(\gamma)$, of length $n = \rho p^s$ whose Tanner graph has girth at least 6. The rate of this code is at least $(\rho - \gamma)/\rho$.

Since no two rows in $\boldsymbol{H}_{GA}(\gamma)$ have more than one 1-component in common and each column of the matrix has weight γ, there are γ rows in $\boldsymbol{H}_{GA}(\gamma)$ that are orthogonal [10] on every code bit of $\mathcal{C}_{GA}(\gamma)$. It follows from *Massey's orthogonality theorem* [10,11] that the minimum distance of $\mathcal{C}_{GA}(\gamma)$ is at least $\gamma + 1$. This lower bound on minimum distance can be improved if the structure of parity check matrix $\boldsymbol{H}_{GA}(\gamma)$ is taken into account. Recall that each symbol location matrix $\boldsymbol{A}_i$ in $\mathcal{A}$ consists of a row of ρ $p^s \times p^s$ permutation matrices. Therefore, the parity check matrix $\boldsymbol{H}_{GA}(\gamma)$ consists of ρ columns of $p^s \times p^s$ permutation matrices. Each column of $\boldsymbol{H}_{GA}(\gamma)$ consists of γ sections, each section consists of a single 1-component. For a set of columns in the parity check matrix $\boldsymbol{H}_{GA}(\gamma)$ to sum to zero, the number of columns in the set must be even. This implies that the minimum distance of $\mathcal{C}_{GA}(\gamma)$ must be even. As a result, for even γ, the minimum distance of $\mathcal{C}_{GA}(\gamma)$ is then at least $\gamma + 2$. Summarizing the above results, we have the following lower bound on the minimum distance $d_{min}(\gamma)$ of $\mathcal{C}_{GA}(\gamma)$:

$$d_{min}(\gamma) \geq \left\{ \begin{array}{l} \gamma + 1, \text{ for odd } \gamma \\ \gamma + 2, \text{ for even } \gamma. \end{array} \right.$$

For a given choice of p, s, and ρ, we can construct a sequence of Gallager-LDPC codes of length $n = \rho p^s$ for $\gamma = 1, 2, \cdots, p^s$. When $\rho = p^s - 1$, the code is *quasi-cyclic*. For a given choice of p, s, and γ, we can construct a sequence of Gallager-LDPC codes of different lengths with minimum distance at least $\gamma + 1$ or $\gamma + 2$ by varying ρ. For the case with $\rho = \gamma = p^s - 1$, the parity check matrix $\boldsymbol{H}_{GA}(\gamma)$ is a $p^s(p^s - 1) \times p^s(p^s - 1)$ square matrix with row and column weights $p^s - 1$.

Since the construction is based on the $(\rho, 2, \rho - 1)$ shortened RS code $\mathcal{C}_b$ over $\mathrm{GF}(p^s)$, we call $\mathcal{C}_b$ and $\mathrm{GF}(p^s)$ the base code and construction field, respectively.

4   Some RS-Based Gallager-LDPC Codes and Their Error Performances

In this section, we present several RS-based LDPC codes and their error performances with iterative decoding using the SPA. For performance computation, we assume BPSK transmission over an AWGN channel.

Example 1. Let $\mathrm{GF}(2^6)$ be the field for code construction. Let $\rho = 32$. Then the base code is the (32,2,31) shortened RS code with two information symbols over $\mathrm{GF}(2^6)$. The location vector of each symbol in $\mathrm{GF}(2^6)$ is a 64-tuple over $\mathrm{GF}(2)$ with a single 1-component. Suppose we set $\gamma = 6$. Then the RS-based Gallager-LDPC code $\mathcal{C}_{GA}(6)$ is a (6,32)-regular (2048,1723) code with rate 0.841 and minimum distance at least 8. The bit and block error performances with the SPA decoding are shown in Fig. 1. At the BER of 10^{-6}, the code performs only 1.55 dB from the Shannon limit and achieves a 6 dB coding gain over the uncoded BPSK. For LDPC codes, the code length of 2048 is considered to be short. For such a short LDPC code, its error performance is very good. For comparison, the performance of the MacKay computer generated code of the same length and rate is also given in Fig. 1. We can see that the two codes have almost the same bit error performance. However, the RS-based Gallager-LDPC code has slightly better block error performance than the MacKay code below the block error rate of 10^{-5}. Fig. 2 depicts the convergence of the SPA decoding for the RS-based Gallager-LDPC code given in this example. We see that at BER of 10^{-4} the performance gap between 5 and 100 iterations is less than 0.4 dB, and the performance gap between 10 and 100 iterations is less than 0.2 dB. Fig. 3 depicts the convergence of the SPA decoding for the MacKay code. We see that this code has slightly slower convergence than the RS-based Gallager-LDPC code.   △△

Example 2. Again we use $\mathrm{GF}(2^6)$ as the construction field. Let $\rho = 63$. Then the base code for construction is the (63,2,62) RS code. Set $\gamma = 60$. The Gallager-LDPC code constructed is a (60,63)-regular (4032,3307) quasi-cyclic code with rate 0.82 and minimum distance at least 62. The bit and block error performances of the code are shown in Fig. 4. At the BER of 10^{-6}, the code performs 1.65 dB from the Shannon limit. Since the code has a very large minimum distance, there should not be any error floor or the error floor occurs at a very low bit error rate.   △△

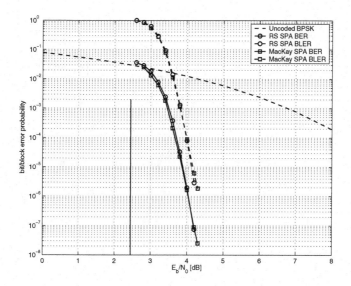

Fig. 1. Error performance of the (2048,1723) RS-based Gallager (6,32)-regular LDPC code with construction field $GF(2^6)$

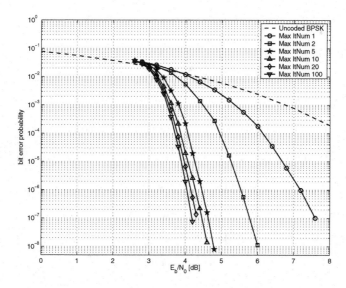

Fig. 2. Convergence of the SPA decoding for the (2048,1723) RS-based Gallager-LDPC code

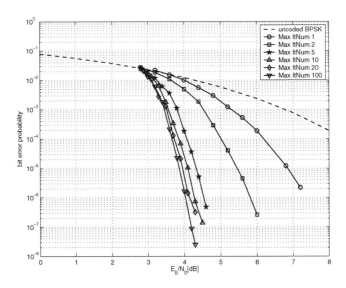

Fig. 3. Convergence of the SPA decoding for the (2048,1723) MacKay computer generated LDPC code

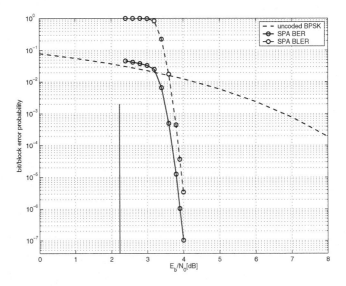

Fig. 4. Error performance of the (4032,3307) RS-based Gallager (60,63)-regular quasi-cyclic LDPC code with construction field $GF(2^6)$

Example 3. Suppose we construct a (32,2,31) shortened RS code $\mathcal{C}_b$ over GF(2^8). Set $\gamma = 6$. Then the Gallager-LDPC code constructed based on the shortened RS code $\mathcal{C}_b$ is a (6,32)-regular (8192,6754) code with rate 0.824 and minimum distance at least 8. The bit and block error performances of the LDPC code are shown in Fig. 5. At the BER of 10^{-6}, the code performs 1.25 dB from the Shannon limit and achieves a 6.7 dB coding gain over the uncoded BPSK. $\triangle\triangle$

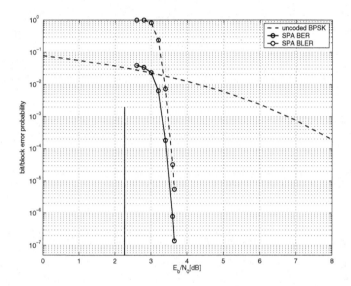

Fig. 5. Error performance of the (8192,6754) RS-based Gallager (6,32)-regular LDPC code with construction field GF(2^8)

Example 4. Again we use GF(2^8) as the construction field. The shortened RS code $\mathcal{C}_b$ used for code construction is the (48,2,47) code over GF(2^8). Set $\gamma = 6$. Then the Gallager-LDPC code constructed is a (6,48)-regular (12288,10845) code with rate 0.8825 and minimum distance at least 8. The bit and block error performances are shown in Fig. 6. At the BER of 10^{-6}, the code performs 1.1 dB from the Shannon limit. For comparison, the performance of the MacKay computer generated code of the same length and rate is also given in Fig. 6. We see that in this case MacKay code is 0.2 dB better than RS-based Gallager-LDPC code. However, the error performance of the RS-based Gallager-LDPC code has larger dropping rate. The performance curves of the two codes may intersect at lower BER. $\triangle\triangle$

5   Conclusion

In this paper, a simple RS-based algebraic method for constructing regular LDPC codes with girth at least 6 has been presented. Construction gives a large

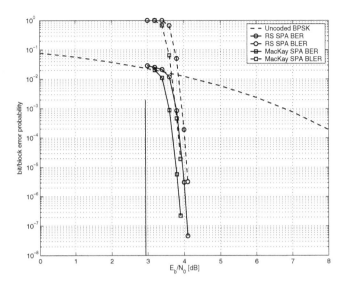

Fig. 6. Error performance of the (12288,10845) RS-based Gallager (6,48)-regular LDPC code with construction field $GF(2^8)$

class of regular LDPC codes in Gallager's orginal form. A possible generalization of this method is to use a RS code with three information symbols as the base code. In this case, there will be cycles of length 4 in the Tanner graph of the code constructed. These short cycles do not necessarily prevent the code from having good error performance with iterative decoding if the code has large minimum distance and good cycle structure in its Tanner graph. It has been shown in [12] that some geometry codes with millions of cycles of length 4 perform very well with iterative decoding due to their large minimum distance and good cycle structure in a ring form. We wonder whether other known low rate codes with symbols from nonbinary fields can be used for constructing LDPC codes. We hope that the humble results in this paper will motivate further research in this direction.

References

1. R.G. Gallager, *Low Density Parity Check Codes*, Cambridge, MA: MIT Press, 1963.
2. D.J.C. MacKay and R.M. Neal, "Near Shannon Limit Performance of Low Density Parity Check Codes," *Electronics Letters* 32 (18): 1645–1646, 1996.
3. M. Sipser and D. Spielman, "Expander Codes," *IEEE Transactions on Information Theory*, vol. 42, pp. 1710–1722, November 1996.
4. D.J.C. MacKay, "Good Error-Correcting Codes Based on Very Sparse Matrices," *IEEE Transactions on Information Theory*, vol. 45, pp. 399–432, March 1999.
5. T.J. Richardson, M.A. Shokrollahi, and R. Urbanke, "Design of Capacity-Approaching Irregular Low-Density Parity-Check Codes," *IEEE Transactions on Information Theory*, vol. 47, pp. 619–637, February 2001.

6. T.J. Richardson, and R. Urbanke, "The Capacity of Low-Density Parity-Check Codes Under Message-Passing Decoding," *IEEE Transactions on Information Theory*, vol. 47, pp. 599–618, February 2001.

7. S.-Y. Chung, G.D. Forney, Jr., T.J. Richardson, and R. Urbanke, "On the Design of Low-Density Parity-Check Codes within 0.0045 dB of the Shannon Limit," *IEEE Commununications Letters*, vol. 5, pp. 58–60, February 2001.

8. Y. Kou, S. Lin, and M. Fossorier, "Low Density Parity Check Codes: A Rediscovery and New Results," *IEEE Transactions on Information Theory*, vol. 47, pp. 2711–2736, November 2001.

9. R.M. Tanner, "A Recursive Approach to Low Complexity Codes," *IEEE Transactions on Information Theory*, vol. IT-27, pp. 533–547, September 1981.

10. S. Lin and D.J. Costello, Jr., *Error Control Coding: Fundamentals and Applications*, Englewood Cliffs, NJ: Prentice Hall, 1983.

11. J.L. Massey, *Threshold Decoding*, Cambridge, MA: MIT Press, 1963.

12. J. Xu, H. Tang, Y. Kou, S. Lin, and K. Abdel-Ghaffar, "A General Class of LDPC Finite Geometry Codes and their Performance," *Proc. IEEE Int. Symp. Information Theory*, Lausanne, Switzerland, June 30–July 5, 2002, p. 309.

Relative Duality in MacWilliams Identity

L.S. Kazarin[1], V.M. Sidelnikov[2], and I.B. Gashkov[3]

[1] Yaroslavl State University, Chair of Algebra and Mathematical Logic
Department of Mathematics, 150000 Yaroslavl, Russia
`kazarin@uniyar.ac.ru`
[2] Chair "Discrete mathematics" Department Mechanics and Mathematics
Moscow State University, 119899 Moscow, Russia
`sid@vertex.inria.msu.ru`
[3] Department of Engineering Sciences,Physics and Mathematics
Karlstad University, 651 88 Karlstad, Sweden
`Igor.Gachkov@kau.se`

Abstract This work considers relations, which connect weight enumerator of a linear code $\mathcal{L} \subset \mathbb{F}_p^n$ and weight enumerator of code $\mathcal{K}$, lying between $\mathcal{L}$ and $\mathbb{F}_p^n$: $\mathcal{L} \subseteq \mathcal{K} \subseteq \mathbb{F}_p^n$. The connection is establish by code $\mathcal{L}^\perp(\mathcal{K})$, which is dual code of code $\mathcal{L}$ in the code $\mathcal{K}$. It has been show, that there is a code $\mathcal{L}^\perp(\mathcal{K})$, which dimension is $k-l$, where l and k are dimensions of $\mathcal{L}$ and $\mathcal{K}$ respectively. These relations for weight enumerators generalise the MacWilliams relation, in which $\mathcal{K} = \mathbb{F}_p^n$.
The new relations are written for linear code of length $2n$ of code $\mathcal{L} = \{(a|a + b),\ a \in \mathcal{L}_1,\ b \in \mathcal{L}_2\}$, where $\mathcal{L}_1, \mathcal{L}_2 \subseteq \mathbb{F}_p^n$, for two different comprehensive codes $\mathcal{K}$. As a result we obtain two different expressions for weight enumerator of code $\mathcal{L}$.
As example we consider a case $\mathcal{L}_1 = RM_{m-2.m}$, $\mathcal{L}_2 = RM_{1.m}$, where $RM_{r.m}$ is the r-th order Reed-Muller code of lenght $n = 2^m$. [1]

1 Introduction

Consider a linear code $\mathcal{K} \subseteq \mathbb{F}_p^n$ of dimension k and subcode $\mathcal{L}$ of dimension l, $l \leq k$.
We define a characteristic function $\varphi_{\mathcal{L},\mathcal{K}}(\boldsymbol{x})$ of subcode $\mathcal{L}$ in the code $\mathcal{K}$ as

$$\varphi_{\mathcal{L},\mathcal{K}}(\boldsymbol{x}) = \begin{cases} 0, & \boldsymbol{x} \notin \mathcal{L},\ \boldsymbol{x} \in \mathcal{K} \\ 1, & \boldsymbol{x} \in \mathcal{L},\ \boldsymbol{x} \in \mathcal{K} \end{cases}. \tag{1}$$

Ordinary characteristic function of subcode $\mathcal{L}$

$$\varphi_{\mathcal{L}}(\boldsymbol{x}) = \begin{cases} 0, & \boldsymbol{x} \notin \mathcal{L},\ \boldsymbol{x} \in \mathbb{F}_p^n, \\ 1, & \boldsymbol{x} \in \mathcal{L},\ \boldsymbol{x} \in \mathbb{F}_p^n, \end{cases}$$

in our notations is written as $\varphi_{\mathcal{L}}(\boldsymbol{x}) = \varphi_{\mathcal{L},\mathbb{F}_p^n}(\boldsymbol{x})$.
Note, that function $\varphi_{\mathcal{L},\mathcal{K}}(\boldsymbol{x})$ for vectors $\boldsymbol{x} \notin \mathcal{K}$ can be equal both 0 and 1.

---

[1] Thise research was support by The Royal Swedesh akademy of Sciences

M. Fossorier, T. Hoeholdt, and A. Poli (Eds.): AAECC 2003, LNCS 2643, pp. 108–118, 2003.

Definition 1. *Let us denote by $\mathcal{L}^{\perp}(\mathcal{K})$ subspace $\mathcal{M}$ of $\mathbb{F}_p^n$. Subspace $\mathcal{M}$ have the following properties:*

(A) *if $\boldsymbol{x} \in \mathcal{L}$, then $\langle \boldsymbol{x}, \boldsymbol{y} \rangle = 0$ for all $\boldsymbol{y} \in \mathcal{M}$;*

(B) *if $\boldsymbol{x} \in \mathcal{K} \setminus \mathcal{L}$, then there exist $\boldsymbol{y} \in \mathcal{M}$, that $\langle \boldsymbol{x}, \boldsymbol{y} \rangle \neq 0$.*

We will call as $\mathcal{L}^{\perp}(\mathcal{K})$ dual subspace of $\mathcal{L}$ in the space $\mathcal{K}$.

Definition 2. *Let $\mathcal{M}$ be subspace of $V = \mathbb{F}^n$, with following properties:*

(a) $\mathcal{M} \subseteq \mathcal{L}^{\perp}$;

(b) $\mathcal{M}^{\perp} \cap \mathcal{K} \subseteq \mathcal{L}$.

Any such subspace let's denote by $\mathcal{L}^{\perp}(\mathcal{K})$.

Note, that subspace $\mathcal{L}^{\perp}(\mathcal{K})$ can be chosen by different methods.

Lemma 1. *Definitions 1 and 2 are equivalent.*

Proof. Obviously.$\square$

Further in the path 2 we will show, that there are subspaces $\mathcal{L}^{\perp}(\mathcal{K})$, which dimensions are $k - l$, and some other properties of $\mathcal{L}^{\perp}(\mathcal{K})$ will be studies.

Definition 1 allows to represent a characteristic function $\varphi_{\mathcal{L},\mathcal{K}}(\boldsymbol{x})$ (see. (1)) in form

$$\varphi_{\mathcal{L},\mathcal{K}}(\boldsymbol{x}) = \frac{1}{|\mathcal{L}^{\perp}(\mathcal{K})|} \sum_{\boldsymbol{y} \in \mathcal{L}^{\perp}(\mathcal{K})} \exp\left(\frac{2\pi i \langle \boldsymbol{x}, \boldsymbol{y} \rangle}{p}\right) \tag{2}$$

The opposite is true as well: If for any subspace $\mathcal{L}^{\perp}(\mathcal{K})$ function (2) is a characteristic function of subcode $\mathcal{L}$ in the code $\mathcal{K}$, then $\mathcal{L}^{\perp}(\mathcal{K})$ is dual subspace of $\mathcal{L}$ in the space $\mathcal{K}$.

From (2) follows, that

$$\sum_{\boldsymbol{x} \in \mathcal{L}} z^{w(\boldsymbol{x})} = \frac{1}{|\mathcal{L}^{\perp}(\mathcal{K})|} \sum_{\boldsymbol{y} \in \mathcal{L}^{\perp}(\mathcal{K})} \sum_{\boldsymbol{x} \in \mathcal{K}} \exp\left(\frac{2\pi i \langle \boldsymbol{x}, \boldsymbol{y} \rangle}{p}\right) z^{w(\boldsymbol{x})}. \tag{3}$$

In case $\mathcal{K} = \mathbb{F}_p^n$ relation (3) is well known MacWilliams Identity [1]

$$\sum_{\boldsymbol{x} \in \mathcal{L}} z^{w(\boldsymbol{x})} = \frac{1}{|\mathcal{L}^{\perp}|} \sum_{\boldsymbol{y} \in \mathcal{L}^{\perp}(\mathbb{F}_p^n)} (1 - z)^{w(\boldsymbol{y})} (1 + (p - 1))^{n - w(\boldsymbol{y})} \tag{4}$$

This follows from as well known relation

$$T(\boldsymbol{y}, \mathbb{F}_p^n) = \sum_{\boldsymbol{x} \in \mathbb{F}_p^n} \exp(\frac{2\pi i \langle \boldsymbol{x}, \boldsymbol{y} \rangle}{p}) z^{w(\boldsymbol{x})} =$$

$$\prod_{j=1}^{n} \sum_{x \in \mathbb{F}_p} \exp\left(\frac{2\pi i x y_j}{p}\right) z^{w(x)} = (1 - z)^{w(\boldsymbol{y})} (1 + (p - 1)z)^{n - w(\boldsymbol{y})}, \tag{5}$$

because

$$\sum_{x\in\mathbb{F}_p} \exp\left(\frac{2\pi ixy}{p}\right) z^{w(x)} = \begin{cases} (1-z), & \text{for } y \neq 0, \\ 1+(p-1)z, & \text{for } y = 0, \end{cases} \tag{6}$$

If we can represent sum in (3)

$$T(\boldsymbol{y},\mathcal{K}) = \sum_{\boldsymbol{x}\in\mathcal{K}} \exp(\frac{2\pi i\langle\boldsymbol{x},\boldsymbol{y}\rangle}{p}) z^{w(\boldsymbol{x})}, \ \boldsymbol{y} \in \mathcal{L}^{\perp}(\mathcal{K}), \tag{7}$$

through any parameters of the vectors $\boldsymbol{y}$, we will get relation, which generalize (4).

One of these obvious methods is representation $T(\boldsymbol{y},\mathcal{K})$ by characteristic function $\varphi_{\mathcal{K}}(\boldsymbol{x})$. Minding (5) this representation will lead to

Theorem 1.

$$\sum_{\boldsymbol{x}\in\mathcal{L}} z^{w(\boldsymbol{x})} = \frac{1}{|\mathcal{L}^{\perp}(\mathcal{K})||\mathcal{K}^{\perp}|} \sum_{\boldsymbol{y}\in\mathcal{L}^{\perp}(\mathcal{K})} \sum_{\boldsymbol{x}\in\mathcal{K}^{\perp}} (1-z)^{w(\boldsymbol{x}+\boldsymbol{y})} (1+(p-1)z)^{n-w(\boldsymbol{x}+\boldsymbol{y})}. \tag{8}$$

In many cases space $\mathcal{L}^{\perp}(\mathcal{K}) \subseteq \mathcal{L}^{\perp}$ has dimension which is a great deal less then dimension of space $\mathcal{L}^{\perp}$. Therefore (8) allows in these cases, using quite obvious ideas simplify calculations of the weight enumerator $\sum_{\boldsymbol{x}\in\mathcal{L}} z^{w(\boldsymbol{x})}$.

Our main task is choosing of a good code $\mathcal{K}$ and $\mathcal{L} \subseteq \mathcal{K}$, for which we can calculate sum (7) in explicit form. In path 3 this is done for some codes $\mathcal{L}$, having a certain structure.

2   Properties of Spaces $\mathrm{L}^{\perp}(\mathrm{K})$

Lemma 2. *Let* $\mathcal{N} = \mathcal{L}^{\perp}(\mathcal{K})$ *and* $\mathcal{N} \subseteq \mathcal{M} \subseteq \mathcal{L}^{\perp}$. *Then* $\mathcal{M}$ *satisfies (a), (b) in the definition 1. In particular,* $\mathcal{M}$ *can we chose as* $\mathcal{L}^{\perp}(\mathcal{K})$.

Proof. Since $\mathcal{N} \subseteq \mathcal{M}$, then $\mathcal{M}^{\perp} \subseteq \mathcal{N}^{\perp}$. Therefore $\mathcal{M}^{\perp}\cap\mathcal{K} \subseteq \mathcal{N}^{\perp}\cap\mathcal{K} \subseteq \mathcal{L}$. $\square$

Lemma 3. *Let* $\mathcal{M} = \mathcal{L}^{\perp}(\mathcal{K})$. *Then* $\dim\mathcal{M} \geqslant \dim\mathcal{K} - \dim\mathcal{L}$.

Proof. Obviously, that $\dim(\mathcal{M}^{\perp} + \mathcal{K}) \leqslant \dim V = n$. Since $\dim(\mathcal{M}^{\perp} + \mathcal{K}) = \dim\mathcal{M}^{\perp} + \dim\mathcal{K} - \dim(\mathcal{M}^{\perp} \cap \mathcal{K}) = \dim V - \dim\mathcal{M} + \dim\mathcal{K} - \dim(\mathcal{M}^{\perp} \cap \mathcal{K}) \leqslant \dim V$, then $\dim\mathcal{K} - \dim\mathcal{M} \leqslant \dim(\mathcal{M}^{\perp} \cap \mathcal{K})$. Since $\mathcal{M}^{\perp}\cap\mathcal{K} \subseteq \mathcal{L}$, then $\dim\mathcal{M} \geqslant \dim\mathcal{K} - \dim\mathcal{L}$. $\square$

Remark 1. Property (b) in Definition 2 indicate, that $\mathcal{L}^{\perp}(\mathcal{K})\cap\mathcal{K} = \mathcal{L}$.

Indeed, $\mathcal{L} \subseteq \mathcal{L}^{\perp}(\mathcal{K})\cap\mathcal{K}$.

Remark 2. In the capacity of $\mathcal{L}^{\perp}$ always can we chose $\mathcal{L}^{\perp}$ so, that set of subspaces $\mathcal{L}^{\perp}(\mathcal{K})$ are not empty.

Remind that vector x is called "isotropic vector" if $\langle x, x \rangle = 0$, and "anti-isotropic vector" if $\langle x, x \rangle \neq 0$.

Lemma 4. *Let $\boldsymbol{x}$ be isotropic vector ($\langle \boldsymbol{x}, \boldsymbol{x} \rangle = 0$). If $\langle \cdot, \cdot \rangle$ is non-singular bilinear form and $char\mathbb{F} \neq 2$, then there exist an isotropic vektor $\boldsymbol{y} \in V$, for which $\{\boldsymbol{x}, \boldsymbol{y}\}$ is hyperbolic pair.*

Proof. Let $\langle \boldsymbol{t}, \boldsymbol{x} \rangle \neq 0$ and $\langle \boldsymbol{t}, \boldsymbol{t} \rangle \neq 0$. Then $V = \langle \boldsymbol{t} \rangle \oplus \langle \boldsymbol{t} \rangle^{\perp}$. Therefore $\boldsymbol{x} = \lambda \boldsymbol{t} + \boldsymbol{z}$ for some $\boldsymbol{z} \in \langle \boldsymbol{t} \rangle^{\perp}$. In addition $\langle \boldsymbol{x}, \boldsymbol{x} \rangle = \lambda^2 \langle \boldsymbol{t}, \boldsymbol{t} \rangle + \langle \boldsymbol{z}, \boldsymbol{z} \rangle = 0$. Obviously, that $\boldsymbol{x} = \lambda \boldsymbol{t} - \boldsymbol{z}$ have the same property. If $\langle \boldsymbol{x}, \boldsymbol{y} \rangle = 0$, then $\lambda^2 \langle \boldsymbol{t}, \boldsymbol{t} \rangle - \langle \boldsymbol{z}, \boldsymbol{z} \rangle = 0$, and then $2\lambda^2 \langle \boldsymbol{t}, \boldsymbol{t} \rangle = \langle \boldsymbol{t}, \boldsymbol{t} \rangle = 0$. $\square$

Lemma 5. *If $char\mathbb{F} \neq 2$, then $\mathcal{L} = \mathcal{L}_1 \oplus \mathcal{L}_0$, where $\mathcal{L}_0 = \mathcal{L} \cap \mathcal{L}^{\perp}$, and $\mathcal{L}_1$ have basis which consist of anti-isotropic vectors.*

Proof. Let x be some any anti-isotropic vector, $x \in \mathcal{L}$. Then $V = \langle x \rangle \oplus \langle x \rangle^{\perp}$. If we continue this process, we obtain, that $\mathcal{L} = \mathcal{L}_1 \oplus \mathcal{L}_0$, where $\mathcal{L}_1$ have basis consisting of anti-isotropic vectors and $\mathcal{L}_0 \subseteq \mathcal{L}^{\perp}$.

Using Lemma 4 $\langle \boldsymbol{x}, \boldsymbol{y} \rangle = 0$ for all $\boldsymbol{x}, \boldsymbol{y} \in \mathcal{L}_0$. Let $\mathcal{M}$ be subspace of $\mathcal{L}$, which have Orthogonal decomposition: $\langle \boldsymbol{x}_1 \rangle \oplus \cdots \oplus \langle \boldsymbol{x}_k \rangle$, where $\boldsymbol{x}_j$ are anti-isotropic vectors. Then $\mathcal{M} \cap \mathcal{M}^{\perp} = \{0\}$. Really, if $\alpha_1 \boldsymbol{x}_1 + \cdots \alpha_k \boldsymbol{x}_k = \boldsymbol{u} \in \mathcal{M} \cap \mathcal{M}^{\perp}$, then $\langle \boldsymbol{u}, \boldsymbol{x}_i \rangle = \alpha_i \langle \boldsymbol{x}_i, \boldsymbol{x}_i \rangle = 0$ and we get $\alpha_i = 0$. It means, that $(\mathcal{L} \cap \mathcal{L}^{\perp}) \cap \mathcal{L}_1 = \{0\}$. Therefore $\mathcal{L} \cap \mathcal{L}^{\perp} = \mathcal{L}_0$. $\square$

Lemma 6. *Let $\mathcal{L} \subseteq \mathcal{K}$. If $\mathcal{L}^{\perp}(\mathcal{K}) \subseteq \mathcal{K}$, then $\mathcal{K}^{\perp} \cap \mathcal{K} \subseteq \mathcal{L}^{\perp} \cap \mathcal{L}$.*

Proof. Let $\mathcal{L}^{\perp}(\mathcal{K}) \subseteq \mathcal{K}$. Then $\mathcal{K}^{\perp} \subseteq \mathcal{L}^{\perp}(\mathcal{K})^{\perp}$. Therefore $\mathcal{K} \cap \mathcal{K}^{\perp} \subseteq \mathcal{L}^{\perp}(\mathcal{K})^{\perp} \cap \mathcal{K}$. Using (b) in Definition 2, we have $\mathcal{K} \cap \mathcal{K}^{\perp} \subseteq \mathcal{L}$. Since $\mathcal{L} \subseteq \mathcal{K}$, then $\mathcal{K}^{\perp} \subseteq \mathcal{L}^{\perp}$ and we obtain, that $\mathcal{K} \cap \mathcal{K}^{\perp} \subseteq \mathcal{L} \cap \mathcal{L}^{\perp}$. $\square$

Theorem 2. *Let $\mathcal{K} \cap \mathcal{K}^{\perp} \subseteq \mathcal{L} \cap \mathcal{L}^{\perp}$. If $(\mathcal{L}^{\perp} \cap \mathcal{K})^{\perp} \subseteq \mathcal{K}^{\perp}$, then $K \cap \mathcal{L}^{\perp} = \mathcal{L}^{\perp}(K)$.*

Proof. We define $\mathcal{L}^{\perp}(\mathcal{K}) = K \cap \mathcal{L}^{\perp}$. Then $\mathcal{L}^{\perp}(\mathcal{K}) \subseteq \mathcal{L}^{\perp}$ $\mathcal{L}^{\perp}(\mathcal{K})^{\perp} \cap \mathcal{K} \subseteq (\mathcal{L}^{\perp} \cap \mathcal{K})^{\perp} \subseteq \mathcal{K}^{\perp} \cap \mathcal{K} \subseteq \mathcal{L}^{\perp} \cap \mathcal{L} \subseteq \mathcal{L}$. $\square$

Lemma 7. *Let $\mathcal{K} = \mathcal{L} \oplus (\mathcal{L}^{\perp} \cap \mathcal{K})$. Then $\mathcal{L}^{\perp} \cap \mathcal{K} = \mathcal{L}^{\perp}(\mathcal{K})$.*

Proof. $(\mathcal{L}^{\perp}(\mathcal{K}))^{\perp} \cap \mathcal{K} = \mathcal{L}$. $\square$

Corollary 1. *Let $\mathcal{L}$ have a basis which consist of anti-isotropic vectors. Then $\mathcal{L}^{\perp} \cap \mathcal{K} = \mathcal{L}^{\perp}(\mathcal{K})$.*

Proof. $(\mathcal{L}^{\perp}(\mathcal{K}))^{\perp} \cap \mathcal{K} = \mathcal{L}$. $\square$

Lemma 8. *(i) For any subspace $\mathcal{M} \subseteq V = \mathbb{F}^n$ there is subspace $\mathcal{N}$, that $V = \mathcal{M} \oplus \mathcal{N} = \mathcal{N} \oplus \mathcal{N}^{\perp}$;*
(ii) If $V = \mathcal{M} \oplus \mathcal{N}$, then $V = \mathcal{M}^{\perp} \oplus \mathcal{N}^{\perp}$.

Proof.

(i) Complete the basis of $\mathcal{M}$ until the basis of V by vectors from canonical orthonormal basis of V. We denote these vectors by $e_1, e_2, \cdots e_k$ and $k = \dim V - \dim \mathcal{M}$. Then $\langle e_1, e_2 \cdots e_k \rangle = \mathcal{N}$ is desired subspace. Remaining basis vectors define a dual subspace $\mathcal{N}^\perp$.

(ii) Let $x \in \mathcal{M}^\perp \cap \mathcal{N}^\perp$. Then $\langle \boldsymbol{x}, \boldsymbol{y} \rangle = 0$ for all $y \in \mathcal{M}$ and $\langle \boldsymbol{x}, \boldsymbol{z} \rangle = 0$ for all $z \in \mathcal{N}$. Therefore $\langle \boldsymbol{x}, \boldsymbol{v} \rangle = 0$ for all $\boldsymbol{v} \in V$. Since $\langle \cdot, \cdot \rangle$ is the non-singular bilinear form, then $\boldsymbol{x} = 0$. So $\mathcal{M}^\perp \cap \mathcal{N}^\perp = \{0\}$. Therefore $\dim \mathcal{M}^\perp = \dim \mathcal{N}$, $\dim \mathcal{N}^\perp = \dim \mathcal{M}$ and $\dim \mathcal{N} + \dim \mathcal{M} = \dim V$, then $V = \mathcal{M}^\perp \oplus \mathcal{N}^\perp$. $\square$

Lemma 9. *Let $V = \mathcal{M} \oplus \mathcal{N}$. If $\mathcal{K} = \mathcal{L} \oplus \mathcal{N}$, where $\mathcal{L} \subseteq \mathcal{M}$, then $L^\perp(\mathcal{K}) = \mathcal{M}^\perp$.*

Proof. By Lemma 8 (ii) $V = \mathcal{M}^\perp \oplus \mathcal{N}^\perp$. Since $\mathcal{M}^\perp \subseteq \mathcal{L}^\perp$, then condition (a) is realized. On the other hand, $(\mathcal{M}^\perp)^\perp = \mathcal{M}$ and $\mathcal{M}^{\perp\perp} \cap \mathcal{K} = \mathcal{M} \cap \mathcal{K} = \mathcal{L}$, i.e. condition (b) is realized also. $\square$

Theorem 3. *For any subspace $\mathcal{L} \subseteq \mathcal{K}$ there is subspace $\mathcal{L}^\perp(\mathcal{K})$ with dimension $\dim \mathcal{K} - \dim \mathcal{L}$.*

Proof. Let $\mathcal{L} \subset \mathcal{K} \subset V$. Then $\mathcal{K} = \mathcal{N} \oplus \mathcal{L}$ for some $\mathcal{N}$ and $V = \mathcal{M} \oplus \mathcal{N}$, where $\mathcal{M} \cap \mathcal{K} = \mathcal{L}$. The by Lemma 8 (ii) $\mathcal{M}^\perp \oplus \mathcal{N}^\perp = V$, and also $\dim \mathcal{M}^\perp = \dim V - \dim \mathcal{N}^\perp = \dim \mathcal{N} = \dim \mathcal{K} - \dim \mathcal{L}$. By Lemma 9 we get conclusion of the theorem. $\square$

Lemma 10. *Let $\mathcal{L} \subseteq \mathcal{K}$ be a linear code with $\dim \mathcal{L}^\perp(\mathcal{K}) = \dim \mathcal{K} - \dim \mathcal{L}$. Then for all subspace $\mathcal{N}$ so, that $\mathcal{K} = \mathcal{L} \oplus \mathcal{N}$ fulfilled, that $\mathcal{L}^\perp(K)^\perp \oplus \mathcal{N} = V$ and $\mathcal{L}^\perp(K)^\perp \cap \mathcal{K} = \mathcal{L}$.*

Proof. Let $(\mathcal{L}^\perp(\mathcal{K}))^\perp = \mathcal{M}$ $\dim \mathcal{N} = \dim \mathcal{K} - \dim \mathcal{L}$. If $\mathcal{K} = \mathcal{L} \oplus \mathcal{N}$, then $\dim \mathcal{N} = \dim \mathcal{M}$. Obviously, that $\mathcal{M}^\perp \cap \mathcal{K} \supseteq \mathcal{L}$. Since $\mathcal{M}^\perp \cap \mathcal{K} \subset \mathcal{L}$, then $\dim V = \dim \mathcal{M}^\perp + \dim \mathcal{N} = \dim(\mathcal{M}^\perp \oplus \mathcal{N})$. Therefore $\mathcal{M}^\perp \cap \mathcal{N} = \{0\}$. Consequently $\mathcal{K} = \mathcal{L} \oplus \mathcal{N}$, where $\mathcal{M} \oplus \mathcal{N} = V$ and $\mathcal{L} \subseteq \mathcal{M}$, then $\mathcal{L}^\perp(\mathcal{K}) = \mathcal{M}^\perp$. $\square$

Lemma 11. *Let $\mathcal{L} \subseteq \mathcal{K} \subseteq V$ and for all $\boldsymbol{x} \in \mathcal{K} \setminus \mathcal{L}$ $\boldsymbol{x}$ is isotropic vector. Then $\mathcal{L}$ and $\mathcal{K}$ are isotropic subspaces.*

Proof. Let $\mathcal{K} = \mathcal{L} \oplus \mathcal{N}$, where $\mathcal{N}$ is an arbitrary complement space. So, that any vector $\boldsymbol{y} \in \mathcal{N}$ is isotropic vector , then $\langle \boldsymbol{y}, \boldsymbol{y} \rangle = 0$ for all $\boldsymbol{y} \in \mathcal{N}$. Since $\langle \boldsymbol{y} + \boldsymbol{x}, \boldsymbol{y} + \boldsymbol{x} \rangle = 0$ for all $\boldsymbol{x} \in \mathcal{L}$, then $\langle \boldsymbol{x}, \boldsymbol{x} \rangle + 2\langle \boldsymbol{x}, \boldsymbol{y} \rangle = 0$ and we get, that $\langle \boldsymbol{x}, \boldsymbol{x} + 2\boldsymbol{y} \rangle = 0$. If characteristic $char \mathbb{F} = 2$, then $\langle \boldsymbol{x}, \boldsymbol{x} \rangle = 0$, Q.E.D.

Let characteristic $\mathbb{F}$ be odd. Since $\langle \boldsymbol{x}, \boldsymbol{x} + 2\boldsymbol{y} \rangle = 0$ for all $\boldsymbol{y} \in \mathcal{N}$, then, this is correctly for $\boldsymbol{y}_1 = 1/2\boldsymbol{y}$. from here follows as well, that $\langle \boldsymbol{x}, \boldsymbol{x} \rangle = 0$ for all $x \in \mathcal{L}$ as before. $\square$

Corollary 2. *If $\mathcal{L}$ contain at least one anti-isotropic vector, the for every $\mathcal{K} \supseteq \mathcal{L}$ there is basis, consisting of anti-isotropic vectors.*

Proof. Let $\mathcal{K}_0 \supseteq \mathcal{L}$ be maximal subspace. The ther is vector $y \in \mathcal{K} \setminus \mathcal{K}_0$, which is not isotropic. Then induction on dimension. $\square$

Let $\mathcal{M}$ be a subspace of the space $\mathbb{F}_p^n$ and $m = \dim \mathcal{M}$. By $B(\mathcal{M})$ we will be denote $m \times n-$ matrix, which rows are basis of subspace $\mathcal{M}$.

By $\mathrm{Com}\mathcal{M}$, where $\dim \mathrm{Com}\mathcal{M} = n - m$, let be denote the complement of $\mathcal{M}$ in $\mathbb{F}_p^n$. The space $\mathrm{Com}\mathcal{M}$ not uniquely defined: if $\{b_1, \dots, b_{n-m}\}$ is a basis of one of the subspsces $\mathrm{Com}\mathcal{M}$, then vectors $\{b_1 + x_1, \dots, b_{n-m} + x_{n-m}\}$, where $x_j \in \mathcal{M}$ are a basis of another subspaces.

If $\mathcal{K} = \mathcal{L} \oplus \mathcal{L}'$, then by $\mathrm{Com}_\mathcal{K}\mathcal{L}$ we will be denote space $\mathcal{L}'$, i.e. complement of $\mathcal{L}$ in $\mathcal{K}$.

If $\mathcal{K} = \mathcal{L} \oplus \mathrm{Com}_\mathcal{K}\mathcal{L}$, then obviously, that $B(\mathcal{L}) \cdot B(\mathcal{L}^\perp(\mathcal{K}))^T = 0$ and $B(\mathrm{Com}_\mathcal{K}\mathcal{L}) \cdot B(\mathcal{L}^\perp(\mathcal{K}))^T$ is $k - l \times t-$ matrix with $k - l$ linear independent rows of length t, where $k = \dim \mathcal{K}, l = \dim \mathcal{L}, t = \dim \mathcal{L}^\perp(\mathcal{K})$. From here, in particular, we can get another proof of Lemma 3.

Note, that if $\mathcal{K} \subseteq \mathcal{K}'$, then space $\mathcal{L}^\perp(\mathcal{K}')$ also is a dual space $\mathcal{L}$ in the space $\mathcal{K}$. In particular, $\mathcal{L}^\perp(\mathcal{K}) \subseteq \mathcal{L}^\perp = \mathcal{L}^\perp(\mathbb{F}_p^n)$.

For application it is more convenient to write Lemma 8 as follows.

Lemma 12. *For every subspace $\mathcal{M} \subseteq \mathbb{F}_p^n$ ther are subspaces $\mathrm{Com}\mathcal{M}$ and $\mathrm{Com}\mathcal{M}^\perp$ so, that*

$$(\mathrm{Com}\mathcal{M})^\perp = \mathrm{Com}\mathcal{M}^\perp. \tag{9}$$

Proof. Sinec $\mathbb{F}_p^n = \mathcal{M}^\perp \oplus \mathrm{Com}\mathcal{M}^\perp$, then it will be enough to proff, that there is subspace $(\mathrm{Com}\mathcal{M})$, that $\mathbb{F}_p^n = \mathcal{M}^\perp \oplus (\mathrm{Com}\mathcal{M})^\perp$.

Let $\{a_1, \dots, a_m\}$, $\{b_1, \dots, b_{n-m}\}$ $\{e_1, \dots, e_m\}$ are bases of subspaces $\mathcal{M}$, $\mathrm{Com}\mathcal{M}$ $\mathrm{Com}\mathcal{M}^\perp$.

We have to show, that there exists basis in the form of $\{b_1 + x_1, \dots, b_{n-m} + x_{n-m}\}$, $x_j \in \mathcal{M}$, one of the space $\mathrm{Com}\mathcal{M}$ and bsis in the form of $e_1 + y_1, \dots, e_m + y_m$, $y_j \in \mathcal{M}^\perp$, one of the space $\mathrm{Com}(\mathcal{M}^\perp)$ so that (9) is true, i.e. $\langle e_i + y_i, b_j + x_j \rangle = \langle e_i, b_j \rangle + \langle e_i, x_j \rangle + \langle y_i, b_j \rangle = 0$, $i = 1, \dots, m, j = 1, \dots, n-m$. If we write $x_j = \sum_{t=1}^{n-m} x_{j,t} a_t$ and $y_i = \sum_{s=1}^m y_{i,s} e_s$, $x_{j,t}, y_{i,s} \in \mathbb{F}_p$, then the last system will become

$$\langle e_i, b_j \rangle + \sum_{t=1}^{n-m} x_{j,t} \langle e_t, a_j \rangle + \sum_{s=1}^m y_{i,s} \langle e_s, b_j \rangle = 0, \quad i = 1, \dots, m, j = 1, \dots, n - m. \tag{10}$$

The matrix $\|\langle e_t, a_i \rangle; \langle e_t, b_j \rangle\|$, $t = 1, \dots, n - m$, $i = 1, \dots, m, j = 1, \dots, n - m$, has rang $n - m$, since vectors $\{e_1, \dots, e_m\}$ and vectors $\{a_1, \dots, a_m, b_1, \dots, b_{n-m}\}$ are linear independent. Thus the system (10) has solution for every fixed i. $\square$

Corollary 3. *Let $\mathcal{M}$ be subspace of $\mathbb{F}_p^n$. Then for some space $\mathrm{Com}\mathcal{M}^\perp$ the space $\mathbb{F}_p^n$ can be represented as direct sum*

$$\mathbb{F}_p^n = \mathcal{M} \oplus (\mathrm{Com}\mathcal{M}^\perp)^\perp \tag{11}$$

Corollary 4. *Let for subspaces* $\mathrm{Com}\mathcal{M}^{\perp}$ *and* $\mathrm{Com}\mathcal{M}$ *we have that* $\mathbb{F}_p^n = \mathcal{M} \oplus (\mathrm{Com}\mathcal{M}^{\perp})^{\perp}$, $\mathbb{F}_p^n = \mathcal{M}^{\perp} \oplus (\mathrm{Com}\mathcal{M})^{\perp}$ *and* $\mathcal{K} = \mathcal{L} \oplus (\mathrm{Com}\mathcal{M}^{\perp})^{\perp}$, *where* $\mathcal{L}$ *is subspace of* $\mathcal{M}$. *Then subspace* $\mathcal{M}^{\perp}$ *is orthogonal of* $\mathcal{L}$ *in* $\mathcal{K}$, *i.e.* $\mathcal{M}^{\perp} = \mathcal{L}^{\perp}(\mathcal{K})$.

Proof. Really, $B(\mathcal{M}^{\perp}) \cdot B(\mathcal{L})^T = 0$ and $B((\mathrm{Com}\mathcal{M}^{\perp})^{\perp}) \cdot B(\mathcal{M}^{\perp})$ is nonsingular matrix of dimension $n - m \times n - m$. $\square$

3   The MacWilliams Identity for Some Codes with Specific Structure

3.1 $(a\,|a + b)$ Construction

Lets start with code $\mathcal{L} = \mathcal{L}(\mathcal{L}_1,\ \mathcal{L}_2)$ of lenght $2n$, then code vecotrs we can write in the form

$$(a|a + b),\ a \in \mathcal{L}_1,\ b \in \mathcal{L}_2, \tag{12}$$

where $\mathcal{L}_1$, $\mathcal{L}_2$ are subspaces of space $\mathbb{F}_p^n$ of dimensions l_1, l_2, respectively.

The dimension of code $\mathcal{L} = \mathcal{L}(\mathcal{L}_1,\ \mathcal{L}_2)$ is $l = l_1 + l_2$. The r-th order Reed-Muller code $RM_{r,m}$ have the same construction in view of the fact that its vectors can we write in the form (12) with $\mathcal{L}_1 = RM_{r,m-1}$ and $\mathcal{L}_2 = RM_{r-1,m-1}$.

3.2 $\mathrm{K} = \mathrm{L}\left(\mathrm{L}_1, \mathbb{F}_p^n\right)$.

As comprehending code $\mathcal{K}$ lets consider a code $\mathcal{K} = \mathcal{L}(\mathcal{L}_1, \mathbb{F}_p^n)$. In this case, the space $\mathcal{L}^{\perp}(\mathcal{K})$, obviously, consist of vectors $(a^{\perp}|a^{\perp})$, where $a^{\perp} \in \mathcal{L}_2^{\perp}$. The space $\mathcal{L}^{\perp}(\mathcal{K})$ has dimension $n - l_2$.

Theorem 4 (The MacWilliams Relation for Code $\mathrm{L}\left(\mathrm{L}_1,\ \mathrm{L}_2\right) \square \mathrm{K} = \mathrm{L}\left(\mathrm{L}_1, \mathbb{F}_p^n\right)$**).** *We have*

$$W_{\mathcal{L}}(z) = \sum_{x \in \mathcal{L}} z^{w(x)}$$

$$= \frac{1}{|\mathcal{L}_2^{\perp}|} \sum_{a \in \mathcal{L}_1} \sum_{c \in \mathcal{L}_2^{\perp}} \exp\left(\frac{-2\pi i \langle a, c \rangle}{p}\right) \times$$

$$z^{w(a)}(1 + (p-1)z)^{n-w(c)}(1-z)^{w(c)} \tag{13}$$

$$= \frac{1}{|\mathcal{L}_2^{\perp}||\mathcal{L}_1^{\perp}|} \sum_{c \in \mathcal{L}_2^{\perp}} \sum_{b \in \mathcal{L}_1^{\perp}} (1 + (p-1)z)^{2n-w(c)-w(b-c)}(1-z)^{w(c)+w(b-c)}.$$

$$\tag{14}$$

Proof. Using the characteristic function $\varphi_{\mathcal{L},\mathcal{K}}(\boldsymbol{x})$ of code $\mathcal{L}$ in the code $\mathcal{K}$ (see. (1)), we get

$$W_{\mathcal{L}}(z) = \sum_{\boldsymbol{a}\in\mathcal{L}_1}\sum_{\boldsymbol{b}\in\mathcal{L}_2} z^{w(\boldsymbol{a}|\boldsymbol{a}+\boldsymbol{b})}$$

$$= \frac{1}{|\mathcal{L}^{\perp}(\mathcal{K})|} \sum_{\boldsymbol{a}\in\mathcal{L}_1}\sum_{\boldsymbol{b}\in\mathbb{F}_p^n}\sum_{\boldsymbol{c}\in\mathcal{L}_2^{\perp}} \exp\left(\frac{2\pi i\langle \boldsymbol{b},\boldsymbol{c}\rangle}{p}\right) z^{w(\boldsymbol{a}|\boldsymbol{a}+\boldsymbol{b})}$$

$$= \frac{1}{|\mathcal{L}_2^{\perp}|} \sum_{\boldsymbol{a}\in\mathcal{L}_1}\sum_{\boldsymbol{c}\in\mathcal{L}_2^{\perp}}\prod_{j=1}^{n}\sum_{x\in\mathbb{F}_p} \exp\left(\frac{2\pi i c_j x}{p}\right) z^{w(a_j)+w(x+a_j)}$$

$$= \frac{1}{|\mathcal{L}_2^{\perp}|} \sum_{\boldsymbol{a}\in\mathcal{L}_1}\sum_{\boldsymbol{c}\in\mathcal{L}_1^{\perp}}\prod_{c,a\in\mathbb{F}_p} (\tau_{c,a}(z))^{\nu_{c,a}(\boldsymbol{c},\boldsymbol{a})}, \tag{15}$$

where $\nu_{c,a}(\boldsymbol{c},\boldsymbol{a})$ is number of Coordinate of vectors $\boldsymbol{c} = (c_1,\ldots,c_n)$ and $\boldsymbol{a} = (a_1,\ldots,a_n)$, for which $c_j = c$, $a_j = a$ and $\tau_{c,a}(z) = \sum_{x\in\mathbb{F}_p}\exp\left(\frac{2\pi i c x}{p}\right) z^{w(a)+w(x+a)}$.

The sum $\tau_{c,b}(z)$ is easy to calculate:

$$\tau_{c,a}(z) = \begin{cases} (1+(p-1)z)z^{w(a)}, & c=0, \\ \exp(\frac{-2\pi i c a}{p})(1-z)z^{w(a)}, & c\neq 0 \end{cases}. \tag{16}$$

This proves the equality (13).

The equality (14) follows from relation

$$\sum_{\boldsymbol{a}\in\mathcal{L}_1}\exp\left(\frac{-2\pi i\,\langle \boldsymbol{a},\boldsymbol{c}\rangle}{p}\right)z^{w(\boldsymbol{a})} = \frac{1}{|\mathcal{L}_1^{\perp}|}\sum_{\boldsymbol{a}\in\mathbb{F}_p^n}\sum_{\boldsymbol{b}\in\mathcal{L}_1^{\perp}}\times$$

$$\exp\left(\frac{-2\pi i(\langle \boldsymbol{a},\boldsymbol{b}\rangle - \langle \boldsymbol{a},\boldsymbol{c}\rangle)}{p}\right)z^{w(\boldsymbol{a})}$$

$$= \frac{1}{|\mathcal{L}_1^{\perp}|}\sum_{\boldsymbol{b}\in\mathcal{L}_1^{\perp}}(1+(p-1)z)^{n-w(\boldsymbol{b}-\boldsymbol{c})}(1-z)^{w(\boldsymbol{b}-\boldsymbol{c})},$$

e where $\frac{1}{|\mathcal{L}_1^{\perp}|}\sum_{\boldsymbol{b}\in\mathcal{L}_1^{\perp}}\exp\left(\frac{2\pi i(\langle \boldsymbol{a},\boldsymbol{b}\rangle}{p}\right)$ is characteristic function of the code $\mathcal{L}_1$ in the space $\mathbb{F}_p^n$. $\square$

3.3   K = L $(\mathbf{F}_p^n, \mathbf{L}_2)$

As comprehending code $\mathcal{K}$ lets consider a code $\mathcal{K} = \mathcal{L}(\mathbb{F}_p^n, \mathcal{L}_2)$. In this case the space $\mathcal{L}^{\perp}(\mathcal{K})$ consist of vectors in the form $(\boldsymbol{a}^{\perp}|0)$, where $\boldsymbol{a}^{\perp}\in\mathcal{L}_1^{\perp}$, and has dimension $n - l_1$.

The characteristic function $\varphi_{\mathcal{L},\mathcal{K}}(\boldsymbol{x})$ of code $\mathcal{L}$ in the code $\mathcal{K}$ in this case looks like follows

$$\varphi_{\mathcal{L},\mathcal{K}}(\boldsymbol{x}) = \frac{1}{|\mathcal{L}^{\perp}(\mathcal{K})|}\sum_{\boldsymbol{y}\in\mathcal{L}^{\perp}(\mathcal{K})}\exp\left(\frac{2\pi i\langle \boldsymbol{x},\boldsymbol{y}\rangle}{p}\right) = \tag{17}$$

$$\frac{1}{|\mathcal{L}_1^{\perp}|} \sum_{c \in \mathcal{L}_1^{\perp}} \exp\left(\frac{2\pi i \langle a, c \rangle}{p}\right), \quad x = (a|a+b), \quad y = (c, 0), (a \in \mathbb{F}_p^n, b \in \mathcal{L}_2, c \in \mathcal{L}_1^{\perp}.$$

(18)

Theorem 5 (Analog of MacWilliams Identity for Code $\mathrm{L}(\mathrm{L}_1, \mathrm{L}_2)$ ▫ $\mathrm{K} = \mathrm{L}(\mathbb{F}_p^n, \mathrm{L}_2))$. *We have*

$$W_{\mathcal{L}}(z) = \sum_{x \in \mathcal{L}} z^{w(x)} = \frac{1}{|\mathcal{L}_1^{\perp}|} \sum_{b \in \mathcal{L}_2} \sum_{c \in \mathcal{L}_1^{\perp}} \prod_{c, b \in \mathbb{F}_p} (\tau_{c,b}(z))^{\nu_{a,b}(c,b)},$$

(19)

where

$$\tau_{c,b}(z) = \begin{cases} 1 + (p-1)z^2, & \text{if } c = 0, b = 0, \\ 2z + (p-2)z^2, & \text{if } c = 0, b \neq 0, \\ (1 - z^2), & \text{if } c \neq 0, b = 0, \\ (1 + \exp(\frac{-2\pi i c b}{p}))z - (1 + \exp(\frac{-2\pi i c b}{p}))z^2, & \text{if } c \neq 0, b \neq 0, \end{cases}$$

(20)

Proof. Using the characteristic function (17), we have

$$W_{\mathcal{L}}(z) = \sum_{a \in \mathcal{L}_1} \sum_{b \in \mathcal{L}_2} z^{w(a|a+b)}$$

$$= \frac{1}{|\mathcal{L}^{\perp}(K)|} \sum_{a \in \mathbb{F}_p^n} \sum_{b \in \mathcal{L}_2} \sum_{c \in \mathcal{L}_1^{\perp}} \exp\left(\frac{2\pi i \langle a, c \rangle}{p}\right) z^{w(a|a+b)}$$

$$= \frac{1}{|\mathcal{L}_1^{\perp}|} \sum_{b \in \mathcal{L}_2} \sum_{c \in \mathcal{L}_1^{\perp}} \prod_{j=1}^{n} \sum_{x \in \mathbb{F}_p} \exp\left(\frac{2\pi i c_j x}{p}\right) z^{w(x)+w(x+b_j)}$$

$$= \frac{1}{|\mathcal{L}_1^{\perp}|} \sum_{b \in \mathcal{L}_2} \sum_{c \in \mathcal{L}_1^{\perp}} \prod_{c, b \in \mathbb{F}_p} (\tau_{c,b}(z))^{\nu_{a,b}(c,b)},$$

where $\nu_{c,b}(c, b)$ is number of coordinate of vectors $c = (c_1, \ldots, c_n)$ and $b = (b_1, \ldots, b_n)$, for which $c_j = c$, $b_j = b$, and $\tau_{c,b}(z) = \sum_{x \in \mathbb{F}_p} \exp\left(\frac{2\pi i c x}{p}\right) z^{w(x)+w(x+b)}$. We can easily see that, $\tau_{c,b}(z)$ looks like (20). □

Note, that for $p = 2$ $\tau_{1,1}(z) = 0$. Thus $\prod_{c,b \in \mathbb{F}_p} (\tau_{c,b}(z))^{\nu_{c,b}(c,b)} = 0$, if the vectors a, b have at least one coordinate $a_j = b_j = 1$, i.e. if $(c_1 \cdot b_1, \ldots, c_n \cdot b_n) = c \cdot b \neq 0$.

With fixed vector c a set of vectors $b \in \mathcal{L}_2$, for which $c \cdot b = 0$, make up a subspace $\mathcal{L}_2(c)$. In particular, $\mathcal{L}_2(0) = \mathcal{L}_2$. Note, that relation $c \cdot b = 0$ is fulfilled if and only if $w(c + b) = w(c) + w(b)$. We have also, if $b \in \mathcal{L}_2(c)$, then $\nu_{0,1}(c, b) = w(b)$, $\nu_{1,0}(c, b) = w(c)$ and $\nu_{0,0}(c, b) = n - w(c) - w(b)$.

This and (19) give

Corollary 5. *Let* $p = 2$. *Then for binary code* $\mathcal{L} = \{(\boldsymbol{x}|\boldsymbol{x} + \boldsymbol{y}); \boldsymbol{x} \in \mathcal{L}_1, \boldsymbol{y} \in \mathcal{L}_2\}$ *we have*

$$W_{\mathcal{L}}(z) = \sum_{\boldsymbol{x} \in \mathcal{L}} z^{w(\boldsymbol{x})}$$

$$= \frac{1}{|\mathcal{L}_1^{\perp}|} \sum_{\boldsymbol{c} \in \mathcal{L}_1^{\perp}} \sum_{\boldsymbol{b} \in \mathcal{L}_2(\boldsymbol{c})} \left(1 + z^2\right)^{n - w(\boldsymbol{b}) - w(\boldsymbol{c})} \left(1 - z^2\right)^{w(\boldsymbol{c})} (2z)^{w(\boldsymbol{b})}$$

$$= \frac{1}{|\mathcal{L}_1^{\perp}|} \sum_{\boldsymbol{b} \in \mathcal{L}_2} \sum_{\boldsymbol{c} \in \mathcal{L}_1^{\perp}(\boldsymbol{b})} \left(1 + z^2\right)^{n - w(\boldsymbol{b}) - w(\boldsymbol{c})} \left(1 - z^2\right)^{w(\boldsymbol{b})} (2z)^{w(\boldsymbol{c})} \quad (21)$$

Corollary 6. *For binary code* $\mathcal{L} = \{(\boldsymbol{x}|\boldsymbol{x} + \boldsymbol{y}); \boldsymbol{x} \in \mathcal{L}_1, \boldsymbol{y} \in \mathcal{L}_2\}$ *we have*

$$W_{\mathcal{L}}(z) = \frac{1}{|\mathcal{L}_2^{\perp}|} \sum_{\boldsymbol{a} \in \mathcal{L}_1} \sum_{\boldsymbol{c} \in \mathcal{L}_2^{\perp}} (-1)^{\langle \boldsymbol{a}, \boldsymbol{c} \rangle} z^{w(\boldsymbol{a})} (1 + z)^{n - w(\boldsymbol{c})} (1 - z)^{w(\boldsymbol{c})} = \quad (22)$$

$$= \frac{1}{|\mathcal{L}_1^{\perp}|} \sum_{\boldsymbol{c} \in \mathcal{L}_1^{\perp}} \sum_{\boldsymbol{b} \in \mathcal{L}_2(\boldsymbol{c})} \left(1 + z^2\right)^{n - w(\boldsymbol{b}) - w(\boldsymbol{c})} \left(1 - z^2\right)^{w(\boldsymbol{c})} (2z)^{w(\boldsymbol{b})}.$$

Example 1. $\mathcal{L} = \{(\boldsymbol{a}|\boldsymbol{a} + \boldsymbol{b}); \boldsymbol{a} \in RM_{m-2,m}, \boldsymbol{b} \in RM_{1,m}\}$.

Let $n = 2^m$, $\{\alpha_1, \dots, \alpha_n\} = \mathbb{F}_2^m$ and $\boldsymbol{c}(l) = (l(\alpha_1), \dots, l(\alpha_N))$ is code vector of $RM_{1,m}$, where $l(\boldsymbol{x}) = a_1 x_1 + \cdots + a_m x_m + \epsilon$ is an affine function. It's known, that weight istribution of $RM_{1,m}$ consists of one vector of weight 0 and one vector of weight n and $2^{m+1} - 2$ vectors of weight $\frac{n}{2} = 2^{m-1}$

From the fact, that $RM_{m-2,m}^{\perp} = RM_{1,m}$ and the fact said above follows, that code $\mathcal{L}_2(\boldsymbol{c}(l))$ for $l \neq \boldsymbol{0}, \boldsymbol{1}$, consists of vectors $\boldsymbol{c}(l + 1)$ and $\boldsymbol{0}$, because, for affine functions l', distinct from $l + 1, 0$, a set of vectors α_j, for which $l'(\alpha_j) = 0$ for all j such as $l(\alpha_j) = 1$, is empty set. If $l = 0$, then, obviously, $RM_{1,m}(\boldsymbol{0}) = RM_{1,m}$. If $l = 1$, i.e. $\boldsymbol{c} = \boldsymbol{1}$, then $RM_{1,m}(\boldsymbol{c}) = \{\boldsymbol{0}\}$.

This and (22) give, that

$$W_{\mathcal{L}}(z) = \sum_{\boldsymbol{x} \in \mathcal{L}} z^{w(\boldsymbol{x})}$$

$$= \frac{1}{|\mathcal{L}^{\perp}(\mathcal{K})|} \left(\sum_{\boldsymbol{b} \in RM_{1,m}} \left(1 + z^2\right)^{n - w(\boldsymbol{b})} (2z)^{w(\boldsymbol{b})} + \left(1 - z^2\right)^{n} + \right.$$

$$\sum_{\boldsymbol{c} \in RM_{1,m}) \backslash \{0,1\}} \sum_{\boldsymbol{b} \in RM_{1,m}(\boldsymbol{c})} \left(1 + z^2\right)^{n - w(\boldsymbol{b}) - w(\boldsymbol{c})} \times$$

$$\left. \left(1 - z^2\right)^{w(\boldsymbol{c})} (2z)^{w(\boldsymbol{b})} \right)$$

$$= \frac{1}{|\mathcal{L}^{\perp}(\mathcal{K})|} \left(\left(1 + z^2\right)^{n} + (2z)^{n} + \left(1 - z^2\right)^{n} + (2^{m+1} - 2) \times \right.$$

$$\left. \left(\left(1 - z^2\right)^{\frac{n}{2}} (2z)^{\frac{n}{2}} + \left(1 + z^2\right)^{\frac{n}{2}} \left(1 - z^2\right)^{\frac{n}{2}}\right) \right), \quad (23)$$

where $|\mathcal{L}^{\perp}(\mathcal{K})| = 2^{n-m-1}$, $n = 2^m$.

References

1. F.W. MacWilliams and N.W.A. Sloane, The Theory of Error-Correcting Codes. North-Holland, Amsterdam (1977).
2. J. Simonis, MacWillams Identities and Coordinate Partitions, Linear Algebra and its Applications, Vol. 216 (1995), pp. 81–91.
3. G. David Forney, Jr. Transforms and Groups, Codes, curves, and signals (Urbana, IL, 1997), 79–97, Kluwer Acad. Publ., Boston. MA. 1998.

Good Expander Graphs and Expander Codes: Parameters and Decoding

H. Janwa*

UPR-Rio Piedras, San Juan, PR00931-3355
`hjanwa@rrpac.upr.clu.edu`
`http://ramanujan.cnnet.clu.edu/~janwa`

Abstract. We present some explicit constructions of families of constant degree expander graphs. We use them to give constructions of expander codes and determine stronger bounds on their parameters. We also show that some of these codes can be decoded very efficiently using the decoding algorithm recently given by Janwa and Lal.

1   Introduction

Ever since the seminal work of Claude Shannon 50 years ago, the research areas of codes and sequence designs have been of fundamental importance to communication science.

Expander Graphs and Ramanujan graphs were first introduced in connection with communication networks, sorting networks and were utilized in the construction of the famous AKS optimal sorting networks. They have been used in theoretical computer science, and in the design of networks for high-performance computing. A recent breakthrough is in the applications of these graphs in the construction of codes (expander codes, low density parity check codes) that have error-correcting capacity close to the theoretical limit, a goal long thought to be unattainable. Conversely, coding theory has led to recent exciting developments in theoretical computer science, such as the design of probabilistically checkable proofs and hardness of approximation problems, and pseudo random generators. Furthermore, the tools and techniques developed in coding theory have become extremely useful in cryptography, image processing, data compression, rate distortion theory, computer science, and several other areas of discrete applied sciences.

In this paper we present some explicit constructions of families of constant degree expander graphs. We use them to give constructions of expander codes and determine stronger bounds on their parameters. We also show that some of these codes can be decoded very efficiently using the decoding algorithm recently given by Janwa and Lal [29].

---

* Research supported in part by the NSF grant number CR:981498

M. Fossorier, T. Hoeholdt, and A. Poli (Eds.): AAECC 2003, LNCS 2643, pp. 119–128, 2003.

2 Expander Graphs and Ramanujan Graphs

Expander graphs are graphs that are sparse but "highly connected" (see Bien [4]) for a survey. First dramatic application of expander graphs was by Ajtai, Komlós and Szemerédi in constructing an $\mathcal{O}(n \log n)$ sorting network on sequential machines and a $c \log n$ sorting network on parallel machines (see [56] for references). Since then they have been used in the construction of pseudo-random graphs, in constructing asymptotically good error-correcting codes (expander-codes, low density parity check (LDPC) codes [58]) (and a profusion of articles by these and other authors), in several areas of network theory, in graph theory, in cryptography, in probability theory and dynamical systems, in group theory, in theoretical computer science (probabilistically checkable proofs), in complexity theory and in cryptography (for example, in construction extractors and pseudo-random graphs). For surveys of these applications, we refer to [4], Sarnak [56], Lubotzky [47], [14], and Janwa and Rangachari [35]. Some applications of the graphs constructed in this article are discussed in the last section.

Ramanujan graphs were defined by Lubotzky, Phillips, and Sarnak [48] as k-regular expander graphs with remarkable eigenvalue properties. Namely, they are precisely those graphs whose Ihara zeta function satisfies the Riemann hypothesis (see [47]). Furthermore, as a family, Ramanujan graphs provide asymptotically optimal expander graphs. We introduce the notion of these graphs via bipartite graphs first.

Let $\mathcal{X} = (V, E)$ be a graph with vertex set $V = \{v_1, v_2, \ldots, v_n\}$ and edge set E. Let $A = (a_{ij})$ be the adjacency matrix of $\mathcal{X}$, where a_{ij} equals the number of edges between the vertex v_i and v_j. If $\mathcal{X}$ has a loop at vertex v_i, then a_{ii} is set equal to 2. The eigenvalues of $\mathcal{X}$ are, by definition, the eigenvalues of A.

For a subset Y of the vertices, let $\partial Y := \{x \in V \setminus Y | d(x, y) = 1$ for some $y \in Y\}$.

Definition 1. *[48]* Let $\mathcal{X} = (V, E)$ be a bipartite graph with kn edges and bipartition given by $V = I \cup O$ so that $|I| = n$ and $|O| = \theta n$. Then $\mathcal{X}$ is called an $(n, \theta, k, \alpha, c)$ *expander graph* (or a *bounded strong concentrator* (bsc)) in the terminology of Gabber and Galil (see [63]), if it satisfies the following property:

$$\text{For } X \subseteq I \text{ with } |X| \leq \alpha n, \quad |\partial X| \geq c|X|. \tag{1}$$

Let $\lambda_0, \lambda_1, \cdots, \lambda_{n-1}$ be the eigenvalues of a k-regular graph $\mathcal{X}$. Define,

$$\mu_1(\mathcal{X}) = \max_{|\lambda_i| \neq k \ 0 \leq i \leq n-1} |\lambda_i| \tag{2}$$

The eigenvalues of A^2 are λ_i^2 ($0 \leq i \leq n - 1$). Thus $\mu_1(\mathcal{X})^2$ is the next-to-largest eigenvalue of A^2. Clearly, $|\lambda_1| \leq \mu_1(\mathcal{X})$. For a bipartite graph, $|\lambda_1| = \mu_1(\mathcal{X})$. Determination of the expansion coefficient was made feasible after Tanner [63] demonstrated the following result.

Proposition 1. *Let $\mathcal{X}$ be a k-regular bipartite graph on $2n$ vertices, with n input nodes and n output nodes, then $\mathcal{X}$ is an $(n, 1, k, \alpha, c(\alpha))$ expander with $c(\alpha) \geq \frac{k^2}{\alpha k^2 + (1-\alpha)\mu_1^2}$. In particular, for $0 < \alpha < 1$, $c(\alpha) > 1$.*

Proposition 1 shows that, for two k-regular bipartite graphs $\mathcal{X}_1$ and $\mathcal{X}_2$, if $\mu_1(\mathcal{X}_1) < \mu_1(\mathcal{X}_2)$, then $\mathcal{X}_2$ is expected to have better expansion coefficients than $\mathcal{X}_1$.

Similarly, from results of Alon and Millman, Chung, and Mohar (see Sarnak [56] or Janwa and Rangachari for references) it is known that graphs with smaller μ_1 are expected to have smaller diameter. Similarly, smaller values of μ_1 yield better upper bounds on diameters (see [12]) and better lower bounds on chromatic, girth, and high vertex connectivity (useful in fault tolerance in networks) (see [56], [47], and [35] for references).

It is also know that as a family, for a fixed regularity k, Ramanujan graphs have asymptotically optimal $\mu_1 = 2\sqrt{k-1}$.

Following the works of Lubotzky, Phillips and Sarnak, and Margulis, many constructions of Ramanujan graphs are known (see the survey articles mentioned above). In Janwa and Moreno [34], a new construction of graphs from codes called *projective coset graphs* is introduced. These new graphs yield many sequences of Ramanujan graphs (see also [32] and [33]). The projective coset graphs connect three very fruitful applications of exponential sums, namely, Coding Theory, Sequence Designs, and Ramanujan Graphs.

Reingold, Vadhan, and Wigderson [54] recently introduced the "zig-zag" graph product construction that, that beginning with a good expander graphs in iterative and recursive manner, provide sequences of new constant-degree expanders. For practical applications their construction requires graphs that can be described in a compact manner so that one can compute their "rotation maps". We use modifications of some excellent expander graphs constructed in Janwa and Moreno [34] to construct families of constant-degree expanders. Since the graphs constructed are based on highly regular objects such as *uniformly packed* codes, they have further desirable properties.

3   The Zig-Zag Graph Product and New Constant-Degree Expanders from Projective Coset Graphs

3.1   The Zig-Zag Graph Product

First we define the zig-zag graph product construction of Reingold, Vadhan, and Wigderson [54].

Definition 2. *[54] For a D-regular undirected multigraph G on N vertices, we assume that the edges incident at each vertex are colored (or labeled). We will also assume that the graphs could be multigraphs The rotation map* Rot_G : $[N] \times [D] ----> [N] \times [D]$ *is defined as follows:*

$\text{Rot}_G(v.i) = (w, j)$ *if the ith edge incident to v leads to w, and this edge is the jth edge incident to w. Let G_1 be an (N_1, D_1, λ_1)-graph, and let G_2 be an (D_1, D_2, λ_2)-graph. The zig-zag product of $G_1 \textcircled{z} G_2$ $(N_1 \times D_1, D_2^2, \lambda)$-graph such that for each edge $e = (v, w)$ of G_1, there are two associated vertices of $G_1 \textcircled{z} G_2$—-(v, w) and (w, l), where e is the kth edge leaving v and lth edge leaving w. Furthermore, $\text{Rot}_{G_1}(v, k) = (w, l)$.*

Theorem 1. [54] Let G_1 and G_2 be as in Definition 2. Then $G_1 ⊘ G_2$ is an $(N_1 \times D_1, D_2^2, f(\lambda_1, \lambda_2))$-graph where $f(\lambda_1, \lambda_2) \leq \lambda_1 + \lambda_2^2$ and $f(\lambda_1, \lambda_2) < 1$ when $\lambda_1, \lambda_2 < 1$. Moreover, $\mathrm{Rot}_{G_1 ⊘ G_2}$ can be computed in time $\mathrm{Poly}(\log N_1, \log D_1, \log D_2)$ with one oracle query to Rot_{G_1} and two oracle queries to Rot_{G_2}.

The following *recursive construction* from [54] will be instrumental in some of our constant-degree expander graphs.

Let H be a (D^8, D, λ)-graph form some D and λ. Let G_1 be H^2 and G_2 be $H \otimes H$. Then, for $t > 2$, G_t is defined recursively by

$$G_t = \left(G_{\lceil (t-1)/2 \rceil} \bigotimes G_{\lfloor (t-1)/2 \rfloor} \right)^2 ⊘ H$$

Theorem 2. [54] For every $t \geq 0$, G_t is a (D^{8t}, D^2, λ_t)-graph with $\lambda_t = \lambda + \mathcal{O}(\lambda^2)$. Moreover, Rot_{G_t} can be computed in time $\mathrm{poly}(t, \log D)$ with $\mathrm{poly}(t)$ oracle queries to Rot_H.

Definition 3. Let $\mathcal{X}$ be a D regular graph on N vertices. If $\lambda = \mu_1(\mathcal{X})/D$ denotes its normalized second highest eigenvalue (in absolute value), then $\mathcal{X}$ is called an (N, D, λ)-graph.

Lemma 1. [54] If $\mathcal{X}_1$ is an (N_1, D_1, λ_1)-graph, then its square, $\mathcal{X}_1^2$, is an $(N_1, D_1^2, \lambda_1^2)$ graph. Furthermore, if $\mathcal{X}_2$ is another (N_2, D_1, λ_1)-graph, then their tensor product $\mathcal{X}_1 \otimes \mathcal{X}_2$ is an $(N_1 \times N_2, D_1 \times D_2, \lambda_1^2)$ graph.

For this recursive construction to yield excellent constant degree expander graphs, we need initial expander graph H that is highly expanding. In the next section, we give such examples.

4   Projective Coset Graphs of Codes

In Janwa and Moreno [34] the following notion of graphs from codes was introduced.

Let C be an $[n, k, d]$ code over $\mathcal{F}_q$. Assume that $d \geq 3$. Let $T = \mathcal{F}_q^n/C$ be the coset space of C. Let $S \subset T$ be the set of cosets of weight 1, where weight of a coset $D := \min\{|x| | x \in D\}$. Let $\mathcal{P}(S) := S/\approx$, where $D \approx E$ if and only if $D = \lambda E$, where $\lambda \in \mathcal{F}_q^*$ (i.e., $\mathcal{P}(S)$ is the projective set associated to S). Since $d \geq 3$, we may take, $\mathcal{P}(S) = \{\S_i + C | i = 1, \ldots, n\}$, where for each i, $|\S_i|$ is the vector of weight 1 with a 1 in the ith coordinate. Consequently, $|\mathcal{P}(S)| = n$.

Definition 4. *(Janwa and Moreno [34])* The graph $\mathcal{X}_{\mathcal{P}}(C)(T, \mathcal{P}(S))$ is defined to be having vertex set T and the edge set labeled by the elements of $\mathcal{P}(S)$, such that two vertices $\S + C$ and $\dagger + C$ in T are connected if and only if $\dagger - \S + C \in \mathcal{P}(S)$ (i.e., this is the Cayley graph associated to the group T and the subset $\mathcal{P}(S)$) (see Biggs [5]). Therefore, $\mathcal{X}\mathcal{P}(C)$ is an n-regular graph on q^{n-k} vertices.

Remark 1. The projective coset graphs that are are attached to linear codes are, in general, directed. To get undirected graphs, one can modify the definition for adjacency by declaring that two vertices $\S + C$ and $\dagger + C$ in T are connected if and only if $\dagger + \S + C \in \mathcal{P}(S)$ see [43] and [12] for motivation. *The eigenvalue expressions for these graphs are the same as those for $\mathcal{X}_P$, and since eigenvalues are what matters in applications considered here, we deal only with the graphs $\mathcal{X}_P$.*

Theorem 3. *(Janwa and Moreno [32]) Let C be an $[n,k,d]$ code defined over $\mathcal{F}_q$, where $q = p^m$. The set of eigenvalues of $\mathcal{X}_P(C)$ is $\{(\sum_{i=1}^{n} e^{2\pi i Tr(c_i)/p})|c \in C^{\perp}\}$, where Tr is the trace mapping from $\mathcal{F}_q$ to $\mathcal{F}_p$.*

4.1 Explicit Construction of Constant-Degree Expanders

We use as our base graph modifications of some excellent expander graphs constructed in Janwa and Moreno [34] as base graphs in conjunction with ziz-zag construction, censoring, squaring, and replacement techniques to give constructions of families of constant degree expander graphs.

Construction 1. Let α be a *primitive* element of $\mathcal{F}_{2^m}$ (i.e., α is a generator for the multiplicative group $\mathcal{F}_{2^m} \setminus \{0\}$) and let $m_r(x)$ denote the minimal polynomial of α^r over $\mathcal{F}_2$. Let $n = 2^m - 1$. Let C_r be the cyclic code of length $2^m - 1$ over $\mathcal{F}_2$ generated by $m_1(x)m_r(x)$ (i.e., the vector space $\frac{x^n-1}{\langle m_1(x)m_r(x)\rangle}$).

Let $r = 2^i + 1$, or $r = 2^{2i} - 2^i + 1$. Then, $C_r^{\perp}$ has three non-zero weights, namely $2^{m-1}, 2^{m-1} \pm 2^{(n+e-2)/2}$, when m/e is odd, where $e = $ G.C.D.(m,i) (see [53] and [49]).

Then it is shown in Janwa and Moreno [34] that the graphs $\mathcal{X}(C_r)$ is a $K = (2^m - 1)$-regular graph on $N = (K+1)^2$ vertices and that $\mu_1(\mathcal{X}(C_r)) = 1 + 2^{e/2}\sqrt{K+1}$.

Consequently, $e = 1$ yields strongly RGs with $\mu_1 \approx \sqrt{2}\sqrt{K-1}$.

However, we need graphs H of the type (D^8, D, γ) with small γ.

Let $e = 1$ and let be odd. Let $r = 2^i + 1$, or $r = 2^{2i} - 2^i + 1$. We modify C_r as follows: Let D_e be its even subcode. Consider the code $C = (D_e, 0) \cup [(D_e, 0) + 1^{n+1}]$. Then the graph $\mathcal{X}(C)$ is a 2^m regular graph on m vertices. From the weight distribution of C, we conclude that $\mathcal{X}^{(1)} = \mathcal{X}(C)$ is a $[2^{2m}, 2^m, \frac{1}{2^{m-1/2}}]$-graph.

Let

$$\mathcal{X}^{(2)} = \mathcal{X}^{(1)} \bigotimes \mathcal{X}^{(1)}$$

. Then $\mathcal{X}^{(2)}$ is a $[2^{4m}, 2^{2m}, \frac{1}{2^{m-1/2}}]$-graph.

Now, for $i \geq 3$, let

$$\mathcal{X}^{(i+1)} = \mathcal{X}^{(i)} \bigotimes \mathcal{X}^{(1)}$$

Lemma 1. *Then $\mathcal{X}^{(i)}$ is a $[q^{2(i+1)}, q^2, \lambda_i]$-graph. Where λ_i is recursively defined by $\lambda_{i+1} = \lambda_i + \lambda_i^2 + \frac{1}{2^{m-1/2}}$.*

In particular $H = \mathcal{X}^{(7)}$ is a $[D^8, D, \lambda]$ graph with $\lambda = \mathcal{O}(\frac{(7\sqrt{2})}{q})$. Now the following Proposition is a consequence of Theorem 2.

Theorem 4. *Let m be odd and let $D = q = 2^m$. Let $H = X^{(7)}$. Let G_t be as defined in Theorem 2. Then for $t \geq 0$, G_t is a (D^{8t}, D^2, λ_t)-graph with $\frac{\lambda = O(7\sqrt{2}}{D)}$. $\lambda_t = \lambda + O(\lambda^2)$. Moreover, Rot_{G_t} can be computed in time $\text{poly}(t, \log D)$ with $\text{poly}(t)$ oracle queries to Rot_H.*

These class of graphs are expected to be highly practical for the following reason. As discussed in [34], for $r = 2^i + 1$ or $r = 2^{2i} - 2^i + 1$, $i \geq 1$, and $e = 1$, $C_r^{\perp}$ $i \geq 1$, and $e = 1$, C_r has minimum distance 5 (see van Lint and Wilson [46] and Janwa and Wilson [36]), Consequently, the graphs $X(C_r)$ have larger girth (=4). Furthermore, the codes C_r are *uniformly packed* (see Janwa and Wilson [37]).

A better construction in many respect is the following:

Construction 2. *Let $q = p^m$, and let B_m be the primitive BCH code of length $p^m + 1$ over $\mathcal{F}_p$. Then $X(B_m)$ is a $[p^{2m}, p^m + 1, \frac{1}{p^{m/2}}]$ graph (see Janwa and Moreno [33]. This graphs are equivalent to some abelian Ramanujan graphs of of Li [43].)*
We modify these graphs to give graphs of G_{bch} of the type $[p^{2m}, p^m, \lambda]$, where $\lambda = \frac{1}{p^{m/2+\epsilon}}$.
Let $\Gamma^{(1)} = X(B_m)$. Let

$$\Gamma^{(2)} = \Gamma^{(1)} \bigotimes \Gamma^{(1)}$$

.

Now, for $i \geq 3$, let

$$\Gamma^{(i+1)} = \Gamma^{(i)} \bigotimes \Gamma^{(1)}$$

.

Then,

Proposition 2. *Γ^i is a $(q^2(i+1), q^2, O(i/\sqrt{q})$-graph.*

5 Some Expander Codes, Their Parameters, and Decoding Algorithms

In Janwa and Lal [29] Ramanujan graphs of Lubotzky, Phillips and Sarnak [48] are tensored with graphs from N-gons to give explicit families of asymptotically good graphs for which decoding algorithms involving $O(n \log n)$ of depth $O(\log n)$ is presented. We use the excellent families of graphs constructed here and tensor them with N-gons to give families of bipartite graphs that lead us to families of asymptotically good expander codes suitable for decoding, in complexity involving $O(n \log n)$ of depth $O(\log n)$, using the algorithm presented in Janwa and Lal [29]. Further details will be presented at the conference.

References

1. N. Alon, "Eigenvalues and expanders," *Combinatorica,* 6 (1986), 83–96.
2. N. Alon, J. Bruck, J. Naor, M. Naor, R.M. Roth, "Construction of asymptotically good low-rate error-correcting codes through pseudo-random graphs," *IEEE Trans. on Inform. Theory,* vol. 38, no. 2, pp. 509–516, 1992.
3. H. Bass, "The Ihara-selberg zeta function of a tree lattice," International Journal of Mathematics," vol. 3, no. 6, pp. 717–797, 1992.
4. Frédéric Bien, "Constructions of telephone networks by group representations," *Notices Amer. Math. Soc.,* vol. 36 (1989), no. 1, 5–22.
5. N. Biggs, *Algebraic Graph Theory.* 2nd edition. Cambridge University Press, 1993.
6. E. Bombieri, "On Exponential Sums in Finite Fields," *Amer. Jour. of Math.,* **88** (1966), 71–105.
7. A.E. Brouwer, A.M. Cohen, A. Neumaier, *Distance Regular Graphs.* Academic Press, London, 1978.
8. R. Calderbank and J.-M. Goethals, "On a pair of dual subschemes of the Hamming scheme $H_n(q)$", *European Journal of Combinatorics,* vol. 6, pp. 133–147, 1985.
9. R. Calderbank and W.M. Kantor, "The geometry of two weight codes," *Bull. London Math. Soc.,* vol. 18, 1986, pp. 97–122.
10. P.J. Cameron and J.H. van Lint, *Designs Graphs and their Links.* London Mathematical Society Student Texts 22, 1991.
11. L. Carlitz and S. Uchiyama, "Bounds on Exponential Sums," *Duke Math. Journal* **24** (1957), pp. 37–41.
12. F.R.K. Chung, "Diameters and eigenvalues," *Journal of AMS,* 2 (1989), 187–200.
13. F.R.K. Chung, "Constructing random like graphs," *Proc. Symp. in Appl. Math.,* vol. 44, 1991, pp. 21–55.
14. F.R.K. Chung, *Spectral graph theory.* CBMS Regional Conference Series in Mathematics, 92. Published for the Conference Board of the Mathematical Sciences, Washington, DC; by the American Mathematical Society, Providence, RI, 1997
15. R.M. Damerell, "On Moore graphs," *Proc. Cambridge Philos. Soc.,* vol. 74, pp. 227–236, 1973.
16. C. Delorme and P. Solé, "Diameters, covering index, covering radius and eigenvalues," *European Journal of Combinatorics,* vol. 12, no. 2, pp. 95–108, 1991.
17. P. Delsarte, "Four fundamental parameters of a code and their combinatorial significance," *Inform. Control.,* vol. 23, pp. 403–438, 1973.
18. P. Delsarte, "An algebraic approach to the association schemes of coding theory," *Phillips Research Reports Suppl.,* **10**, 1973.
19. P. Delsarte, "On Subfield Subcodes of Reed-Solomon codes," *IEEE Trans. Info. Theory,* **21** (1975), 575–576.
20. P. Delsarte and J.-M. Goethals, "Alternating bilinear forms over $GF(q)$," *J. Combinatorial Theory (A),* vol. 19, pp. 26–50, 1975.
21. H. Janwa , O. Moreno, P.V. Kumar, and T. Helleseth,"Ramanujan graphs from codes over Galois Rings," in preparation.
22. Joel Friedman (ed.), *Expanding Graphs.* DIMACS Series in Discrete Mathematics nd Theoretical Computer Science, vol. 10, AMS, 1993.
23. Joel Friedman, "Some graphs with small second eigenvalue," *Combinatorica,* vol. 15, no. 1, pp. 31–42, 1995.
24. H. Janwa, *Relations among various parameters of codes.* Ph.D. Dissertation, Syracuse University, 1986.

25. H. Janwa, "Covering radius of codes, expander graphs, and Waring's problem over finite fields," in preparation.

26. H. Janwa, "A graph theoretic proof of the Delsarte bound on the covering radius of linear codes," preprint.

27. H. Janwa, "Explicit families of constant degree expander graphs," in preparation.

28. H. Janwa and A.K. Lal, "On Expander Graphs: Parameters and Applications" submitted, January 2001.

29. H. Janwa and A.K. Lal, "On Tanner Codes: Parameters and Decoding," *Journal of Applicable Algebra, Algebraic Algorithms, and Error-Correcting Codes,* **Springer Journal**, vol. 13, pp. 335–347, 2003.

30. H. Janwa and H.F. Mattson, Jr., "The Projective Hypercube and Its Properties and Applications" (with H.F. Mattson, Jr.), *Proceedings of the 2001 International Symposium on Information Theory, (ISIT-2001)* June 2001, Washington, USA.

31. H. Janwa and H.F. Mattson, Jr., "Covering radii of even subcodes of t-dense codes," *Lecture Notes in CS,* vol. 229, pp. 120–130, Springer-Verlag, 1986.

32. H. Janwa and O. Moreno, "Elementary constructions of some Ramanujan graphs," *CONGRESSUS NUMERANTIUM,* December 1995.

33. H. Janwa and O. Moreno, "Coding theoretic constructions of some number theoretic Ramanujan graphs," *CONGRESSUS NUMERANTIUM,* Vol. 130, pp. 63–76, 1998.

34. H. Janwa and O. Moreno, "Expander Graphs, Ramanujan Graphs, Codes, Exponential Sums, and Sequences," *IEEE Transactions on Information Theory,* (to be submitted).

35. H. Janwa and S.S. Rangachari, *Ramanujan Graphs.* Lecture Notes. Preliminary Version, July 1994.

36. H. Janwa and R.M. Wilson, "Hyperplane sections of Fermat varieties in IP^3 in char. 2 and some applications to cyclic codes," *Proceedings of the 10th International Conference on Algebraic Algorithms and Error-Correcting Codes. Springer-Verlag Lecture Notes in Computer Science,* No. 673, 1993, pp. 180–194.

37. H. Janwa and R.M. Wilson, "On the minimum distance of the cyclic codes $\langle m_1(x) m_t(x) \rangle$," in preparation.

38. P. Vijay Kumar and O. Moreno, "Prime-phase sequences with periodic correlation properties better than binary sequences," *IEEE Transactions on Information Theory,* vol. IT-37, pp. 603–616, 1991.

39. G. Lachaud and J. Wolfmann, "The Weights of the Orthogonals of the Extended Quadratic Binary Goppa Codes," *IEEE Trans. on Information Theory*, vol. 36, no. 3, pp. 686–692, 1990.

40. G. Lachaud, "Artin-Schreier curves, exponential sums, and the Carlitz-Uchiyama bound for geometric Goppa codes," *J. Number Theory,* vol. 39, no. 1, 1991.

41. W.-C.W. Li and P. Sole, "Spectra of regular graphs and hypergraphs and orthogonal polynomials," *Europ. J. Combinatorics,* vol. 17, pp. 461–477, 1996.

42. F. Thomson Leighton and Bruce Maggs, INTRODUCTION TO PARALLEL ALGORITHMS AND ARCHITECTURES: EXPANDERS • PRAMS • VLSI. Morgan and Kaufmann, Pub.. San Mateo, California. *To appear.*

43. Wen-Ching Winnie Li, "Character sums and abelian Ramanujan graphs," *Journal of Number Theory,* vol. 41, pp. 199–217, 1992.

44. Wen-Ching Winnie Li, "Number theoretic constructions of Ramanujan graphs," *COLUMBIA UNIVERSITY NUMBER THEORY SEMINAR,* Astérisques, no. 228, 1995, pp. 101–120.

45. J.H. van Lint and R.M. Wilson, *A Course in Combinatorics.* Cambridge University Press, 1992.

46. J.H. van Lint and R.M. Wilson, "On the minimum distance of cyclic codes," *IEEE Transactions on Information Theory,* vol. IT-32, no. 1, pp. 23–40, 1986.

47. A. Lubotzky, *Discrete Groups, Expanding Graphs and Invariant Measures.* Birkhauser, 1994.

48. A. Lubotzky, R. Phillips and P. Sarnak, "Ramanujan Graphs," *Combinatorica,* 8 (1988), 261–277.

49. F.J. MacWilliams and N.J.A. Sloane, *The Theory of Error-Correcting Codes.* North-Holland, Amsterdam, 1977.

50. G.A. Margulis, "Explicit group theoretic constructions of combinatorial schemes and their applications for construction of expanders and super-concentrators," *Journal of Problems of Information Transformation,* 1988, 39–46.

51. C. J. Moreno, *Algebraic Curves over Finite Fields, Tracks in Mathematics* **97**, Cambridge University Press, 1991.

52. Moreno, O. and Moreno, C., "A *p*-adic Serre bound," *Finite Fields Appl.* vol. 4, 1998, no.3 201–217.

53. Niho, *Ph.D. Thesis.* University of Southern California, 197?.

54. O. Reingold, S. Vadhan and A. Wigderson, "Entropy waves, the zig-zag graph product, and new constant-degree expanders," *Ann. of Math.* (2), **155** (2002), no. 1, 157–187.

55. Saints, Keith; Heegard, Chris Algebraic-geometric codes and multidimensional cyclic codes: a unified theory and algorithms for decoding using Grübner bases. Special issue on algebraic geometry codes. IEEE Trans. Inform. Theory 41 (1995), no. 6, part 1, 1733–1751.

56. Peter Sarnak, *Some Applications of Modular Forms.* Cambridge University Press, 1990.

57. H. Schellwat, "Highly expanding graphs obtained from dihedral groups," in [22], pp. 117-125.

58. Sipser, Michael; Spielman, Daniel A. Expander codes. Codes and complexity. IEEE Trans. Inform. Theory 42 (1996), no. 6, part 1, 1710–1722.

59. A.N. Skorobogatov, "On the parameters of subcodes of algebraic-geometric codes over prime subfields. Applied algebra, algebraic algorithms, and error-correcting codes," *Discrete Appl. Math.* 33 (1991), no. 1-3, 205–214; MR 93e:94017

60. P. Solé, "Packing radius, covering radius and dual distance," *IEEE Trans. on Inform. Theory,* vol. 41, pp. 268–272, 1995.

61. D.A. Spielman, "Linear-time encodable and decodable error-correcting codes," *IEEE Trans. on Inform. Theory,* Vol. 42, No. 6, pp. 1710–1722, 1996.

62. H. Stichtenoth, "Which extended Goppa codes are cyclic?" *Journal of Combinatorial Theory, Series A* vol. 51,pp. 205–220, 1989.

63. R.M. Tanner, "Explicit concentrators from generalized *n*-gons", *SIAM J. of Discrete and Applied Mathematics,* 5 (1984), 287–29.

64. A. Thomason, "Dense Expanders and Pseudo-Random Bipartite Graphs," *Discrete Mathematics,* vol. 75, pp. 381–386, 1987.

65. A. A. Tietaväinen, "An Asymptotic Bound on the Covering Radii of Binary BCH Codes, *IEEE Trans. on Information Theory,* vol. 36, no. 1, Jan. 1990, pp. 211–213.

66. H.C.A. van Tilborg, *Uniformly Packed Codes.* Ph.D. Dissertation, Eindhoven Tech. Univ., 1976.

67. M.A. Tsfasman and S.G. Vladut, *Algebraic Geometric Codes.* Kluwer Academic Press, 1991.

68. S.G. Vlèduts and A.N. Skorobogatov, "On the spectra of subcodes over a subfield of algebro-geometric codes," *Problemy Peredachi Informatsii* **27** (1991), no. 1, 24–36; translation in *Problems Inform. Transmission* **27** (1991), no. 1, 19–29 ; MR 95i:94012

69. van der Vlugt, Marcel, "The true dimension of certain binary Goppa codes. *IEEE Trans. Inform. Theory,*" 36 (1990), no. 2, 397–398. 94B27 (11T71 14G15 94B05)

70. J. Wolfmann, "Résultats sur les paramèters des codes linéaires," *Revue du Cethedec,* no. 66, pp. 25–33, 1975.

71. A. Weil, "On Some Exponential Sums," *Proc. National Academy of Sciences,* USA, vol. 34 (1948), pp. 204–207.

On the Covering Radius of Certain Cyclic Codes

Oscar Moreno and Francis N. Castro

University of Puerto Rico, Rio Piedras, PO Box 23355, SJ, PR, 00931-3355
o_moreno@UPR1.UPR.clu.edu
fcastro@goliath.cnnet.clu.edu

Abstract. In this paper we apply divisibility techniques to obtain new results on the covering radius of certain cyclic codes.

1 Introduction

In Sect. 2 we give a summary of the divisibility results about the number of solutions of a system of polynomials equations. We review results of Chevalley-Warning, Ax-Katz and Moreno-Moreno.

In Sect. 3, we discuss the tightness of Moreno-Moreno's improvement of the Chevalley-Warning's Theorem.

In Sect.4 we review our results of [14], where we introduced a new method to obtain the covering radius of codes and in particular to prove quasi-perfection in codes. Our techniques combine divisibility results of Ax-Katz and Moreno-Moreno as well as coding theoretic methods. We answered a problem posed in [5] about the covering radius for BCH codes.

In the present paper we generalize the methods of [14], and in Sect. 5 we give our main results. In the first part of Sect. 5 we give a criterion to obtain the covering radius of codes with zeroes α^{d_1} and α^{d_2} and minimum distance 5 over $\mathbf{F}_{2^f}$ (see Theorem 10). This result provides an elementary proof that: (1) the code with zeroes $\alpha^{2^{t-1}+1}, \alpha^{2^t+1}$ over $\mathbf{F}_{2^{2t+1}}$ is quasi-perfect and (2) the code with zeroes α, $\alpha^{2^{2i}-2^i+1}$ over $\mathbf{F}_{2^f}$ is quasi-perfect, whenever $(f, i) = 1$.

2 On the Divisibility of Number of Solutions of a System of Polynomial Equations

In this section we give a brief exposition of the present day methods to obtain divisibility properties of a system of polynomials equations. In 1935, Chevalley proved a conjectured by E. Artin(see [4]).

Theorem 1 (Chevalley). *Let* $F(X_1, \dots, X_n)$ *be a polynomial over a finite field* $\mathbf{F}_q$ *having* q *elements and suppose that*
- $F(0, \dots, 0) = 0$
- $n > d = deg(F)$.

Then F *has at least two zeroes.*

M. Fossorier, T. Hoeholdt, and A. Poli (Eds.): AAECC 2003, LNCS 2643, pp. 129–138, 2003.

Let $F_1, \ldots, F_t$ be polynomials in n variables with coefficients in $\mathbf{F}_q$ of total degrees $d_1, \ldots, d_t$, respectively, and let $N(F_1, \ldots, F_t)$ be the number of simultaneous solutions of $F_1, \ldots, F_t$.

Warning obtained the following archimedean lower bound:

Theorem 2 (Warning). *Let $F_1, \ldots, F_t$ be polynomials in n variables with coefficients in $\mathbf{F}_q$ of total degrees $d_1, \ldots, d_t$, respectively. If $N(F_1, \ldots, F_t) \geq 1$, then*

$$N(F_1, \ldots, F_t) \geq q^{n - \sum_{i=1}^t d_i}.$$

For the proof see [22].

Ax obtained an improvement of Chevalley's theorem(see [2]).

Theorem 3 (Ax). *With the notation of Theorem 1. If μ is equal to $\lceil n/d \rceil - 1$, where $\lceil a \rceil$ is the smallest integer larger or equal to a, then the number of zeroes of F is divisible by q^μ.*

In 1971, N. Katz improved Ax's theorem(see [11]) as follows:

Theorem 4 (Katz). *Let $F_1, \ldots, F_t$ be polynomials in n variables with coefficients in F_q of total degrees $d_1, \ldots, d_t$. Let μ be the least integer that satisfies*

$$\mu \geq \frac{n - \sum_{i=1}^t d_i}{max_i \, d_i}.$$

Then q^μ divides $N(F_1, \ldots, F_t)$.

Moreno and Moreno improved the Ax-Katz's theorem and their result is stated following the definition provided below.

Definition 1. *Let $k = p$ or $k = q$. For each integer n with k-expansion*

$$n = a_0 + a_1 k \cdots + a_s k^s \text{ where } 0 \leq a_i < k,$$

we denote its k-weight by $\sigma_k(n) = \sum_{i=0}^s a_i$. The k-weight degree of a monomial $X^d = X_1^{d_1} \cdots X_n^{d_n}$ is $w_k(X^d) = \sigma_k(d_1) + \cdots + \sigma_k(d_n)$. The k-weight degree of a polynomial $F(X_1, \ldots, X_n) = \sum_d a_d X^d$ is $w_k(F) = max_{X^d, a_d \neq 0} w_k(X^d)$

The Moreno-Moreno's result(see [13]) is the following:

Theorem 5 (Moreno-Moreno). *Let $F_1, \ldots, F_t$ be polynomials in n variables with coefficients in $\mathbf{F}_q$, a finite field with $q = p^f$ elements. Let l_i be the p-weight degree of F_i and let μ be the smallest integer such that*

$$\mu \geq f \left(\frac{n - \sum_{i=1}^t l_i}{max_i \, l_i} \right).$$

Then p^μ divides $N(F_1, \ldots, F_t)$.

Remark 1. Moreno-Moreno is a total improvement of the Chevalley's result since the p-weight degree $\leq$ total degree. Theorem 5 improves Theorem 4 whenever the characteristic is small with comparison to the degrees, i.e., we need, say $p < \max_i d_i$ in order for an improvement to occur.

In [15], we improved Theorems 4 and 5.

Theorem 6. *Let $F_1, \ldots, F_t$ be polynomials in n variables with coefficients in $\mathbf{F}_{q^s}$, a finite field with q^s elements. Let $w_q(F_i)$ be the q-weight degree of F_i and let μ be the smallest integer such that*

$$\mu \geq s\left(\frac{n - \sum_{i=1}^{t} w_q(F_i)}{max_i \, w_q(F_i)}\right).$$

Then q^μ divides $N(F_1, \ldots, F_t)$.

Example 1. Consider the following system of diagonal equations:

$$F_1(X_1, \ldots, X_4) = X_1 + X_2 + X_3 + \beta_1 X_4$$
$$F_2(X_1, \ldots, X_4) = X_1^{q+1} + X_2^{q+1} + X_3^{q+1} + \beta_2 X_4^{q+1}$$

over $\mathbf{F}_{q^s}$. We have that $w_q(F_1) = 1$, and $w_q(F_2) = 2$, hence $q^{\lceil \frac{s}{2} \rceil}$ divides $N(F_1, F_2)$. Note that the theorem of Ax-Katz does not give any information about $N(F_1, F_2)$ and Moreno-Moreno's result gives that $p^{\lceil \frac{sf}{2} \rceil}$ divides $N(F_1, F_2)$. In particular, if $q = 2^{13}$ and $s = 3$, our theorem implies that 2^{26} divides $N(F_1, F_2)$ and Moreno-Moreno's result implies that 2^{20} divides $N(F_1, F_2)$.

3   On the Tightness of the Result of Chevalley-Warning

In this section we substitute the total degree in Theorem 2 by the p-weight degree.

Theorem 7. *Let $F_1, \ldots, F_t$ be polynomials in n variables with coefficients in $\mathbf{F}_q$ of p-weight degrees $w_p(F_1), \ldots, w_p(F_t)$, respectively. If $N(F_1, \ldots, F_t) \geq 1$, then*

$$N(F_1, \ldots, F_t) \geq q^{(n - \sum_{i=1}^{t} w_p(F_i))}$$

where $N(F_1, \ldots, F_t)$ is the number of simultaneous solutions of $F_1 = 0, \ldots, F_t = 0$.

Proof. Let $q = p^f$ and let $N(F_1, \ldots, F_t)$ be the number of solutions of the following system of polynomials equations

$$F_1(X_1, \ldots, X_n) = 0$$

$$\vdots \qquad \vdots$$

$$F_t(X_1, \ldots, X_n) = 0$$

We have

$$N(F_1,\dots,F_t) = \frac{1}{q^t} \sum_{\substack{X_1,\dots,X_n \in \mathbf{F}_q \\ Y_1,\dots,Y_t \in \mathbf{F}_q}} \psi(\sum_{i=1}^{t} Y_i F_i(X_1,\dots,X_n)),$$

where $\psi(X) = e^{2\pi i Tr_{\mathbf{F}_q/\mathbf{F}_p}(X)}$.

For the sake of completeness we describe the reduction to the ground field method for a finite field of characteristic 2. Given a monomial $X_1^{d_1}\cdots X_n^{d_n}$ over $\mathbf{F}_{2^f}$, we can choose a $\mathbf{F}_2$-basis $\alpha_1,\dots,\alpha_f$ of $\mathbf{F}_{2^f}$ such that

$$X_j = \sum_{i=1}^{f} x_{ji}\alpha_i \text{ for } j = 1,\dots,n.$$

If we substitute the above in $X_1^{d_1}\cdots X_n^{d_n}$, we get

$$\begin{aligned}
X_1^{d_1}\cdots X_n^{d_n} &= (\sum_{i=1}^{f} x_{1i}\alpha_i)^{d_1}\cdots(\sum_{i=1}^{f} x_{ni}\alpha_i)^{d_n} \\
&= (\sum_{i=1}^{f} x_{1i}\alpha_i)^{\sum_{j=1}^{k_1} a_{1j}2^{j_{1j}}}\cdots(\sum_{i=1}^{f} x_{ni}\alpha_i)^{\sum_{j=1}^{k_n} a_{nj}2^{j_{nj}}} \\
&= (\sum_{i=1}^{f} x_{1i}\alpha_i)^{a_{11}2^{j_{11}}}(\sum_{i=1}^{f} x_{1i}\alpha_i)^{a_{12}2^{j_{12}}}\cdots \\
&= (\sum_{i=1}^{f} x_{1i}\alpha_i^{2^{j_{11}}})^{a_{11}}(\sum_{i=1}^{f} x_{1i}\alpha_i^{2^{j_{21}}})^{a_{12}}\cdots \quad (1)
\end{aligned}$$

where $d_i = a_{i1}2^{j_{i1}} + a_{i2}2^{j_{i2}} + \cdots$. Applying to (1) the trace map from $\mathbf{F}_{2^f}$ to $\mathbf{F}_2$, we get a polynomial in fn variables over $\mathbf{F}_2$ and degree less than or equal to $w_2(X_1^{d_1}\cdots X_n^{d_n})$. Hence we can transform any polynomial over $\mathbf{F}_{2^f}$ by using the reduction to the ground field method to a polynomial over $\mathbf{F}_2$ since any polynomial is a sum of monomials and the trace map is linear.

We now apply the reduction to the ground field method to $y_i F_i$ for $i = 1,\dots,t$ (see [13]). We obtain

$$Tr_{\mathbf{F}_q/\mathbf{F}_p}(Y_i F_i) = \sum_{j=1}^{f} y_{ij} F'_{ij}(x_{i1},\dots,x_{nf}).$$

Note that the degree of $F'_{ij} \leq w_p(F_i)$ for $j = 1, \ldots, f$, and $y_{ij} F'_{ij}$ is polynomial over $\mathbf{F}_p$. We get

$$N(F_1, \ldots, F_t) = \frac{1}{q^t} \sum_{\substack{X_1, \ldots, X_n \in \mathbf{F}_q \\ Y_1, \ldots, Y_t \in \mathbf{F}_q}} \psi(\sum_{i=1}^{t} Y_i F_i(X_1, \ldots, X_n))$$

$$= \frac{1}{q^t} \sum_{\substack{x_{11}, \ldots, x_{nf} \in \mathbf{F}_q \\ y_{11}, \ldots, y_{tf} \in \mathbf{F}_q}} \psi'(\sum_{i=1}^{t} \sum_{j=1}^{f} y_{ij} F'_{ij}(x_{11}, \ldots, x_{nf})) = N(F'_{11}, \ldots, F'_{tf})$$

Now we can apply Warning's theorem(Theorem 2) to the following system of polynomials equations:

$$F'_{11} = 0$$
$$F'_{12} = 0$$
$$\vdots$$
$$F'_{tf} = 0,$$

and we obtain that

$$N(F'_{11}, \ldots, F'_{tf}) \geq p^{nf - \sum_{i=1}^{t} \sum_{j=1}^{f} w_p(F'_{ij})} = p^{f(n - \sum_{i=1}^{t} w_p(F_i))}.$$

Remark 2. Theorem 7 follows from the results in [13]. We decided to include it here since it is not stated in [13]. Note that Theorem 7 is a total improvement of Theorem 2.

An immediate consequence of Theorems 5 and 7 is the following corollary.

Corollary 1 ([13]). *Let F be polynomial in n variables over $\mathbf{F}_q$ and suppose that*
- $F(0, \ldots, 0) = 0$
- $n > w_p(F)$.
Then F has a nontrivial solution.

We now discuss when Corollary 1 is tight. Consider the diagonal equation

$$X_1^d + \cdots + X_{n-1}^d + \beta X_n^d = 0 \tag{2}$$

over $\mathbf{F}_q$. If $(d, q - 1) = 1$, then equation (2) has q^{n-1} solutions. From now on, we assume that d divides $q - 1$. We want to compute the minimum number of variables such that (2) has a nontrivial solution. Let us define:

$$\delta(d, q) = \min\{n \mid X_1^d + \cdots + X_{n-1}^d + \beta X_n^d \text{ has a nontrivial solution}\}.$$

- **Corollary 1 is tight, when $\sigma_p(d) = 2$ for equations of type (2).** If $d = p^j + 1$ and $n = 2$, then equation (2) becomes $X_1^{p^j+1} + \beta X_2^{p^j+1} = 0$. If $-\beta$ is not a $(p^j + 1)$th power in $\mathbf{F}_q$, then $X_1^{p^j+1} + \beta X_2^{p^j+1}$ only has the trivial solution. Hence $\delta(p^j + 1, q) > 2$(see [15]). Using Corollary 1, we have that the equation $X_1^d + \cdots + X_{n-1}^d + \beta X_n^d$ has a nontrivial solution for any $n \geq 3$. Hence $\delta(p^j + 1, q) = 3$. In this case Corollary 1 is tight.

- **Corollary 1 is not tight, when $\sigma_p(d) > 2$, for equations of type (2)** . We now assume that $\sigma_p(d) > 2$. Then $\delta(d, q) \leq 3$ whenever $q > (d-1)^4$. This is true since $X_1^d + X_2^d = -\beta$ has a solution for every $\beta \in \mathbf{F}_q$ whenever $q > (d-1)^4$ (see [18]). In this case Theorem 7 implies that we need more than three variables to guarantee a nontrivial solution. Consequently Corollary 1 is not tight.
We can conclude that the Corollary 1 is tight, for equations of type (2), only when the exponent d has p-weight 2.

4   Review of Applications of Divisibility to Covering Radius

In this section we will state the main results of [14]. In the following section we generalize the results below.

In [14], we computed the covering radius of certain cyclic codes. This solved a question posed in [5]. The question was to give an direct proof of the computation of the covering radius for $BCH(3)$(see [5], [12]).

If a code C has minimum distance $2e+1$ and all the coset leaders have weight $\leq e + 1$ then the code is called quasi-perfect (A coset leader of a coset $\alpha + C$ is a vector of smallest weight in its coset). The covering radius is the weight of a coset leader with maximum weight or equivalently, the least r such that the spheres of radius r around the code words cover $\mathbf{F}_q^n$ (see [20], [7]). Our results were:

Theorem 8. *Let α be a primitive root of $\mathbf{F}_{2^f}$ and let C be the code of length $n = 2^f - 1$ with zeroes α, α^d over $\mathbf{F}_{2^f}$, where $d = 2^i + 1$. If $(i, f) = 1$, then C is a quasi-perfect code.*

Theorem 9. *Let α be a primitive root of $\mathbf{F}_{2^f}$. The code C with zeroes $\alpha, \alpha^d, \alpha^{d'}$ and minimum distance 7, where $d = 2^i + 1$, and $d' = 2^j + 1$, has covering radius 5 for $f > 8$.*

Theorem 9 provided an elementary proof for BCH(3), as well as the Non-BCH triple error correcting codes of Sect. 9.11 in [12]. Notice that the computation of the covering radius of BCH(3) required 3 papers (see [1], [6], and [19]). The first paper by J.A. van der Horst and T. Berger is elementary; the second paper by E.F. Assmus and H.F. Mattson used the Delsarte's bound, and the final paper by Helleseth invokes the Weil-Carlitz-Uchiyama bound and many people would consider it non-elementary.

An immediate consequence of the above theorem is the calculation of the covering radius of the Non-BCH triple correcting code of Sect. 9.11 in [12].

Corollary 2. Let $f = 2t + 1$ and α be a primitive root of $\mathbf{F}_{2^f}$. The code C with zeroes $\alpha, \alpha^d, \alpha^{d'}$, where $d = 2^{t-1} + 1$, and $d' = 2^t + 1$ has covering radius 5.

5 Main Results

Following the techniques of [14] in this section we will compute the covering radius of certain cyclic codes(see [12], [5]). Let d_1, d_2 be distinct natural numbers. Let $N((d_1, d_2), 4, \mathbf{F}_q)$ be the number of solutions over $\mathbf{F}_q$ of the following system of polynomials equations:

$$
\begin{aligned}
x_1^{d_1} + x_2^{d_1} + x_3^{d_1} &= \beta_1 x_4^{d_1} \\
x_1^{d_2} + x_2^{d_2} + x_3^{d_2} &= \beta_2 x_4^{d_2}
\end{aligned}
\tag{3}
$$

The following theorem gives a way to prove quasi-perfection.

Theorem 10. Let α be a primitive root of $\mathbf{F}_{2^f}$ $(f \geq 2)$ and let C be the code of length $n = 2^f - 1$ with zeroes $\alpha^{d_1}, \alpha^{d_2}$ over $\mathbf{F}_{2^f}$. We assume that the minimum distance of C is 5. Then C is a quasi-perfect code whenever 4 divides $N((d_1, d_2), 4, \mathbf{F}_{2^f})$ (see (3)).

Proof. C is a quasi-perfect code if its minimum distance is 5 and its covering radius is 3. Hence we need to prove that C has covering radius 3. Equivalently, we need to prove that the following system of equations has a solution:

$$
\begin{aligned}
x_1^{d_1} + x_2^{d_1} + x_3^{d_1} &= \beta_1 \\
x_1^{d_2} + x_2^{d_2} + x_3^{d_2} &= \beta_2
\end{aligned}
\tag{4}
$$

for any $(\beta_1, \beta_2) \in \mathbf{F}_{2^f}^2$.

The proof consists of three steps:

Step 1. By hypothesis, the number of solutions of the following system of equations

$$
\begin{aligned}
x_1^{d_1} + x_2^{d_1} + x_3^{d_1} &= \beta_1 x_4^{d_1} \\
x_1^{d_2} + x_2^{d_2} + x_3^{d_2} &= \beta_2 x_4^{d_2}
\end{aligned}
\tag{5}
$$

is divisible by 4, i.e., 4 divides $N((d_1, d_2), 4, \mathbf{F}_{2^f})$.

Step 2. We will prove that system (5) has a solution with $x_4 \neq 0$. If the system (5) does not have a solution with $x_4 \neq 0$, then

$$
\begin{aligned}
x_1^{d_1} + x_2^{d_1} + x_3^{d_1} &= 0 \\
x_1^{d_2} + x_2^{d_2} + x_3^{d_2} &= 0.
\end{aligned}
\tag{6}
$$

and system (5) have the same number of solutions.

We say that a solution (x_1, x_2, x_3) of (6) is non-trivial if all the x_i's are distinct. The system (6) has only trivial solutions since the minimum distance of C is 5, i.e., $x_{i_1} = 0$ and $x_{i_2} = x_{i_3}$. Using the inclusion-exclusion principle it is easy to obtain that number of solutions of system (6) is $2(3 \times 2^{f-1} - 1)$. But this is a contradiction since 4 has to divide the number of solutions of (6). Hence, system (5) has at least one solution with $x_4 \neq 0$. Therefore, the system (4) has at least one solution for any $(\beta_1, \beta_2) \in \mathbf{F}_{2^f}$. Hence, we can conclude that the covering radius of C is 3. That implies that C is a quasi-perfect code.

The above theorem generalizes Theorem 8.

We now combine the techniques of [16] and Theorem 10 to obtain the following theorem.

Theorem 11. *Let α be a primitive root of $\mathbf{F}_{2^{2t+1}}$, and let C be the code of length $n = 2^{2t+1} - 1$ with zeroes $\alpha^{2^i+1}, \alpha^{2^j+1}$. If C has minimum distance 5, then C is quasi-perfect.*

Proof. Following the techniques of [16, Lemma 24], we associate to the following system of equations

$$x_1^{2^i+1} + \cdots + x_3^{2^i+1} = \beta x_4^{2^i+1}$$
$$x_1^{2^j+1} + \cdots + x_3^{2^j+1} = \beta x_4^{2^j+1}$$

the following system of modular equations:

$$\begin{aligned}
(2^i + 1)j_1 + (2^j + 1)l_1 &\equiv 0 \bmod 2^f - 1 \\
(2^i + 1)j_2 + (2^j + 1)l_2 &\equiv 0 \bmod 2^f - 1 \\
(2^i + 1)j_3 + (2^j + 1)l_3 &\equiv 0 \bmod 2^f - 1 \qquad (7) \\
(2^i + 1)j_4 + (2^j + 1)l_4 &\equiv 0 \bmod 2^f - 1 \\
j_1 + j_2 + j_3 + j_4 &\equiv 0 \bmod 2^f - 1. \\
l_1 + l_2 + l_3 + l_4 &\equiv 0 \bmod 2^f - 1
\end{aligned}$$

The solutions of the modular system (7) determine the 2-divisibility of $N((2^i + 1, 2^j + 1), 4, \mathbf{F}_{2^{2t+1}})$, i.e., if

$$\mu = \min_{\substack{(j_1, \ldots, j_4, l_1, \ldots, l_4) \\ \text{is solution of } (7)}} \{\sigma_2(j_1) + \cdots + \sigma_2(j_4) + \sigma_2(l_1) + \cdots + \sigma_2(l_4)\} - 2(2t + 1),$$

then 2^μ divides $N((2^i + 1, 2^j + 1), 4, \mathbf{F}_{2^{2t+1}})$. Hence we need to prove that $\mu \geq 2$.

Applying the function σ_2 to the modular system (7) and dividing by 2, we obtain

$$\begin{aligned}
\sigma_2(j_1) + \sigma_2(l_1) &\geq t + 1 \\
\sigma_2(j_2) + \sigma_2(l_2) &\geq t + 1 \\
\sigma_2(j_3) + \sigma_2(l_3) &\geq t + 1 \\
\sigma_2(j_4) + \sigma_2(l_4) &\geq t + 1.
\end{aligned}$$

Hence $\sigma_2(j_1) + \cdots + \sigma_2(l_4) \geq 4t + 4 - (4t + 2) = 2$. Hence 4 divides $N((2^i + 1, 2^j + 1), 4, \mathbf{F}_{2^{2t+1}})$. This proves the theorem.

In the following corollary we use Theorem 10 to compute the covering radius of certain cyclic codes.

Corollary 3. *Let α be a primitive root of $\mathbf{F}_{2^f}$.*

1. *Let $f = 2t + 1$ and let C be the code of length $n = 2^{2t+1} - 1$ with zeroes $\alpha^{2^{t-1}+1}, \alpha^{2^t+1}$ over $\mathbf{F}_{2^{2t+1}}$, then C is a quasi-perfect code.*

2. *Let f be an odd number and let C be the code of length $n = 2^f - 1$ with zeroes $\alpha, \alpha^{2^{2i}-2^i+1}$ over $\mathbf{F}_{2^f}$, then C is a quasi-perfect code whenever $(i, f) = 1$.*

Proof. If C is the code with zeroes $\alpha^{2^{t-1}+1}, \alpha^{2^t+1}$ over $\mathbf{F}_{2^{2t+1}}$, then C has minimum distance 5(see [3] or [17]). Hence, applying Corollary 11, we obtain that C is quasi-perfect code. If C is the code with zeroes $\alpha^{2^i+1}, \alpha^{2^{3i}+1}$ over $\mathbf{F}_{2^f}$, then C has minimum distance 5 whenever $(i, f) = 1$ (see [21]). Applying Corollary 11, we obtain that C is quasi-perfect. If we change the primitive element, we obtain the code with zeroes α and $\alpha^{2^{2i}-2^i+1}$ (see [8]). Hence the code C with zeroes α and $\alpha^{2^{2i}-2^i+1}$ over $\mathbf{F}_{2^f}$ is quasi-perfect whenever $(f, i) = 1$.

Remark 3. Note that the dual of the code C with zeroes α and $\alpha^{2^{2i}-2^i+1}$ over $\mathbf{F}_{2^f}$ for $f/(f, i)$ odd has three nonzero weights (Kasami code, see [9], [10]) and using a result of Delsarte(see [12]) gives that the covering radius is 3.

Acknowledgments. We thank the referees for valuable suggestions, which have made the paper clearer.

References

1. E.F. Assmus, Jr. and H.F. Mattson Jr., Some 3-Error Correcting BCH Codes Have Covering Radius 5, *IEEE Trans. Inform. Th*, **22**, pp. 348–349, 1976.
2. J. Ax, Zeros of Polynomials over Finite Fields, *Am. J. Math.*, **86**, 1964, pp. 255–261.
3. R.D. Baker, J.H. van Lint, and R.M. Wilson, On the Preparata and Goethals Codes, *IEEE Trans. on Inform Theory*, **29**, no. 1, pp. 342–345, 1983.
4. C. Chevalley, Demonstration d'une hypothese de M. Artin, *Abhandlungen aus dem Mathematischen Seminar der Universität Hamburg*, **11**, pp. 73–75, 1936.
5. G. Cohen, L. Honkala, S. Litsyn, and A. Lobstein, *Covering Radius*, North-Holland Mathematical, Library **54**, 1997.
6. T. Helleseth, All Binary 3-Error-Correcting BCH Codes of Length $2^m - 1$ Have Covering Radius 5, *IEEE Trans. Inform. Th.*, **24**, pp. 257–258, 1978.
7. T. Helleseth, On the Covering Radius of Cyclic Linear Codes and Arithmetic Codes, *Discrete Applied Mathematics*, **11**, pp. 157–173 (1985).
8. H. Janwa and R. M. Wilson, *Hyperplane Sections of Fermat Varieties in $\mathbf{P}^3$ in Char. 2 and Applications to Cyclic Codes*, Applied Algebra, Algebraic Algorithms and Error-Correcting Codes, Proceedings AAECC-10 (G. Cohen, T. Mora, and O. Moreno Eds.), Lecture Notes in Computer Science, Springer-Verlag, New York/Berlin 1993.

9. T. Kasami, Weight Distribution of Bose-Chaudhuri-Hocquenghen Codes, *Combinatorial Math. and its Applications* (R.C. Bose and T.A. Dowling, eds.), Univ. of North Carolina Press, Chapel Hill, NC 1969.

10. T. Kasami, Weight Enumerators of several Classes of Subcodes of the and Order Binary Reed-Muller Codes, *Info. and Control*, **18** (1971), pp. 369–394.

11. N.M. Katz, On a Theorem of Ax, *Am. J. Math.*, **93**, 1971, pp. 485–499.

12. F.J. MacWilliams and N.J.A. Sloane, *Theory of Error-Correcting Codes*, North-Holland Publ. Comp., 1977.

13. O. Moreno and C.J. Moreno, Improvement of Chevalley-Warning and the Ax-Katz Theorems, *Am. J. Math.*, **117**, no. 1, pp. 241–244, 1995.

14. O. Moreno and F.N. Castro, Divisibility Properties for Covering Radius of Certain Cyclic Codes (accepted).

15. O. Moreno and F.N. Castro, Improvement of Ax-Katz and Moreno-Moreno Results and Applications (submitted).

16. O. Moreno, W. Shum, F.N. Castro, and P. Vijay Kumar, An Improved Tight Bound for the Divisibility of Exponential Sums (submitted).

17. C. Roos, A New Lower Bound for the Minimum Distance of a Cyclic Code, *IEEE Trans. on Inform Theory*, **29**, no. 3, pp. 330–332, 1983.

18. C. Small, Sums of Powers in Large Finite Fields, *Proc. Amer. Soc.* **65** (1977), pp. 35–36.

19. J.A. van der Horst and T. Berger, Complete Decoding of Triple-Error-Correcting Binary BCH Codes, *IEEE Trans. Inform. Theory*, **22**, pp. 138–147.

20. J.H. van Lint, *Introduction to Coding Theory*, GTM, Springer-Verlag 1992.

21. J.H. van Lint and R.M. Wilson, On the Minimum Distance of Cyclic Codes, *IEEE Trans. Inform. Theory*, **32**, pp. 23–40.

22. E. Warning, "Bemerkung zur vorstehenden Arbeit von Herrn Chevalley" *Abhandlungen aus dem Mathematischen Seminar der Universität Hamburg*, **11** (1936), pp. 76–83.

Unitary Error Bases: Constructions, Equivalence, and Applications

Andreas Klappenecker[1] and Martin Rötteler[2]

[1] Department of Computer Science
Texas A&M University, College Station, TX 77843-3112, USA
klappi@cs.tamu.edu
[2] Department of Combinatorics and Optimization
University of Waterloo, Waterloo, Ontario, Canada, N2L 3G1
mroetteler@uwaterloo.ca

Abstract. Unitary error bases are fundamental primitives in quantum computing, which are instrumental for quantum error-correcting codes and the design of teleportation and super-dense coding schemes. There are two prominent constructions of such bases: an algebraic construction using projective representations of finite groups and a combinatorial construction using Latin squares and Hadamard matrices. An open problem posed by Schlingemann and Werner relates these two constructions, and asks whether each algebraic construction is equivalent to a combinatorial construction. We answer this question by giving an explicit counterexample in dimension 165 which has been constructed with the help of a computer algebra system.

Keywords: quantum codes, unitary error bases, monomial representations, Hadamard matrices, Latin squares.

1 Introduction

Orthonormal bases of unitary matrices are a recurring theme in quantum information processing. The most well-known example is provided by the Pauli basis

$$\mathcal{P} = \left\{ \begin{pmatrix} 1 & 0 \\ 0 & 1 \end{pmatrix}, \begin{pmatrix} 0 & 1 \\ 1 & 0 \end{pmatrix}, \begin{pmatrix} 1 & 0 \\ 0 & -1 \end{pmatrix}, \begin{pmatrix} 0 & -i \\ i & 0 \end{pmatrix} \right\},$$

which is a unitary basis for the vector space $\mathbb{C}^{2 \times 2}$. More generally, a set $\mathcal{E}$ of d^2 unitary $d \times d$ matrices is called a *unitary error basis* if and only if $\mathcal{E}$ is orthonormal with respect to the inner product $\langle A, B \rangle = \mathrm{tr}(A^\dagger B)/d$. These bases of $\mathbb{C}^{d \times d}$ play a fundamental role for quantum error-correcting codes and other typical applications of quantum information processing, like e.g. super-dense coding and teleportation schemes.

Unitary error bases have been studied in numerous works, even before the advent of quantum computing [6,8,9,10,12], yet surprisingly little is known about their general structure. Currently, there are two fundamentally different constructions of unitary error bases: An algebraic construction due to Knill [9],

M. Fossorier, T. Hoeholdt, and A. Poli (Eds.): AAECC 2003, LNCS 2643, pp. 139–149, 2003.
© Springer-Verlag Berlin Heidelberg 2003

which yields so-called nice error bases, and a combinatorial construction due to Werner [14], which yields so-called shift-and-multiply bases. Schlingemann and Werner posed the following problem [11]:

Is every nice error basis equivalent to a basis of shift-and-multiply type?

We solve this problem.

2   Two Constructions of Unitary Error Bases

We begin by giving a simple characterization of unitary error bases. Suppose that U is a unitary matrix of dimension d^2. Let $E_i(U) = \sqrt{d}(U_{i,nd+m})_{n,m=0,\ldots,d-1}$ be the matrix obtained by arranging the ith row of U into a matrix. We call a matrix $U \in \mathcal{U}(d^2)$ a *UEB-matrix* if and only if $E_i(U)$ is a unitary matrix for all i in the range $0 \le i < d^2$. The UEB-matrices characterize the unitary error bases:

Lemma 1. *If U is a UEB-matrix, then $\mathcal{E} = \{E_i \mid 0 \le i < d^2\}$ is a unitary error basis. Conversely, if $\mathcal{E}$ is a unitary error basis in dimension d, then the matrix $\frac{1}{\sqrt{d}}(\mathrm{vec}(E))_{E \in \mathcal{E}} \in \mathcal{U}(d^2)$ is a UEB-matrix.*

Here we denote by $\mathrm{vec}(M)$ the row vector $(\mathrm{vec}(M))_\mu = M_{(\mu \, \mathrm{div} \, d, \, \mu \, \mathrm{mod} \, d)}$ for any $d \times d$ matrix M. The proof of the lemma is a direct consequence of the definitions. If U is a UEB-matrix, then $\mathrm{diag}(c_E : E \in \mathcal{E})U(A^t \otimes B)$ is a UEB-matrix for all $A, B \in \mathcal{U}(d)$. This motivates the following notion of equivalence of unitary error bases given in the next paragraph.

Equivalence. Let $\mathcal{E}$ and $\mathcal{E}'$ be two unitary error bases in d dimensions. We say that $\mathcal{E}$ and $\mathcal{E}'$ are equivalent, in signs $\mathcal{E} \equiv \mathcal{E}'$, if and only if there exist unitary matrices $A, B \in \mathcal{U}(d)$ and constants $c_E \in \mathcal{U}(1)$, $E \in \mathcal{E}$, such that

$$\mathcal{E}' = \{c_E AEB : E \in \mathcal{E}\}.$$

One readily checks that $\equiv$ is an equivalence relation.

Lemma 2. *Any unitary error basis in dimension 2 is equivalent to the Pauli basis.*

Proof: Let $\mathcal{A} = \{A_1, A_2, A_3, A_4\}$ be the unitary error basis. This basis is equivalent to a basis of the form $\{1_2, \mathrm{diag}(1, -1), B_3, B_4\}$. The diagonal elements of B_3 and B_4 are necessarily zero, because of the trace orthogonality relations. We may assume that B_3 and B_4 are of the form $B_3 = \mathrm{antidiag}(1, a)$ and $B_4 = \mathrm{antidiag}(1, -a)$, where $a = \exp(i\phi)$ for some $\phi \in \mathbb{R}$, since we are allowed to multiply the matrices with scalars. Conjugating the basis elements with the matrix $\mathrm{diag}(1, \exp(-i\phi/2))$ yields the matrices of Pauli basis up to scalar multiples, hence $\mathcal{A} \equiv \mathcal{P}$. □

In higher dimensions, we have an ubiquity of nonequivalent unitary error bases. In what follows we give a brief description of the two known constructions of unitary error bases.

Nice Error Bases. Let G be a group of order d^2 with identity element 1. A nice error basis in d dimensions is given by a set $\mathcal{E} = \{\rho(g) \in \mathcal{U}(n) \,|\, g \in G\}$ of unitary matrices such that

(i) $\rho(1)$ is the identity matrix,

(ii) $\operatorname{tr}\rho(g) = 0$ for all $g \in G \setminus \{1\}$,

(iii) $\rho(g)\rho(h) = \omega(g,h)\,\rho(gh)$ for all $g, h \in G$,

where $\omega(g,h)$ is a phase factor. Conditions (i) and (iii) state that ρ is a projective representation of the group G.

Lemma 3 (Knill). *A nice error basis is a unitary error basis.*

Proof: Let $\mathcal{E} = \{\rho(g) \in \mathcal{U}(n) \,|\, g \in G\}$ be a nice error basis. Notice that $\rho(g)^\dagger = \omega(g^{-1}, g)^{-1}\rho(g^{-1})$. Assume that g, h are distinct elements of G, then $g^{-1}h \neq 1$, hence $\operatorname{tr}(\rho(g)^\dagger \rho(h)) = \omega(g^{-1}, g)^{-1}\omega(g^{-1}, h)\operatorname{tr}(\rho(g^{-1}h)) = 0$ by property (ii) of a nice error basis. □

The next example shows that nice error bases exist in arbitrary dimensions:

Example 1. Let $d \geq 2$ be a integer, $\omega = \exp(2\pi i/d)$. Let X_d be the shift $X_d\,|x\rangle = |x - 1 \bmod d\rangle$, and let Z_d be the diagonal matrix $\operatorname{diag}(1, \omega, \omega^2, \ldots, \omega^{d-1})$. Then $\mathcal{E}_d := \{X_d^i Z_d^j \,|\, (i,j) \in \mathbb{Z}_d \times \mathbb{Z}_d\}$ is a nice error basis. This has been shown by an explicit calculation in [2].

Shift-and-Multiply Bases. Recall that a Latin square of order d is a $d \times d$ matrix such that each element of the set $\mathbb{Z}_d$ is contained exactly once in each row and in each column. A complex Hadamard matrix H of order d is a matrix in $\mathrm{GL}(d, \mathbb{C})$ such that $H_{ik} \in \mathcal{U}(1)$, $0 \leq i, k < d$, and $H^\dagger H = d\mathbf{1}$.

Let $\mathbf{H} = (H^{(i)} : 0 \leq i < d)$ be a sequence of complex Hadamard matrices, and let L be a Latin square L of order d. A shift-and-multiply basis $\mathcal{E}$ associated with $L, \mathbf{H}$ is given by the unitary matrices

$$E_{ij} = P_j \operatorname{diag}(H_{ik}^{(j)} : 0 \leq k < d), \quad i, j \in \mathbb{Z}_d, \tag{1}$$

where P_j denotes the permutation matrix with entries defined by $P_j(L(j,k), k) = 1$, for $0 \leq k < d$, and 0 otherwise. In short, E_{ij} is determined by the ith row of the jth Hadamard matrix $H^{(j)}$, and by the entries of the jth row of the Latin square L; briefly $E_{ij}\,|k\rangle = H_{ik}^{(j)}\,|L(j,k)\rangle$.

If all matrices in $\mathbf{H}$ are equal to a single Hadamard matrix H, then we refer to this basis as the shift-and-multiply basis associated with L, H.

Lemma 4 (Werner). *A shift-and-multiply basis is a unitary error basis.*

Proof: We have to show that $\operatorname{tr}(E_{ij}^\dagger E_{kl}) = 0$ when $(i, j) \neq (k, l)$. If $j \neq l$, then the matrix $P_j^\dagger P_l$ has a vanishing diagonal, whence $\operatorname{tr}(E_{ij}^\dagger E_{kl}) = 0$ for any choice of i and k. If $j = l$ and $i \neq k$, then $\operatorname{tr}(E_{ij}^\dagger E_{kj})$ is equal to the inner product of the ith and kth row of the complex Hadamard matrix $H^{(j)}$, hence $\operatorname{tr}(E_{ij}^\dagger E_{kj}) = 0$.
□

Example 2. The nice error basis $\mathcal{E}_d$ in Example 1 is a shift-and-multiply basis. Indeed, choose the Latin square $L = (j - i \bmod d)_{i,j \in \mathbb{Z}_d}$ and the complex Hadamard matrix $H = (\omega^{k\ell})_{k,\ell \in \mathbb{Z}_d}$, with $\omega = \exp(2\pi i/d)$. For example, if $d = 3$, then

$$L := \begin{pmatrix} 0\ 1\ 2 \\ 2\ 0\ 1 \\ 1\ 2\ 0 \end{pmatrix}, \; H = \begin{pmatrix} 1\ 1\ \ 1 \\ 1\ \omega\ \ \omega^2 \\ 1\ \omega^2\ \omega \end{pmatrix},$$

where $\omega = \exp(2\pi i/3)$. According to equation (1), the basis matrices E_{01} and E_{12} are respectively given by

$$E_{01} = \begin{pmatrix} 0\ 1\ 0 \\ 0\ 0\ 1 \\ 1\ 0\ 0 \end{pmatrix}, \; E_{12} = \begin{pmatrix} 0\ 0\ \omega^2 \\ 1\ 0\ 0 \\ 0\ \omega\ 0 \end{pmatrix}.$$

The entries of the middle row of the Latin square L and the first row of the complex Hadamard matrix H determine the matrix E_{01}.

Abstract Error Groups. Let $\mathcal{E} = \{\rho(g) : g \in G\}$ be a nice error basis. A group H isomorphic to the group generated by the matrices $\rho(g)$ is called an abstract error group of $\mathcal{E}$.

The group H is not necessarily finite. However, if we multiply the representing matrices $\rho(g)$ by scalars c_g such that $c_g\rho(g)$ has determinant 1, then the resulting nice error basis $\mathcal{E}' = \{c_g\rho(g) : g \in G\}$ is equivalent to $\mathcal{E}$, and its abstract error group H' is finite.

Thus, if we consider a nice error basis up to equivalence, then we may assume without loss of generality that the associated abstract error group is finite.

Example 3. The abstract error group H_d associated with the nice error basis $\mathcal{E}_d$ from Example 1 is by definition isomorphic to the group generated by X_d and Z_d. An element of the group $\langle X_d, Z_d \rangle$ is of the form $\omega^z Z_d^y X_d^x$, because $X_d Z_d = \omega Z_d X_d$. Notice that H_d is isomorphic to the unitriangular subgroup of $\mathrm{GL}(3, \mathbb{Z}_d)$ given by

$$H_d \cong \left\{ \begin{pmatrix} 1\ x\ z \\ 0\ 1\ y \\ 0\ 0\ 1 \end{pmatrix} : x, y, z \in \mathbb{Z}_d \right\}.$$

We prefer to describe the group H_d abstractly by the set of elements $(x, y, z) \in \mathbb{Z}_d^3$ with composition given by $(x, y, z) \circ (x', y', z') = (x + x', y + y, z + z' + xy')$, where all operations are modulo d.

Recall that a finite group H which has an irreducible representation of large degree $d = \sqrt{(H : Z(H))}$ is called a group of central type. It has been shown in [8] that a finite group H is an abstract error group if and only if it is a group of central type with cyclic center. A somewhat surprising consequence is that an abstract error group has to be a solvable group [8].

3   An Application of UEBs: Teleportation Schemes

Teleportation is an effect predicted by quantum mechanics, which allows to communicate the state of a quantum system from one location to another [3]. Teleportation has been demonstrated in experiments for binary quantum systems [5]. We give a description of teleportation schemes systems in *arbitrary dimensions* as an application of unitary error bases.

Let $\mathcal{E} = \{U_{i,j} : i, j = 0, \ldots d - 1\}$ be a unitary error basis in d dimensions. Denote by $|\varepsilon\rangle$ the quantum state $|\varepsilon\rangle := \sum_{i=0}^{d-1} |i\rangle |i\rangle \in \mathbb{C}^d \otimes \mathbb{C}^d$. This state is a maximally entangled, meaning that the partial traces satisfy $\mathrm{tr}_1 |\varepsilon\rangle \langle \varepsilon| = \mathrm{tr}_2 |\varepsilon\rangle \langle \varepsilon| = \frac{1}{d} \mathbf{1}_d$. Therefore, a local measurement of one of the subsystems of $|\varepsilon\rangle$ does not yield any information about the global state.

Next, we construct a unitary operator $U_{\mathcal{E}}$ starting from the unitary error basis $\mathcal{E}$. We denote by $U_{i,j}[n, m]$ the matrix entry at position (n, m) of the matrix $U_{i,j} \in \mathcal{E}$. Then $U_{\mathcal{E}}$ is defined as the UEB matrix which has the flattened and transposed matrices contained in $\mathcal{E}$ as rows: $U_{\mathcal{E}} = \frac{1}{\sqrt{d}}(\mathrm{vec}(E))_{E \in \mathcal{E}} \in \mathcal{U}(d^2)$. As mentioned in the introduction this matrix is unitary. Alternatively, we can write $U_{\mathcal{E}}$ in the form $U_{\mathcal{E}} = \sum_{i,j} \sum_{m,n} U_{i,j}[n, m] |i, j\rangle \langle m, n|$. The following lemma is the essential ingredient for the teleportation circuit.

Lemma 5. *Let $|\varepsilon\rangle \in \mathbb{C}^d \otimes \mathbb{C}^d$ be the maximally entangled state defined above and let $|\psi\rangle \in \mathbb{C}^d$ a quantum state. Furthermore let $U_{\mathcal{E}}$ be the unitary operation corresponding to the error basis $\mathcal{E} = \{U_{i,j}\}$. Then the identity $(U_{\mathcal{E}} \otimes \mathbf{1}_d) |\psi\rangle |\varepsilon\rangle = \sum_{i,j} |i, j\rangle U_{i,j} |\psi\rangle$. holds.*

Proof: Let $|\psi\rangle = \sum_{\mu} \alpha_{\mu} |\mu\rangle$. We make the following computation which is a mere juggling with the bra and ket notation.

$$(U_{\mathcal{E}} \otimes \mathbf{1}_d) |\psi\rangle |\varepsilon\rangle = \left(\sum_{i,j} \sum_{m,n} U_{i,j}[n, m] |i, j, k\rangle \langle m, n, k| \right) \sum_{\mu,\nu} \alpha_{\mu} |\mu\rangle |\nu\rangle |\nu\rangle$$

$$= \sum_{i,j} |i, j\rangle \left(\sum_{m,n} U_{i,j}[n, m] |n\rangle \langle m| \right) \left(\sum_{\mu} \alpha_{\mu} |\mu\rangle \right)$$

$$= \sum_{i,j} |i, j\rangle U_{i,j} |\psi\rangle . \square$$

We obtain the following theorem which generally is called "teleportation" of an unknown quantum state. The context in which this theorem applies is a two party protocol. We call the parties "Alice" and "Bob" and assume that they share an EPR pair, i.e., they both have one part of the state $|\varepsilon\rangle$, but they might be separated from each other. In particular we assume that Alice and Bob cannot perform joint operations on the state $|\varepsilon\rangle$, however they are allowed to perform local operations and to send classical messages.

Theorem 1. *Let $|\psi\rangle \in \mathbb{C}^d$ be an unknown quantum state, let $|\varepsilon\rangle = \sum_i |i\rangle |i\rangle$ be a maximally entangled state, and let U_ε be the unitary transformation corresponding to the unitary error basis $\mathcal{E}$. Then the following protocol will transfer the state $|\psi\rangle$ from Alice to Bob:*

1. *Alice applies the operation $U_\varepsilon \in \mathcal{U}(d^2)$ to $|\psi\rangle$ and her part of $|\varepsilon\rangle$.*
2. *Alice performs a measurement in the standard basis. Let (i_0, j_0) with $i_0, j_0 \in \{0, \ldots, d-1\}$ be the outcome of this measurement.*
3. *Alice sends the classical information (i_0, j_0) to Bob. This can be encoded using $2\log_2 d$ bits.*
4. *Bob performs $U_{i_0,j_0}^\dagger$ on his part of the state $|\varepsilon\rangle$ and retrieves the state $|\psi\rangle$.*

Proof: The hard part was already shown in Lemma 5. The operation affected by Alice prior to the measurement yield the state $(U_\varepsilon \otimes 1_d)|\psi\rangle |\varepsilon\rangle = \sum_{i,j} |i, j\rangle U_{i,j} |\psi\rangle$. A measurement of the first two registers, which are in Alice's possession, gives a random outcome (i_0, j_0). However, by this measurement the entangled quantum state $\sum_{i,j} |i, j\rangle U_{i,j} |\psi\rangle$ will collapse into just the term $U_{i_0,j_0} |\psi\rangle$ of this superposition. After Alice has send the information (i_0, j_0) about the outcome of the measurement to Bob, he in turn is able to apply $U_{i_0,j_0}^\dagger$ to recoup the original quantum state. $\square$

Remark 1. It takes a potentially infinite amount of information to completely characterize the state $|\psi\rangle$ since it involve complex numbers in the description of the amplitudes. A remarkable feature of Theorem 1 is still it is possible to transmit $|\psi\rangle$ to a remote location by sending $2\log_2 d$ bits of classical information.

4   Error Bases That Are Not Nice

A unitary error basis, which is not equivalent to a nice error basis, is called wicked. We show now that there exist an abundance of wicked shift-and-multiply bases.

Theorem 2. *Let $\mathcal{E}_\alpha$ be the shift-and-multiply basis associated with L, H_α, where*

$$
L = \begin{pmatrix} 0\,1\,2\,3 \\ 3\,0\,1\,2 \\ 2\,3\,0\,1 \\ 1\,2\,3\,0 \end{pmatrix}, \quad
H_\alpha = \begin{pmatrix} 1 & 1 & 1 & 1 \\ 1 & 1 & -1 & -1 \\ 1 & -1 & e^{i\alpha} & -e^{i\alpha} \\ 1 & -1 & -e^{i\alpha} & e^{i\alpha} \end{pmatrix}.
$$

If $\alpha \in \mathbb{Q}^\times$, then $\mathcal{E}_\alpha$ is not equivalent to a nice error basis.

Proof: Suppose there exist $A, B \in \mathcal{U}(4)$, and scalars c_{ij} such that the set $\{c_{ij} A U_{ij} B : i, j = 1, \ldots, 4\}$ is a nice error basis. Without loss of generality, we may assume that the group G generated by the matrices $c_{ij} A U_{ij} B$ is finite.

Notice that $\mathcal{E}_\alpha$ contains the matrices $M_1 = \text{diag}(1, -1, e^{i\alpha}, -e^{i\alpha})$, and $M_2 = 1_4$. Consequently, $(c_1 A M_1 B)(c_2 A M_2 B)^{-1} = c_1 c_2^{-1} A M_1 M_2^\dagger A^\dagger = c_1 c_2^{-1} A M_1 A^{-1}$ is an element of the group G. Since G is finite, it follows that $c_1 c_2^{-1} A M_1 A^{-1}$ and hence also that $c_1 c_2^{-1} M_1$ is of finite order. Looking at the entries of this matrix this implies that $c_1 c_2^{-1}$ and $c_1 c_2^{-1} e^{i\alpha}$ are roots of unity. It follows that $e^{i\alpha}$ would have to be a root of unity as well, contradicting the assumption $\alpha \in \mathbb{Q}^\times$. $\square$

5   Nonmonomial Abstract Error Groups

In this section, we will answer the question raised by Schlingemann and Werner:

Theorem 3. *There exist nice error bases which are not equivalent to bases of shift-and-multiply type.*

We will prove this result with the help of abstract error groups. The following result will play a key role in our proof:

Theorem 4 (Dade, Isaacs). *There exists a group H of central type with cyclic center, with nonmonomial irreducible character χ of degree $\chi(1) = \sqrt{(H:Z(H))}$.*

Proof of Theorem 3: We actually show a stronger statement, namely that there are nice error bases which are not equivalent to monomial bases. By Theorem 4, there exists an abstract error group H that has a non-monomial irreducible unitary representation ρ of degree $d = \sqrt{(H:Z(H))}$. Denote by $\mathcal{E}$ a nice error basis associated with ρ, that is, $\mathcal{E} = \{\rho(t) \mid t \in T\}$, where T is a transversal of H modulo $Z(H)$, with $1 \in T$. Since ρ is nonmonomial, it is impossible to find a base change A such that $A\rho(t)A^\dagger$ is monomial for all $t \in T$. We show next that this property is even preserved with respect to $\equiv$.

Seeking a contradiction, we suppose that there exist unitary matrices A, B and scalars c_t such that $c_t A\rho(t)B$ is a monomial unitary error basis. Since the identity matrix $1_d = \rho(1)$ is part of the nice error basis, we can conclude that the matrix $C = c_1 AB$ is monomial. But $c_t A\rho(t)B = c_t A\rho(t)(A^\dagger A)B = c_t/c_1 A\rho(t)A^\dagger C$ shows that the resulting equivalent error basis is nonmonomial. Indeed, among the matrices $A\rho(t)A^\dagger$ is at least one nonmonomial matrix U. Multiplying U with the monomial matrix C and the scalar prefactor c_i/c_1 cannot result in a monomial matrix, leading to a contradiction.   □

In the next section, we want to construct an explicit example of a nice error basis that is not equivalent to a shift-and-multiply basis. We will need an explicit example of a group H satisfying the assumptions of Theorem 4 for that purpose. Theorem 4 was independently proved by Everett Dade and by Martin Isaacs, but unfortunately their results remained unpublished. We construct H, following Dade's approach, and use Magma [4] to construct an explicit counterexample.

6   An Explicit Counterexample

We construct a nice error basis in dimension $165 = 3 \cdot 5 \cdot 11$, which is not equivalent to a shift-and-multiply basis. First, we recall some basic facts on semi-direct products and the representations and automorphism groups of the finite Heisenberg groups.

Semidirect Products. Let N, H be finite groups, and let φ be a group homomorphism from H to $\mathrm{Aut}(N)$. Recall that the (outer) semidirect product $G = N \rtimes_\varphi H$ is a group defined on the set $N \times H$, with composition given by $(n_1, h_1)(n_2, h_2) = (n_1\,\varphi(h_1)(n_2), h_1 h_2)$. If the center of H acts trivially on N, then the center of the semidirect product is given by $Z(G) = Z(N) \times Z(H)$. Recall that a group G is said to be an *inner* semidirect product of the two subgroups H and N if and only if N is a normal subgroup of G such that $HN = G$ and $H \cap N = \{1_G\}$. A detailed discussion of semidirect products can be found for instance in [1].

An Irreducible Representation of H_p. Let p be an odd prime. The Heisenberg group H_p has an irreducible representation $\rho_p \colon H_p \to \mathcal{U}(p)$, which associates to an element $(x, y, z) \in H_p$ the matrix $\rho_p((x, y, z)) = \omega^z Z_p^y X_p^x$. Here ω denotes the primitive root of unity $\omega = \exp(2\pi i/p)$, and X_p and Z_p denote the generalized Pauli matrices, as defined in Example 1. We will derive a faithful irreducible matrix representation of degree 165 of the group $G = (H_5 \times H_{11}) \rtimes_\varphi H_3$ by a suitable composition of the representations ρ_3, ρ_5, and ρ_{11}.

Automorphisms of the Heisenberg Group H_p. Let p be an odd prime. The Heisenberg group H_p defined in Example 3 has p^3 elements. Recall that the special linear group $\mathrm{SL}(2, \mathbb{F}_p)$ is a matrix group of order $(p+1)p(p-1)$, which is generated by the matrices

$$\alpha = \begin{pmatrix} 0 & -1 \\ 1 & 0 \end{pmatrix} \quad \text{and} \quad \beta = \begin{pmatrix} 1 & 0 \\ 1 & 1 \end{pmatrix}.$$

The special linear group acts as an automorphisms group on the Heisenberg group H_p. Indeed, define $\alpha(x, y, z) = (-y, x, z - xy)$ and $\beta(x, y, z) = (x, x + y, z + \frac{(p+1)}{2} x^2)$. It is straightforward to check that $\alpha, \beta \in \mathrm{Aut}(H_p)$. We will construct the abstract error group by semidirect products of Heisenberg groups. We will need some more detailed knowledge about these automorphisms to tailor this construction to our needs.

Making the Automorphisms of H_5 and H_{11} Explicit. The matrix group representing $G = (H_5 \times H_{11}) \rtimes_\varphi H_3$ is an inner semidirect product. This means that the action of the automorphism is realized by a conjugation with a matrix. It suffices to find matrices which realize the action of the generators α and β of $\mathrm{SL}(2, \mathbb{F}_p)$. Recall that $\alpha(1, 0, 0) = (0, 1, 0)$ and $\alpha(0, 1, 0) = (-1, 0, 0)$. This means we need to find a matrix $A \in \mathcal{U}(p)$ such that

$$X_p^A = Z_p \quad \text{and} \quad Z_p^A = X_p^{-1}.$$

The action of the automorphism β is determined by $\beta(1, 0, 0) = (1, 1, (p+1)/2)$ and $\beta(0, 1, 0) = (0, 1, 0)$. Hence we need to find a matrix $B \in \mathcal{U}(p)$ such that

$$X_p^B = \omega^{(p+1)/2} Z_p X_p \quad \text{and} \quad Z_p^B = Z_p.$$

We can choose A to be the discrete Fourier transform (DFT) of length p, i.e., $F_p = \frac{1}{\sqrt{p}} (\omega^{k\ell})_{k,\ell=0,\ldots,p-1}$, with $\omega = \exp(2\pi i/p)$. Notice that the diagonal matrix $D_p = \text{diag}(\omega^{i(i-1)/2} : 0 \leq i < p)$ satisfies $X_p^{D_p} = Z_p X_p$ and $Z_p^{D_p} = Z_p$. It follows that the matrix B can be chosen to be $B = D_p Z_p^{(p+1)/2}$.

A Nonmonomial Error Basis in Dimension 165. We need an element of order 3 of $\text{SL}(2, \mathbb{F}_p)$ to specify the action of H_3 on $H_5 \times H_{11}$. We can choose for instance the element $\gamma = \beta\alpha\beta\alpha$ in $\text{SL}(2, \mathbb{F}_p)$, i.e.,

$$\gamma = \begin{pmatrix} 1 & 0 \\ 1 & 1 \end{pmatrix} \begin{pmatrix} 0 & -1 \\ 1 & 0 \end{pmatrix} \begin{pmatrix} 1 & 0 \\ 1 & 1 \end{pmatrix} \begin{pmatrix} 0 & -1 \\ 1 & 0 \end{pmatrix}.$$

The action of γ on the matrix representation $\rho_5(H_5)$ of the Heisenberg group H_5 is realized by conjugation with the matrix $R_5 = D_5 Z_5^3 \cdot F_5 \cdot D_5 Z_5^3 \cdot F_5 \in \mathcal{U}(5)$. Similarly, the action of γ on the matrix version of H_{11} is given by conjugation with $R_{11} = D_{11} Z_{11}^6 \cdot F_{11} \cdot D_{11} Z_{11}^6 \cdot F_{11} \in \mathcal{U}(11)$.

Recall that H_3 is generated by the two elements $(1,0,0)$ and $(0,1,0)$. According to the construction of Sect. 5, the action of H_3 is chosen such that the generator $(1,0,0)$ acts with γ on H_5 and trivially on H_{11}, and the generator $(0,1,0)$ acts trivially on H_5 and by γ on H_{11}. Explicitly, we obtain the matrix group:

$$G = \langle 1_3 \otimes X_5 \otimes 1_{11}, \ 1_3 \otimes Z_5 \otimes 1_{11}, \ 1_3 \otimes 1_5 \otimes X_{11},$$
$$1_3 \otimes 1_5 \otimes Z_{11}, \ X_3 \otimes R_5 \otimes 1_{11}, \ Z_3 \otimes 1_5 \otimes R_{11}\rangle.$$

The group G is an inner semidirect product of the form $H_3 \ltimes (H_5 \times H_{11})$. Indeed, the subgroup generated by the first four generators is isomorphic to $N = H_5 \times H_{11}$ since the irreducible representations of a direct product are given by tensor products of the irreducible representations of the factors. We have $N \triangleleft G$ because of the choice of R_5 and R_{11}. The complement H of N is given by the group generated by the to remaining generators. Obviously the intersection $H \cap N$ is trivial and $G = HN$ by definition. This shows that $G = H \ltimes N$.

This also shows that $|G| = 4492125 = 3^3 5^3 11^3$, that the centre is of order $\zeta(G) = 165 = 3 \cdot 5 \cdot 11$ and that G itself defines an irreducible representation. This can also be obtained by an enumeration of G using Magma [4]. Starting from this a pc-group presentation [13] of G can be computed and we obtain

$$G = \langle a_1, \ldots, a_9 | a_1^3 = a_9^9, \ a_2^3 = a_8^3, \ a_3^3 = a_4^5 = a_5^5 = a_6^{11} = a_7^{11} = a_8^5 = a_9^{11} = 1,$$
$$[a_2, a_1] = a_3^2, \ [a_4, a_2] = a_5, [a_5, a_2] = a_5^{-1} a_4^2 a_5^3 a_8^3, [a_5, a_4] = a_8,$$
$$[a_6, a_1] = a_7 a_9^{10}, \ [a_7, a_1] = a_7^{-1} a_6^8 a_7^9 a_8^8, \ [a_7, a_6] = a_9^{10}\rangle.$$

Once a finite group is given as a pc-group many computational tasks can be accomplished easily. For instance using computer algebra systems like Gap [7] or Magma [4] we can check that G and $\zeta(G)$ indeed have the desired orders. Using the command CharacterDegrees in Gap we can check that G has irreducible representations of degree 165. One of these representations is given by the generators above. The fact that this representation of degree 165 is nonmonomial can be verified by computing the whole subgroup lattice of G. It turns out that G does not have subgroups of index 165 (in Gap this can be checked using the command Lattice(G)). Since each monomial representation χ is induced by a linear character from a subgroup H of index $\chi(1) = [G : H]$, and no such subgroup exists in the subgroup lattice of G, we can conclude that G is nonmonomial.

We obtain a non-monomial nice error basis by choosing a transversal of $Z(G)$ in G. This nice error basis is in particular not equivalent to a shift-and-multiply basis.

7   Conclusions

We have studied unitary error basis for d dimensional systems, for $d \geq 2$. We have shown that all such bases are equivalent in dimension 2. This changes dramatically in higher dimensions. We have shown that nice error bases are in general not equivalent to shift-and-multiply bases, and vice versa. This answers an open problem posed by Schlingemann and Werner. Also the use of unitary error bases for the construction of general nonbinary teleportation schemes has been described.

Acknowledgments. This research was supported by NSF grant EIA 0218582, Texas A&M TITF, the EC project Q-ACTA (IST-1999-10596), CSE, and MI-TACS. We thank Everett Dade and Martin Isaacs for discussions about groups of central types with nonmonomial representations. The construction of the non-monomial character of extremal degree given in Sect. 5 is due to Everett Dade. We thank Thomas Beth for supporting this work.

References

1. J.L. Alperin and R.B. Bell. *Groups and representations*, volume 162 of *Graduate texts in mathematics*. Springer, 1995.
2. A. Ashikhmin and E. Knill. Nonbinary quantum stabilizer codes. *IEEE Trans. Inform. Theory*, 47(7):3065–3072, 2001.
3. Charles H. Bennett, Gilles Brassard, Claude Crépeau, Richard Jozsa, Asher Peres, and William K. Wootters. Teleporting an Unknown Quantum State via Dual Classical and Einstein-Podolsky-Rosen Channels. *Physical Review Letters*, 70(13):1895–1899, 1993.
4. W. Bosma, J.J. Cannon, and C. Playoust. The Magma algebra system I: The user language. *J. Symb. Comp.*, 24:235–266, 1997.
5. D. Bouwmeester, J. Pan, K. Mattle, M. Eibl, H. Weinfurter, and A. Zeilinger. Experimental quantum teleportation. *Nature*, 390:575–579, 1997.
6. R. Frucht. Über die Darstellung endlicher Abelscher Gruppen durch Kollineationen. *J. Reine Angew. Math.*, 166:16–28, 1931.
7. The GAP Team, Lehrstuhl D für Mathematik, RWTH Aachen, Germany and School of Mathematical and Computational Sciences, U. St. Andrews, Scotland. GAP – *Groups, Algorithms, and Programming, Version 4*, 1997.
8. A. Klappenecker and M. Rötteler. Beyond Stabilizer Codes I: Nice Error Bases. *IEEE Trans. Inform. Theory*, 48(8):2392–2395, 2002.
9. E. Knill. Group representations, error bases and quantum codes. Los Alamos National Laboratory Report LAUR-96-2807, 1996.

10. J. Patera and H. Zassenhaus. The Pauli matrices in n dimensions and finest gradings of simple Lie algebras of type A_{n-1}. *J. Math. Phys.*, 29(3):665–673, 1988.
11. D. Schlingemann. Problem 6 in Open Problems in Quantum Information Theory. http://www.imaph.tu-bs.de/qi/problems/
12. J. Schwinger. Unitary operator bases. *Proc. Nat. Acad. Sci.*, 46:570–579, 1960.
13. C. Sims. *Computation with finitely presented groups.* Cambridge University Press, 1994.
14. R. Werner. All teleportation and dense coding schemes. *J. Phys. A*, 34:7081–7094, 2001.

Differentially 2-Uniform Cocycles – The Binary Case

K.J. Horadam

RMIT University, Melbourne, VIC 3001, Australia
horadam@rmit.edu.au

Abstract. There is a differential operator ∂ mapping 1D functions $\phi : G \to C$ to 2D functions $\partial\phi : G \times G \to C$ which are coboundaries, the simplest form of cocycle. Differentially k-uniform 1D functions determine coboundaries with the same distribution. Extending the idea of differential uniformity to cocycles gives a unified perspective from which to approach existence and construction problems for highly nonlinear functions, sought for their resistance to differential cryptanalysis. We describe two constructions of 2D differentially 2-uniform (APN) cocycles over GF(2^a), of which one gives 1D binary APN functions.

1 Introduction

Let G and C be finite abelian groups and $f : G \to C$ be a mapping. Let $\Delta_f = \max \#\{h \in G : f(g + h) - f(h) = c : g \in G, g \neq 0, c \in C\}$, where $\#X$ denotes the cardinality of set X. Nyberg [8, p.15] defined f to be differentially k-uniform if $\Delta_f = k$. This concept is of interest in cryptography since differential and linear cryptanalysis exploit weaknesses of S-box functions. If f is an S-box function, its susceptibility to differential cryptanalysis is minimised if Δ_f is as small as possible.

Suppose hereafter that G is a finite group (not necessarily abelian) of order v, written multiplicatively, and C is a finite abelian group of order w, written additively. The left and right *derivatives of ϕ in direction h* are the functions $(\Delta\phi)_h$ and $(\nabla\phi)_h : G \to C$ defined by $(\Delta\phi)_h(g) = \phi(hg) - \phi(g)$ and $(\nabla\phi)_h(g) = \phi(gh) - \phi(g)$, respectively. The function ϕ is *differentially k-uniform* if $\max \#\{h \in G : (\Delta\phi)_g(h) = c : g \in G, g \neq 1, c \in C\} = k$.

Nyberg's original *perfect nonlinear* (PN) functions [7, Def. 3.1] have $G = \mathbb{Z}_n^a$ and $C = \mathbb{Z}_n^b$, $a \geq b$, and when $n = 2$ they are precisely the *(vectorial) bent* functions (cf. Chabaud and Vaudenay [2, p.358]). When $a = b$, the concepts of PN, differentially 1-uniform and planar functions all coincide. We know examples of such PN functions exist when n is an odd prime p, but cannot exist when $p = 2$. When $a = b$, a differentially 2-uniform function is also termed *almost perfect nonlinear* (APN) and when, additionally, n is even, it is a *semi-planar* function [3, §2]. When n is odd, any maximally nonlinear function is APN. The best that can be expected when $p = 2$ is an APN function. For primes p, the quest for PN and APN functions $\mathbb{Z}_p^a \to \mathbb{Z}_p^a$ focusses on polynomial power functions in the Galois Field GF(p^a).

M. Fossorier, T. Hoeholdt, and A. Poli (Eds.): AAECC 2003, LNCS 2643, pp. 150–157, 2003.

There is a natural differential operator ∂ which maps 1D functions ϕ to 2D functions $\partial\phi$ called coboundaries, which are the simplest form of 2D cocycles. The purpose of this paper is to explore the case of differential 2-uniform cocycles, generalising the APN case. We describe two constructions of differential 2-uniform cocycles. The first gives new infinite classes of genuinely 2D APN cocycles (ie. which are not coboundaries) and the second gives new APN coboundaries, and consequently 1D APN functions.

2   Coboundaries and Differential Uniformity

The function ϕ is *normalised* if $\phi(1) = 0$. The set of normalised mappings $\phi : G \to C$ forms an abelian group $C^1(G, C)$ under pointwise addition. Given $\phi \in C^1(G, C)$, for each $g \in G$ and $c \in C$ we set $n_\phi(g, c) = \#\{h \in G : \phi(gh) - \phi(h) = c\}$ and say the *distribution* of ϕ is the *multiset* of frequencies $\mathcal{D}(\phi) = \{n_\phi(g, c) : g \in G, c \in C\}$. Then ϕ is differentially Δ_ϕ-uniform if $\max\{n_\phi(g, c) : g \in G, c \in C, g \neq 1\} = \Delta_\phi$.

We are interested in locating functions with the same distributions. There is a differential action of G on $C^1(G, C)$ which preserves distributions.

Definition 1. The *shift action* of G on $C^1(G, C)$ is defined for $k \in G$ and $\phi \in C^1(G, C)$ by $\phi \cdot k : g \mapsto (\nabla\phi)_g(k) = \phi(kg) - \phi(k), \ g \in G$.

Note $\phi \cdot 1 = \phi$ and $\phi \cdot (kh) = (\phi \cdot k) \cdot h$, so shift action is an automorphism action on $C^1(G, C)$. Since $n_{(\phi \cdot k)}(g, c) = n_\phi(kgk^{-1}, c)$ we obtain the following lemma.

Lemma 1. *Let* $\phi \in C^1(G, C)$. *Then for any* $\gamma \in \mathrm{Aut}(C)$, $k \in G$, $\theta \in \mathrm{Aut}(G)$, $\mathcal{D}(\phi) = \mathcal{D}\left(\gamma \circ (\phi \cdot k) \circ \theta\right)$ *and* $\Delta_\phi = \Delta_{\gamma \circ (\phi \cdot k) \circ \theta}$. $\square$

We extend these ideas to two dimensions. Let $C^2(G, C)$ be the set of two-dimensional functions $\Phi : G \times G \to C$ which satisfy $\Phi(1, g) = \Phi(g, 1) = 0$, $g \in G$. Under pointwise addition $C^2(G, C)$ is an abelian group.

Definition 2. For $\Phi \in C^2(G, C)$,

1. the *left first partial derivative* $(\Delta_1\Phi)_k : G \times G \to C$ of Φ in direction k is $(\Delta_1\Phi)_k(g, h) = \Phi(kg, h) - \Phi(g, h)$ and the *right first partial derivative* $(\nabla_1\Phi)_k : G \times G \to C$ of Φ in direction k is $(\nabla_1\Phi)_k(g, h) = \Phi(gk, h) - \Phi(g, h)$.
2. Define $N_\Phi(g, c) = \#\{h \in G : \Phi(g, h) = c\}$, $g \in G, c \in C$. The *distribution* of Φ is the *multiset* of frequencies $\mathcal{D}(\Phi) = \{N_\Phi(g, c) : g \in G, c \in C\}$. Define $\Delta_\Phi = \max\{N_\Phi(g, c) : g \in G, c \in C, g \neq 1\}$. Then we say Φ is *differentially* Δ_Φ-*uniform*.
3. The *shift action* of G on $C^2(G, C)$ is defined for $k \in G$ and $\Phi \in C^2(G, C)$ to be

$$(\Phi \cdot k)(g, h) = (\nabla_1\Phi)_g(k, h) = \Phi(kg, h) - \Phi(k, h), \ g, h \in G.$$

There is a natural mapping $\partial : C^1(G, C) \to C^2(G, C)$ which preserves distribution properties. For $\phi \in C^1(G, C)$, define $\partial \phi : G \times G \to C$ to be

$$\partial \phi(g, h) = \phi(gh) - \phi(g) - \phi(h), \quad g, h \in G. \tag{1}$$

Then $\partial \phi(1, g) = \partial \phi(g, 1) = 0$, $g \in G$. A function $\partial \phi$ satisfying (1) is called a *coboundary*, and measures the amount by which ϕ differs from a homomorphism of groups.

Observe that for each $g \in G$ and $c \in C$, $n_\phi(g, c) = N_{\partial \phi}(g, c - \phi(g))$. Since for each $g \in G$, the mapping $c \mapsto c - \phi(g)$ is a bijection on C, the multisets of frequencies $\{n_\phi(g, c), \ c \in C\}$ and $\{N_{\partial \phi}(g, c), \ c \in C\}$ are the same.

Lemma 2. *Let $\phi \in C^1(G, C)$. Then $\mathcal{D}(\phi) = \mathcal{D}(\partial \phi)$ and $\Delta_\phi = \Delta_{\partial \phi}$.* $\square$

We will focus most attention on those elements of $C^2(G, C)$, including the coboundaries $\partial \phi$, which satisfy a further condition and are known as cocycles. A normalised (2-dimensional) *cocycle* is a mapping $\psi : G \times G \to C$ satisfying $\psi(1, 1) = 0$ and

$$\psi(g, h) + \psi(gh, k) = \psi(g, hk) + \psi(h, k), \quad \forall g, h, k \in G. \tag{2}$$

We can usefully represent a cocycle ψ as a matrix $[\psi(g, h)_{g, h \in G}]$ with rows and columns indexed by the elements of G in some fixed ordering (beginning with the identity 1).

For fixed G and C, the set of cocycles forms a subgroup $Z^2(G, C)$ of $C^2(G, C)$ and the coboundaries form a subgroup $B^2(G, C)$ of $Z^2(G, C)$. The coboundary operator $\partial : \phi \mapsto \partial \phi$ is a homomorphism

$$\partial : C^1(G, C) \to Z^2(G, C) \tag{3}$$

of abelian groups. Its kernel $\ker \partial$ is the group of homomorphisms $\mathrm{Hom}(G, C)$ from G to C.

The *diagonal* $D\Phi$ of $\Phi \in C^2(G, C)$ is $D\Phi(g) = \Phi(g, g)$, $g \in G$. The *diagonal operator* $D : \Phi \mapsto D\Phi$ is a homomorphism of abelian groups (in the reverse direction from ∂)

$$D : C^2(G, C) \to C^1(G, C). \tag{4}$$

The advantages of using cocycles are manifold: firstly, coboundaries are cocycles so the extension from 1D to 2D functions is very natural; secondly, a lot is known about the group of cocycles so the search for highly nonlinear 2D functions can be undertaken in a structured way rather than by exhaustive search; and thirdly, the distribution of a cocycle is preserved under shift action.

Lemma 3. *The* bundle $\mathcal{B}(\psi) \subset Z^2(G, C)$ *containing* $\psi \in Z^2(G, C)$ *is* $\mathcal{B}(\psi) = \{\gamma \circ (\psi \cdot k) \circ (\theta \times \theta), \ \gamma \in \mathrm{Aut}(C), \ k \in G, \ \theta \in \mathrm{Aut}(G)\}$. *Then*
    (i) $\mathcal{B}(\partial \phi) = \{ \ \partial(\gamma \circ (\phi \cdot k) \circ \theta), \ \gamma \in \mathrm{Aut}(C), \ k \in G, \ \theta \in \mathrm{Aut}(G)\}$ *for* $\phi \in C^1(G, C)$;
    (ii) [5, Theorem 3] $\mathcal{D}(\psi) = \mathcal{D}(\varphi)$ *for all* $\varphi \in \mathcal{B}(\psi)$.

Proof. For (i), $((\partial\phi)\cdot k)(g,h) = \phi(kgh) - \phi(kg) - \phi(kh) + \phi(k) = \partial(\phi\cdot k)(g,h)$ and the remainder of the proof is straightforward. $\square$

A mapping $\Phi \in C^2(G,C)$ is *symmetric* if $\Phi(g,h) = \Phi(h,g)$ always. For example, if G is abelian, any coboundary $\partial\phi$ is symmetric. The *symmetrisation* $\widehat{\Phi} \in C^2(G,C)$ of Φ is

$$\widehat{\Phi}(g,h) = \Phi(g,h) + \Phi(h,g), \quad g,h \in G.$$

The symmetrisation of a cocycle $\psi \in Z^2(G,C)$ is not necessarily a cocycle, but if G is abelian or ψ is symmetric, $\widehat{\psi}$ is a cocycle.

3   Differential 2-Uniformity

In this section, we describe properties of differentially 2-uniform cocycles. We will also use the term APN (*almost perfect nonlinear*) for differentially 2-uniform functions when $G = C$ is an elementary abelian p-group.

Firstly consider the case $G = C = (\mathrm{GF}(2^a),+)$ and $\phi \in C^1(G,G)$. We know $(\Delta\phi)_g(h) = (\Delta\phi)_g(h+g)$, so $\partial\phi(g,h) = \partial\phi(g,h+g)$, and $\Delta_{\partial\phi}$ is even. The value 0 is necessarily taken at least twice, as $\partial\phi(g,0)$ and $\partial\phi(g,g)$. Hence, by Lemma 2, $\phi \in C^1(G,G)$ is APN if and only if for each $g \neq 0$, half the elements of G appear twice in $\{\partial\phi(g,h), h \in G\}$ and the other half do not appear.

For example, for $G = C = \mathbb{Z}_2^2$ and a,c distinct nonzero elements of $\mathbb{Z}_2^2$, the differentially 2-uniform cocycles with matrices

$$
\begin{bmatrix} 0 & 0 & 0 & 0 \\ 0 & 0 & a & a \\ 0 & a & 0 & a \\ 0 & a & a & 0 \end{bmatrix},
\begin{bmatrix} 0 & 0 & 0 & 0 \\ 0 & a & c & a+c \\ 0 & c & 0 & c \\ 0 & a+c & c & a \end{bmatrix},
\begin{bmatrix} 0 & 0 & 0 & 0 \\ 0 & a & c & a+c \\ 0 & c & a+c & a \\ 0 & a+c & a & c \end{bmatrix}
$$

correspond to bundles 1.2, 2.3 and 3.2 of [5, Example 8], respectively. The first equals $\partial\phi$ for some 1D APN function ϕ, but the other two cannot. The third is differentially 1-uniform (PN), a situation which cannot arise in the 1D case.

However, even if for each $g \neq 0$, half the elements of G appear twice in $\{\psi(g,h), h \in G\}$ and the other half do not appear, ψ need not be a coboundary. Bundles 2.2, 5.2 and 6.1 of [5, Example 8] represent cocycles with this property which are not coboundaries.

Canteaut's characterisation [1] of binary 1D APN functions in terms of 'second derivatives', namely: ϕ is APN if and only if, for all non-zero $g \neq h \in G$, $\Delta((\Delta\phi)_h)_g(k) \neq 0$, $\forall k \in G$, is simply restated as follows, noting that $\Delta((\Delta\phi)_h)_g(k) = \phi(k+g+h) - \phi(k+g) - \phi(k+h) + \phi(k) = ((\partial\phi)\cdot k)(g,h)$.

Corollary 1. (Canteaut) *Let* $G = (\mathrm{GF}(2^a),+)$. *Then* $\phi \in C^1(G,G)$ *is APN if and only if, for all* $g \neq 0$, $h \neq 0 \in G$ *and all* $k \in G$, $((\partial\phi)\cdot k)(g,h) \neq 0$ *unless* $h = g$. $\square$

This characterisation applies not just to binary APN coboundaries but to arbitrary differentially 2-uniform cocycles: no more than two entries in any non-initial row of the matrix corresponding to any shift of the cocycle can be 0.

Theorem 1. $\psi \in Z^2(G,C)$ satisfies $\Delta_\psi \leq 2$ if and only if, for all $g \neq 1$, $h \neq 1 \in G$ and all $k \in G$, $\#\{h \in G : (\psi \cdot k)(g,h) = 0\} \leq 1$.

Proof. If $\Delta_\psi \leq 2$ then by Lemma 3, $\Delta_{\psi \cdot k} \leq 2$ for any $k \in G$ and the row condition holds. Conversely, if $\Delta_\psi \geq 3$ then for some $g' \neq 1 \in G$, there exist three distinct elements $x, y, z \in G$ such that $\psi(g', x) = \psi(g', y) = \psi(g', z)$. Write $y = xu$, $z = xv$ and $g = x^{-1}g'x$. By Equation (2), $\psi(g', xu) - \psi(g', x) = (\psi \cdot x)(g, u) = 0$ and similarly $(\psi \cdot x)(g, v) = 0$, giving three values 0 for $(\psi \cdot x)(g, h)$. □

We now give a construction of 2-uniform cocycles from 1-uniform cocycles.

Lemma 4 (Construction A). Let $\psi \in Z^2(G,C)$ and let $f \in \mathrm{Hom}(C,C)$. If $|G| = |C|$ and $\Delta_\psi = 1$ then $\Delta_{f \circ \psi} = |\ker f|$. In particular, if $f(C)$ has index 2 in C then $\Delta_{f \circ \psi} = 2$ and $N_{f \circ \psi}(g, c)$ is 2 or 0 for every $c \in C, g \neq 1 \in G$.

Proof. For every $g \neq 1 \in G$ and $c \in C$, $N_\psi(g, c) = 1$, so $N_{f \circ \psi}(g, c) = |\ker f|$ if $c \in f(C)$ and 0 if $c \notin f(C)$. □

This lemma allows us to discover new families of differentially 2-uniform functions and identify known ones from a novel perspective: given any group G and any abelian group C of the same size which possesses an index 2 subgroup, any differentially 1-uniform cocycle in $Z^2(G,C)$ maps to a differentially 2-uniform cocycle in $Z^2(G,C)$. Such cocycles exist even for the binary case when G and C are the same elementary abelian 2-group.

This effect is illustrated in the smallest case when $G = C = \mathbb{Z}_2^2 = \{0, a, c, a+c\} \cong (\mathrm{GF}(4), +)$. The differentially 1-uniform cocycle $\mu_1(g, h) = g^2 h$ corresponding to bundle 6.4 of [5, Example 8], and its images under the homomorphisms $f_1 : C \to C$ given by $f_1(a) = 0, f_1(c) = a$, and $f_2 : C \to C$ given by $f_2(a) = a, f_2(c) = 0$, have matrices

$$\begin{bmatrix} 0 & 0 & 0 & 0 \\ 0 & a & c & a+c \\ 0 & a+c & a & c \\ 0 & c & a+c & a \end{bmatrix}, \begin{bmatrix} 0 & 0 & 0 & 0 \\ 0 & 0 & a & a \\ 0 & a & 0 & a \\ 0 & a & a & 0 \end{bmatrix} \text{ and } \begin{bmatrix} 0 & 0 & 0 & 0 \\ 0 & a & 0 & a \\ 0 & a & a & 0 \\ 0 & 0 & a & a \end{bmatrix}$$

respectively. The second is the APN coboundary corresponding to bundle 1.2 listed earlier and the third is the APN non-coboundary cocycle corresponding to bundle 6.1. Similarly the differentially 1-uniform cocycle $\mu(g, h) = gh$ corresponding to bundle 3.2 listed earlier determines the APN non-coboundary cocycle $f \circ \mu$ corresponding to bundle 2.2 where $f : C \to C$ is given by $f(a) = a, f(c) = 0$.

The differentially 1-uniform cocycles given above are the smallest examples of a general class of differentially 1-uniform binary cocycles described in [5, Theorem 5]. If $(F, +, *)$ is a finite presemifield of order 2^a, set $G = C = (F, +) \cong \mathbb{Z}_2^a$, and let λ, ρ be automorphisms (ie. linearised permutation polynomials (LPP)) of G. Then the generalised multiplication $\mu_{\lambda,\rho} : G \times G \to G$ defined by $\mu_{\lambda,\rho}(g, h) = \lambda(g) * \rho(h)$, $g, h \in G$, is differentially 1-uniform. For example, when $F = \mathrm{GF}(2^a)$, λ is a Frobenius automorphism and $\rho = 1$, we obtain the field multiplication cocycle $\mu(g, h) = gh$ and the *power* cocycles $\mu_i(g, h) = g^{2^i} h$.

Theorem 2. *Let $(F, +, *)$ be a finite presemifield of order 2^a, let $G = C = (F, +)$, let λ and ρ be LPP of G and let $\mu_{\lambda, \rho} : G \times G \to G$ be defined by $\mu_{\lambda, \rho}(g, h) = \lambda(g) * \rho(h)$, $g, h \in G$. Then for any $f \in \mathrm{Hom}(G, G)$ with $|\ker f| = 2$, the cocycle $f \circ \mu_{\lambda, \rho}$ is differentially 2-uniform.* $\square$

If we know what binary differentially 1-uniform cocycles ψ can be mapped by homomorphisms f to a coboundary $\partial\phi$ we know which 1D binary APN functions arise from this construction. A necessary condition is easily derived.

Lemma 5. *Let $G = C = \mathbb{Z}_2^a$ and let $f \in \mathrm{Hom}(G, G)$ with $|\ker f| = 2$. Suppose $\Delta_\psi = 1$, and $f \circ \psi = \partial\phi$. Then there is a unique $x \neq 0 \in G$ such that $\psi(g, g) = x$ for all $g \neq 0 \in G$.*

Proof. For any $\phi \in C^1(\mathbb{Z}_2^a, \mathbb{Z}_2^a)$, $\partial\phi(g, g) = f \circ \psi(g, g) = 0$, $\forall g \in G$. But f maps only one nonzero element of G to 0. $\square$

This condition holds for some instances of Theorem 2. For example, let $(F, +, *)$ be an Albert non-commutative presemifield of order 2^a; that is, the elements of $\mathrm{GF}(2^a)$ are given the multiplication $\mu(g, h) = g * h = gh^{2^b} + cg^{2^b} h$, for some fixed $1 \leq b < a$ and c, where $c \neq x^{2^b - 1}$ for any $x \in \mathrm{GF}(2^a)$. Then $\mu(g, g) = \mu(1, 1) = 1 + c$ for all $g \neq 0$ if and only if $g^{2^b + 1} = 1$ for all $g \neq 0$ if and only if $(2^b + 1, 2^a - 1) = 1$, and this last condition holds, for instance, when a is odd and $b = (a - 1)/2$. It still remains to see if $f \circ \mu = \partial\phi$ for any f, ϕ.

However, if $*$ is field multiplication, none of the instances of Theorem 2 map to coboundaries, with a single exception. By Lemma 3 $\mathcal{B}(\mu_{\lambda, \rho}) = \mathcal{B}(\mu_{\lambda \circ \rho^{-1}, 1})$ so we need only check cocycles of the form $\psi(g, h) = \lambda(g)h$ for λ an LPP. Suppose $\partial\phi = f \circ \psi$, so that for all $g \neq 0 \in \mathrm{GF}(2^a)$, $\lambda(g)g = \lambda(1)$. Let α be a primitive element of $\mathrm{GF}(2^a)$ and let $m_\alpha(x) = \sum_{i=0}^{a} m_i x^i$ be the minimal polynomial for α. Then $\lambda(m_\alpha(\alpha)) = \sum_{i=0}^{a} m_i \lambda(\alpha^i) = \lambda(1) \sum_{i=0}^{a} m_i \alpha^{-i} = 0$, so $m_\alpha(\alpha^{-1}) = 0$. Therefore α^{-1} is in the cyclotomic coset generated by α, hence $2^{a-1} = 2^a - 2$, $a = 2$ and ψ is a scalar multiple of $\mu_1(g, h) = g^2 h$, the example with matrix illustrated earlier, corresponding to bundle 6.4.

Consequently the construction of Lemma 4 gives infinite classes of new (non-coboundary) APN cocycles. They are genuinely 2D, in the sense that they are not the images of 1D APN functions. It may also give classes of 1D APN functions.

4   New 2D APN Cocycles and Coboundaries

We note that to locate 2D APN coboundaries (and thus, by Lemma 2, 1D APN functions) we require symmetric cocycles. Further, it is clear that if ϕ is a power function on $GF(2^a)$, say $\phi(g) = g^d$, and d is odd, then $\partial\phi = \widehat{\psi}$, where $\psi(g,h) = \sum_{i=1}^{(d-1)/2} \binom{d}{i} g^{d-i} h^i$, while if d is even then $\partial\phi = \widehat{\psi}' + \binom{d}{d/2}\mu^{d/2}$, where $\psi'(g,h) = \sum_{i=1}^{d/2-1} \binom{d}{i} g^{d-i} h^i$. Consequently, $D\psi = 0$ or ϕ and $D(\psi' + \binom{d}{d/2}\mu^{d/2}) = 0$ or ϕ. That is, in many cases, ϕ may be recovered by taking the diagonal of a suitable asymmetric 2D function. A natural question to ask is: when is the symmetrisation of a cocycle an APN cocycle?

For a partial answer, there is a set of cocycles with useful properties. A cocycle is *multiplicative* if it is a homomorphism of G on either coordinate (and hence, by (2), on both). For example, any bilinear form is a multiplicative cocycle. All the generalised multiplications on presemifields described in the previous section are multiplicative. The set of multiplicative cocycles forms a subgroup of $Z^2(G,C)$.

Lemma 6. *Suppose* $\psi \in Z^2(G,C)$ *is multiplicative. Then*

(i) $\psi \cdot k = \psi$, *for all* $k \in G$;
(ii) *if* $\Delta_\psi < |G|$, G *is abelian*;
(iii) $\mathcal{D}(D\psi) = \mathcal{D}(\widehat{\psi})$;
(iv) $\partial D(\psi) = \widehat{\psi}$ *(so* $\widehat{\psi}$ *is a cocycle)*;
(v) *when* ψ *is symmetric*, $\partial D(\psi) = 2\psi$;
(vi) *when* $\psi = \partial\phi$ *is a symmetric coboundary, there exists* $f \in \text{Hom}(G,C)$ *such that* $D\partial(\phi) = 2\phi + f$.

Proof. (i) follows by definition. For (ii), the argument of Chen (see [6, 2.11]) applies, since for any $g,k \in G$, $\psi([g,k],h) = 0$ for all $h \in G$. Since $\Delta_\psi < |G|$, for any $x \neq 1$ there exist nonzero $\psi(x,h)$, hence $[g,k] = 1$. For (iii), since ψ is multiplicative, $D\psi(ag) - D\psi(g) = \psi(ag, ag) - \psi(g,g) = \psi(a,a) + \widehat{\psi}(a,g)$. Thus $n_{D\psi}(g,c) = N_{\widehat{\psi}}(g, c - D\psi(g))$. Parts (iv) and (v) follow by definition and (vi) follows from (v) since $\partial(D(\partial\phi)) = \partial(2\phi)$ so $D(\partial\phi) - 2\phi \in \ker \partial$. $\square$

This leads to our second construction of 2D binary APN cocycles.

Theorem 3 (Construction B). *Let* $G = C = \mathbb{Z}_2^a$ *and let* $\psi \in Z^2(G,G)$ *be multiplicative. If, for all* $g \neq 0$, $h \neq 0$ *and* $h \neq g$, $\psi(g,h) \neq \psi(h,g)$, *then* $\widehat{\psi}$ *is APN, and there exists an APN function* $\phi \in C^1(G,G)$ *such that* $\partial\phi = \widehat{\psi}$.

Proof. Since $\widehat{\psi}$ is itself multiplicative, by Lemma 6.i, $\widehat{\psi} \cdot k = \widehat{\psi}$ for all $k \in G$. For every $g \in G$, $\widehat{\psi}(g,g) = 0$. By Theorem 1, $\Delta_{\widehat{\psi}} \leq 2$ if and only if for all $g \neq 0$, $h \neq 0$ and $h \neq g$, $\widehat{\psi}(g,h) \neq 0$. By Lemma 6.iv, $\partial(D\psi) = \widehat{\psi}$ and by Lemma 2, $D\psi$ is APN. $\square$

Examples of such ψ may be found in the generalised multiplications on finite presemifields described earlier, which are all multiplicative cocycles which are differentially 1-uniform. Note that if $(F, +, *)$ is a semifield it has a 1, so that $1*h = h*1 = h$ for all $h \in F$, and we cannot take ψ to be semifield multiplication. Consequently we must search amongst generalised multiplications on semifields, or look for presemifields which are not semifields. We illustrate each type.

For field multiplication, consider the power cocycles $\mu_i(g, h) = g^{2^i} h$. If $g \neq 0$, $h \neq 0$ and $g \neq h$, then $g^{2^i} h = h^{2^i} g$ if and only if $(2^a - 1, 2^i - 1) > 1$ if and only if $(a, i) > 1$. Consequently if $(a, i) = 1$, $\widehat{\mu_i}(g, h) = g^{2^i} h + h^{2^i} g$ and $D\mu_i(g) = g^{2^i + 1}$ is a known 1D binary APN power function (case 2 of [4, Theorem 2]).

For presemifield multiplication, consider the Albert presemifields, with $\mu(g, h) = gh^{2^b} + cg^{2^b} h$, for some fixed $1 \leq b < a$ and c, where $c \neq x^{2^b - 1}$ for any $x \in \mathrm{GF}(2^a)$. Then $\widehat{\mu}(g, h) = (1 + c)(g^{2^b} h + gh^{2^b})$ and $D\mu(g) = (1 + c)g^{2^b + 1}$. So, when $(a, b) = 1$, we obtain a scalar multiple of the APN function just described. There is a huge number of nonisomorphic binary presemifields, so again we have a wealth of choice for construction of APN coboundaries and the corresponding APN functions. This search is currently underway.

However, not all known 1D binary APN functions determine multiplicative coboundaries. For example, the Welch power functions $\phi(g) = g^{2^m + 3}$ for $a = 2m + 1$ and m odd, (case 4 of [4, Theorem 2]) determine 2D APN coboundaries $\partial\phi(g, h) = g^{2^m + 2} h + gh^{2^m + 2} = \widehat{\psi}(g, h)$ where $\psi(g, h) = g^{2^m + 2} h$ and $D\psi = \phi$. We note the first component $g^{2^m + 2}$ of ψ is a Dembowski-Ostrom monomial (cf. [3, p.23]). Note also that $\psi \in C^2(\mathbb{Z}_2^a, \mathbb{Z}_2^a)$ is not a cocycle, even though $\widehat{\psi}$ is a cocycle (because it is a coboundary). Nonetheless, ψ does satisfy $\Delta_\psi = 1$. We collect this evidence in the following conjecture.

Conjecture 1. If $\psi \in C^2(\mathbb{Z}_2^a, \mathbb{Z}_2^a)$ satisfies $\Delta_\psi = 1$ and $\widehat{\psi} = \partial(D\psi)$ then $\Delta_{\widehat{\psi}} = 2$ (and hence $D\psi$ is a 1D APN function).

References

1. A. Canteaut, Cryptographic functions and design criteria for block ciphers, IN-DOCRYPT 2001, eds. C. Pandu Rangan, C. Ding, LNCS 2247, Springer 2001, 1–16.
2. F. Chabaud and S. Vaudenay, Links between linear and differential cryptanalysis, EUROCRYPT-94, LCNS 950, Springer, New York, 1995, 356–365.
3. R.S. Coulter and M. Henderson, A class of functions and their application in constructing semi-biplanes and association schemes, *Discrete Math.* **202** (1999) 21–31.
4. T. Helleseth, C. Rong and D. Sandberg, New families of almost perfect nonlinear power mappings, *IEEE Trans. Inform. Theory* **45** (1999) 475–485.
5. K.J. Horadam, Sequences from cocycles, AAECC-13, LNCS 1719, Springer, Berlin 1999, 121–130.
6. K.J. Horadam and P. Udaya, A New Construction of Central Relative $(p^a, p^a, p^a, 1)$-Difference Sets, *Des., Codes and Cryptogr.* **27** (2002) 281–295.
7. K. Nyberg, Perfect nonlinear S-boxes, EUROCRYPT-91, LCNS 547, Springer, New York, 1991, 378–385.
8. K. Nyberg, Differentially uniform mappings for cryptography, EUROCRYPT-93, LCNS 765, Springer-Verlag, New York, 1994, 55–64.

The Second and Third Generalized Hamming Weights of Algebraic Geometry Codes

Domingo Ramirez-Alzola

University of Basque Country, Department of Mathematics
Apartado 644, 48080 Bilbao, Spain
mtpraald@lg.ehu.es

Abstract. Motivated by cryptographical applications, the second and third generalized Hamming weights of AG codes arising from general curves are studied. This construction includes some of the most important AG codes: Hermitian Codes, elliptic codes, etc.

Keywords: linear codes, generalized Hamming weights.

1 Introduction

The *generalized Hamming weights* of linear codes are the generalization of the minimum distance. They were first motivated by applications from cryptography, namely the wire-tap channel of type II and the t-resilient functions, see [9]. The generalized Hamming weights completely characterize the performance of a linear code when it is used on the type II wire-tap channel. It is also useful in trellis coding (lower bounding the number of trellis states and in truncating a linear block code, see [3]. Later on, another apparently different concept, the Dimension/length Profiles (DLP) of a linear code, was proved to be equivalent to its generalized Hamming weights, see [2]. The close and deep connections between these two concepts make this topic more interesting and active.

Let $\mathbf{F}_q$ be the finite field with q elements. The *support* of a linear code $\mathcal{C}$ over $\mathbf{F}_q$ is defined as $\mathrm{supp}(\mathcal{C}) = \{i \mid x_i \neq 0 \text{ for some } \mathbf{x} \in \mathcal{C}\}$.

If $\mathcal{C}$ is an $[n, k]$ code, for $1 \leq r \leq k$, the rth *generalized Hamming weight* of $\mathcal{C}$ is defined by

$$d_r(\mathcal{C}) = \min\{\#\mathrm{supp}(D) \mid D \text{ is a linear subcode of } \mathcal{C} \text{ with } \dim(D) = r\}.$$

The sequence of generalized Hamming weights, or *weight hierarchy* of $\mathcal{C}$, was introduced by Helleseth, Kløve and Mykkleveit, [4], and rediscovered by Wei, [9]. Nowadays, the weight hierarchy plays a central role in coding theory, and much is known about it for several classes of codes: Hamming codes, Golay codes, Reed-Muller codes, Algebraic-Geometric codes, etc.

Recently, Heijnen and Pellikaan, [6], have given a new bound for higher Hamming weights. This so called *order bound*, which is a generalization of the Feng-Rao bound on the minimum distance, allowed the determination of the complete

M. Fossorier, T. Hoeholdt, and A. Poli (Eds.): AAECC 2003, LNCS 2643, pp. 158–168, 2003.

hierarchy of q-ary Reed-Muller codes. In this paper we study the weight hierarchy of general Algebraic-Geometric codes using this bound. Mainly, we deal with codes arising from curves (curve means plane, non-singular and absolutely irreducible curve) with only one infinity point. Many of most importan codes are a particular case of this one (rationals, elliptics, Hermitian, of Fermat, etc.)

The organization of the paper is as follows: Algebraic geometry and Weierstrass semigroups are studied in Sect. 2. In Sect. 3, we describe the Algebraic-Geometric codes and in Sect. 4 we get some results about their weight hierarchy. The order bound is explained in Sect. 4. In the two last section we applied the order bound for obtaining some values of the second and the third generalized Hamming weights.

2   Algebraic Geometry. Weierstrass Semigroups

Let K be an algebraically closed field.

Let $I\!\!P^n$ be the n-dimensional projective space. A subset $V \subseteq I\!\!P^n$ is a projective algebraic set if it is the zero-set of homogeneous polynomials. $V \subseteq I\!\!P^n$ is a projective variety iff $I(V)$ (the ideal associated to V) is a homogeneous prime ideal in $K[X_0, ..., X_n]$.

Given a non-empty variety $V \subseteq I\!\!P^n$, we define its homogeneous coordinate ring by $\Gamma_h(V) = K[X_0, ..., X_n]/I(V)$. This is an integral domain containing K. The function field of V is defined as

$$K(V) = \{\frac{g}{h} \quad | \quad g, h \in \Gamma_h(V) \text{ are forms of the same degree and } h \neq 0\}$$

which is a subfield of $\mathrm{Quot}(\Gamma_h)$, the quotient field of $\Gamma_h(V)$.

The dimension of V is the transcendence degree of $K(V)$ over K. In particular, when the dimension is one we call it a projective algebraic curve V. Moreover, We say that is plane when $n = 2$. It can be proved that a projective algebraic plane curve is the zero-set of an irreducible homogeneous polynomial $H \in K[X_0, X_1, X_2]$. If $P = (0 : x_1 : x_2) \in V$, we call it infinity point of the curve and $X_0 = 0$ the infinity line. The curve V is called non-singular if all points $P \in V$ are non-singular.

From now on, we assume that k is a perfect field (for example $k = \mathbf{F}_q$ finite field). Let $\overline{k} \supseteq k$ be the algebraic closure of k.

A projective variety $V \subseteq I\!\!P^n$ is defined over k if its ideal can be generated by homogeneous polynomials $F_1, ..., F_r \in k[X_0, ..., X_n]$. We define the ideal $I(V/k) = I(V) \cap k[X_1, ..., X_n]$ and the residue class ring $\Gamma(V/k)$. The quotient field $k(V) = \mathrm{Quot}(\Gamma(V/k)) \subseteq \overline{k}(V)$ is the field of k-rational functions of V. In particular, if χ is a projective curve defined over k, the field $k(\chi)$ is an algebraic function field of one variable over k.

We call $D = \sum_{P \in \chi} n_p P$ a divisor. The functions $f \in k(\chi)$ have also associated a divisor, $\mathrm{div}(f)$ or (f),

$$\mathrm{div}(f) = \sum_{\rho \in P_{k(\chi)}} v_\rho(f)\rho$$

where $P_{k(\chi)}$ denote the set of maximal ideals of the valuation rings of $k(V\chi)/k$. The positive part is called divisor of zeros of f, and we denote by $(f)_0$, and the negative part, with the sign changed, divisor of poles of f, we denote by $(f)_\infty$.

A divisor D is said to be positive, or effective, if $n_\rho \geq 0$ for all $\rho \in P_{k(\chi)}$. We define the genus, g, of the curve by the relationship

$$\inf_{D \in \mathcal{D}iv(P_{k(\chi)})} \{l(D) - deg(D)\} = l(0) - g.$$

We call canonical divisor to the divisor associated to a differential of Weil.

2.1   Weierstrass Semigroups

Let $\mathcal{X}$ be a plane non-singular curve defined on $\mathbf{F}_q$, of degree d with only one infinity point, Q.

Definition 1. *Let $n \in \mathbf{N}_0$. n is said to be a pole number on Q if there exists a function $f \in \mathbf{F}_q(\mathcal{X})$ with one only pole at Q of order n, that is to say, if there exists $f \in \mathbf{F}_q(\mathcal{X})$ such that $(f)_\infty = nQ$. In other case, we call n a gap at Q.*

Obviously, the sum of pole number is a pole number. Let $p_1 < p_2 < ... < p_r < ...$ be the sequence of pole numbers on Q. On the other hand, $p_r \geq \gamma_r$, being $\gamma_r = \min\{deg(A) \mid A \in Div(\mathcal{X}/\mathbf{F}_q), l(A) \geq r\}$ the rth gonality of $\mathcal{X}$. The set $S(Q) = S = \{p_1 = 0, p_2, ...\}$ is a group with respect to the sum of number poles and it is called *the Weierstrass semigroup* of $\mathcal{X}$ at the point Q.

We say that a semigroup S is symmetric if it verifies the following condition: for all n, $n \in S$ if and only if $(2g - 1) - -n \notin S$, where $g = |\mathbf{N}_0| - |S|$ is called the genus of the semigroup. Obviously, if $S = S(Q)$, g is the genus of the curve.

It is well known that there are exactly g gap numbers $i_1, ..., i_g$. If $i_g = 2g - 1$, $S(Q)$ is symmetric and $g > 0$, $(2g - 2)Q$ is a canonical divisor of $\mathcal{X}$.

On the other hand, the semigroup S is generated by a and b, $1 < a < b$, and we write $S =< a, b >$, if for all $n \in S$ there exist $x, y \in \mathbf{N}_0$ such that $n = xa + yb$ (we suppose that $\gcd(a,b)=1$). Every two-generated Weierstrass semigroup in Q associated to the curve χ is symmetric.

Proposition 1. *Let $\mathcal{X}$ be a plane, non-singular curve with only one infinity point Q. If d is the degree of $\mathcal{X}$, the Weierstrass semigroup on Q is*

$$S =< d - 1, d >= \{x(d - 1) + yd \mid 0 \leq x, 0 \leq y \leq d - 2\}.$$

Proof. Let r be the infinite line, $r_1 \neq r$ a line passing across Q and r_2 another line that does not pass across Q. Every line cuts the curve in exactly d points. Moreover, the three divisors associated to these lines are linearly equivalent. In particular, $(r) \sim (r_1)$, so $dQ \sim Q$+(sum of affine points) and $d - 1$ is a pole number on Q. A similar argument proves that d is also a pole number in Q.

Under the previous conditions, it is easy to prove that $p_r = \gamma_r$ for all $r \geq 1$.

3 The Construction of the Algebraic-Geometric Codes

Let $\mathcal{X}$ be a plane, non-singular curve defined on $\mathbf{F}_q$ with degree d and with only one infinity point, Q.

We consider the pencil lines passing through Q and we use the lines which are non-tangent to $\mathcal{X}$ in any affine point. Let $L_1, \ldots, L_t$ be such lines. Their divisors are

$$(L_1) \;=\; P_1^1 + \ldots + P_{d-1}^1 + Q - dQ,$$

$$\vdots$$

$$(L_t) \;=\; P_1^t + \ldots + P_{d-1}^t + Q - dQ,$$

being $P_1^1, \ldots, P_{d-1}^1, \ldots, P_1^t, \ldots, P_{d-1}^t$ distinct points that we suppose rational.

Let $D = P_1^1 + \ldots + P_{d-1}^1 + \ldots + P_1^t + \ldots + P_{d-1}^t$ and $G = mQ$. We construct the Algebraic-Geometric code $\mathcal{C}_{\mathcal{L}}(D, G)$. For the sake of simplicity, we shall write $\mathcal{C}(m)$ instead of $\mathcal{C}_{\mathcal{L}}(D, G)$ and we denote the rth generalized Hamming weight of that code by $d_r(m)$. In the sequel we shall assume that m is a pole number. Otherwise we can replace m by the largest pole number not exceeding m.

4 First Results about the Weight Hierarchy of $\mathcal{C}(m)$

Munuera, [7], gave a geometric interpretation of the generalized Hamming weights.

Proposition 2. *Let $\mathcal{C}(m)$ be a code with dimension k and abundance α. For every $1 \leq r \leq k$, we have*

$$
\begin{aligned}
d_r(m) &= \min\{\deg(D') \mid 0 \leq D' \leq D, l(D' - (n-m)Q) \geq r + \alpha\} \\
&= n - \max\{\deg(D'') \mid 0 \leq D'' \leq D, l(mQ - D'') \geq r + \alpha\}.
\end{aligned}
$$

In particular, $d_r(m) \geq n - m + p_{r+\alpha}$. Moreover, if $r + \alpha > g$ then $d_r(m) = n - k + r$.

Definition 2. *We say that an integer $m \in \mathbf{N}_0$ verifies the property (*) if $m \leq n$ and $m \in (d-1)\mathbf{Z}$.*

If m verifies the property (*) then there exists a divisor $D' \leq D$ such that $D' \sim mQ$. As a consequence, we have the following result.

Theorem 1. *Let $\mathcal{C}(m)$ be a code with abundance α. If $m - p_{r+\alpha}$ or $n - m + p_{r+\alpha}$ verify the property (*), then*

$$d_r(m) = n - m + p_{r+\alpha}.$$

Proof. If $m - p_{r+\alpha}$ satisfies the property (*) then there exists an effective divisor $D' \leq D$ linearly equivalent to $(m - p_{r+\alpha})Q$. We can now apply proposition 2. The proof in the other case is similar.

Proposition 3 (Chain Property). *Let $\mathcal{C}$ and $\mathcal{C}'$ be linear codes on $\mathbf{F}_q{}^n$ such that $\mathcal{C}' \subseteq \mathcal{C}$. Let $k' = \dim \mathcal{C}'$, $k = \dim \mathcal{C}$ and $s = k - k'$. Then, for every $1 \leq r \leq k'$,*

$$d_r(\mathcal{C}) \leq d_r(\mathcal{C}') \leq d_{r+s}(\mathcal{C}).$$

The proof of this result can be found in [7].

4.1   Case $m < n$

Let $\mathcal{C}(m)$ be a code with $m < n$. Using property (*) and proposition 2 we can compute some generalized Hamming weights. We restrict our analysis to the second and third generalized Hamming weights.

Proposition 4. *Let $x \geq 2$. Then,*

$$n - (x - 2)d - 1 \leq d_3(x(d - 1)) \leq n - (x - 2)d + x - 3.$$

Proof.

$$d_2((x - 1)(d - 1) + x - 1) \leq d_3(x(d - 1)) \leq d_4(x(d - 1)) - 1.$$

4.2   Case $m \,\square\, n$

Lemma 1. *Let $m \geq n$.*
(i) If $m = n + (r - 1)(d - 1)$ then $d_r(m) = d - 1$.
(ii) If $m = n + (r - 1)(d - 1) + r$ then $d_{r+1}(m) = d - 1$.

Proof. If α is the abundance of $\mathcal{C}(m)$, $p_{r+\alpha} = r(d - 1)$ and $n - m + p_{r+\alpha}$ verifies property (*). To prove (ii), the abundance is $\alpha = l(((r - 1)(d - 1) + r - 1)Q)$ so that $p_{r+\alpha+1} = r(d - 1) + r$ and then $n - m + p_{r+\alpha+1}$ satisfies property (*).

We can generalize those results.

Theorem 2. *Let $n \leq m \leq n + 2g - 2$, and let us write $m = n + a(d - 1) + b$ with $0 \leq a, -1 \leq b \leq d - 3$. Then,*
(i) $d_r(n + a(d - 1) + b) = d - 1 - a + r - 1$, if $-1 \leq b \leq a$ and $1 \leq r \leq a + 1$.
(ii) $d_r(n + a(d - 1) + b) = d - 1 - a + r - 2$, if $a < b \leq d - 3$ and $1 \leq r \leq a + 2$.

Proof. (i), we observe $\mathcal{C}(n + a(d - 1) - 1) = \mathcal{C}(n + a(d - 1)) = \ldots$
$\ldots = \mathcal{C}(n + a(d - 1) + a)$. From lemma 1, $d_{a+1}(n + a(d - 1)) = d - 1$. On the other hand, $d_1(n + a(d - 1)) = d_1(n + a(d - 1) + a) \geq d - a - 1$ and by the monotonicity $d_1(n + a(d - 1)) + a \leq d_{a+1}(n + a(d - 1))$.
To prove (ii), we consider

$$\mathcal{C}(n + a(d - 1) + a + 1) \subseteq \mathcal{C}(n + a(d - 1) + b) \subseteq \mathcal{C}(n + a(d - 1) + d - 3).$$

First of all, $d_{a+2}(n + a(d - 1) + a + 1) = d - 1$. On the other hand, the first distance is great of equal to $d - 1 - a - 1$ and by monotonicity $d_1(n + a(d - 1) + a + 1) + a + 1 \leq d_{a+2}(n + a(d - 1) + a + 1)$, so $d_r(n + a(d - 1) + a + 1) = d - 1 - a + r - 2$.

To compute the other values we shall use the order bound for generalized Hamming weights.

5   The Order Bound for the Generalized Hamming Weights

The order bound for the generalized Hamming weights has been introduced by P. Heijnen and R. Pellikaan in [6]. It is a generalization of the Feng-Rao bound on the minimum distance. For the reader's convenience we shall explain it briefly in this section.

Let $\mathcal{L} = \cup_{t=0}^{\infty} \mathcal{L}(tQ)$ and let $\{f_1, f_2, \cdots\}$ be a basis of $\mathcal{L}$ such that $-v_Q(f_i) < -v_Q(f_{i+1})$ for all i, where v_Q is the valuation at Q. Note that $\{-v_Q(f) \mid f \in \mathcal{L}\} = S = \langle d-1, d \rangle$.

For every positive integer l let us consider the linear code

$$C_l = \langle ev(f_1), \cdots, ev(f_l) \rangle^{\perp}$$

and the set $A(l) = \{p_i \in S \mid p_i + p_j = p_{l+1} \text{ for some } p_j \in S\}$.

Furthermore, for $l_1 < \cdots < l_r$, let us define the set $A(l_1, \cdots, l_r)$ as the union of the previous sets.

Definition 3. *Let l be a positive integer. The number*

$$d_r^{ORD}(l) = \min\{\#A(l_1, \cdots, l_r) \mid l \le l_1 < \cdots < l_r\}$$

is called the order bound *for the rth generalized weight of C_l. The number*

$$D_r^{ORD}(l) = \min\{\#A(l_1, ..., l_r) \mid l \le l_1 < ... < l_r \text{ y } C_{l_i} \ne C_{l_i+1}$$
$$\text{for all } i = 1, \ldots, r\}.$$

is called the strong order bound *for the rth generalized weight of C_l.*

The relationship between the order bound and the generalized Hamming weights is the next one, [6].

Theorem 3. *For $1 \le l \le n+g$, we have*

$$d_r(C_l) \ge d_r^{ORD}(l) \ge D_r^{ORD}(l).$$

Note that $C_l = (0)$ for $l > n+g$.

Definition 4. *Let $m = x(d-1) + yd$, $0 \le x, 0 \le y < d-1$, be a pole number. We say m is a* type I pole *if $x \le d-1$. If $x > d-1$ then a* type II pole.

If $p_{l+1} = x(d-1) + yd$, with $0 \le x, 0 \le y < d-1$, then

$$\#A(l) = \begin{cases} (x+1)(y+1) & \text{for type I poles;} \\ (x+1)(y+1) + (x-d+1)(d-y-2) & \text{for type II poles.} \end{cases}$$

$S(Q)$ is a semigroup of genus g so if $l > g$ then $A(l) \subseteq A(l+p)$ for every $p \in S$.

We study the sets C_m so that we get the next result.

Proposition 5. *Let $m > g$, then $C_m \neq C_{m+1}$ if and only if $n + g - m - 1$ is a pole at Q.*

Proof. $C_m = < ev(f_1), \ldots, ev(f_m) >^{\perp} = C(D, p_m Q)^{\perp} = C(D, (n + 2g - 2 - p_m)Q)$.

As a corollary we can rewrite the strong order bound when $l > g$. Let

$$\mathcal{P}(l) = \{(l_1, l_2, \ldots, l_r) \mid l \leq l_1 < l_2 < \ldots < l_r; n + g - l_i - 1 \text{ is a pole } i = 1, \ldots, r\}.$$

Then $D_r^{ORD}(l) = min\{\#A(l_1, \ldots, l_r) \mid (l_1, \ldots, l_r) \in \mathcal{P}(l)\}$.

6   The Second Distance in the Case $m < n$

Let us consider $\mathcal{C}(m)$ be with $m < n$. We can write $m = n - a(d - 1) + b$ with $1 \leq a$ and $1 \leq b \leq d - 2$. In the case $b = 1$ the second distance is known, $d_2(m) = (a + 1)(d - 1) - 1$. For the general case, $d_2(m) \leq (a + 1)(d - 1) - 1$. We restrict the study to $1 \leq b \leq d - a - 2$. Then

$$d_2(n - a(d - 1) + d - a - 2) \leq d_2(n - a(d - 1) + b) \leq (a + 1)(d - 1) - 1.$$

Using the order bound, we shall prove that $d_2(n - a(d - 1) + d - a - 2) \geq (a+1)(d-1)-1$. From the order bound, we have to compute $d_2^{ORD}(g+(a-1)d+1)$. First of all, we simplify the order bound for $r = 2$.

Lemma 2 (Reduction Formula). *If $l > g$ then,*

$$d_2^{ORD}(l) = min(\{\#A(l_1, l_2) \mid l \leq l_1 < l_2 \leq l + d - 2\} \cup \{\#A(l + d - 1)\}).e$$

The cardinality of $A(g + ad)$ is $(a + 1)(d - 1)$. We compute now $\#A(l_1, l_2)$ with $g + (a - 1)d + 1 \leq l_1 < l_2 \leq g + ad - 1$. We shall distinguish several cases depending on the type of pole numbers.

Proposition 6.
(i) If $g + (a - 1)d + 1 \leq l_1 < l_2 \leq g + (a - 1)d + d - a - 1$, then $\#A(l_1, l_2) \geq (a + 1)(d - 1) - 1$.
(ii) If $g + ad - a \leq l_1 < l_2 \leq g + ad - 2$, then $\#A(l_1, l_2) \geq (a + 1)d - a$.
(iii) If $l_1 = g + (a - 1)d + r$, $1 \leq r \leq d - a - 1$, and $l_2 = g + a(d - 1) + s$, $0 \leq s \leq a - 2$, then $\#A(l_1, l_2) \geq (a + 1)(d - 1)$.

Proof. (i), the poles are of type I, so $p_{l_1+1} = (d - 3 - r)(d - 1) + (a + r)d$ and $p_{l_2+1} = (d - 3 - s)(d - 1) + (a + s)d$, with $0 \leq r < s \leq d - a - 2$. If $r = d - a - 3, s = d - a - 2$ then

$$\#A(l_1, l_2) = (a + 1)(d - 1) - 1.$$

We can suppose that $0 \leq r < s \leq d - a - 3$, so that $\#A(l_1, l_2) \geq (a + 1)(d - 1) + r$.

To prove (ii), the poles are the type II. We can write $p_{l_1+1} = (d + a - 2 - r)(d-1) + rd$ and $p_{l_2+1} = (d+a-2-s)(d-1) + sd$ with $0 \le r < s \le a - 2$. Hence, $\#A(l_1, l_2) \ge (a+1)d - a + r$.

(iii), we observe that $m = x(d-1) + yd \in A(l_1) \cap A(l_2)$ if and only if $x \le d - r - 2$ and $y \le s$ or $x \le a - s - 2$ and $s + 1 \le y \le a + r - 1$. So, $\#A(l_1, l_2) - (a+1)(d-1) \ge rs \ge 0$.

The other cases are analyzed in much the same manner so we omit them. Then, for $1 \le a \le d - 1$, $d_2^{ORD}(g + (a-1)d + 1) = (a+1)(d-1) - 1$. We summarize our results.

Theorem 4. *Let $\mathcal{C}(m)$ be an AG code with $m = n - a(d-1) + b$. For $1 \le b \le d - a - 2$ we get*

$$d_2(n - a(d-1) + b) = (a+1)(d-1) - 1.$$

7   The Third Distance

We distinguish two cases.

7.1   Case $m \square n$

We observe that $\mathcal{C}(n-1) = \mathcal{C}(n)$ and hence $d_3(n-1) = d_3(n) = p_4$. We compute now $d_3(n+b)$, for $2 \le b \le d - 3$. From the chain property, $d_3(n + d - 3) \le d_3(n + b) \le d_3(n+2)$.

Using the monotonicity, $d_3(n+2) \le 2d - 4$. Our objective is to prove that $d_3(n + d - 3) \ge 2d - 4$.

Lemma 3 (Reduction Formula).
If $l > g$ then, $d_3^{ORD}(l) = \min(\{\#A(l_1, l_2, l_3) \mid l \le l_1 < l_2 < l_3 \le l + d - 1\} \cup \{\#A(l_3) \mid l + d \le l_3 \le l + 2d - 3\}).$

Theorem 5. *For $2 \le b \le d - 3$, $d_3(n+b) = 2d - 4$.*

Proof. From theorem 3, $d_3(n + d - 3) \ge d_3^{ORD}(g - d + 3)$. If we take the values $l_1 = g-2$, $l_2 = g-1$ and $l_3 = g+d-2$ we get $\#A(l_1, l_2, l_3) = 2d-4$. Furthermore, $\#A(l) \ge 2d - 4$ except for l equal to $g + d - 3, g - 2, g - 1, g, g + d - 2$ and $g + d$ and there is no combination of three of these values giving $A(\cdot, \cdot, \cdot)$ with smaller cardinality.

7.2   Case $m < n$

First of all, we compute $d_3(a(d-1))$. From proposition 4, this value is in the interval $[n - (a-2)d - 1, n - (a-2)d + a - 3]$. We can consider $a \ge 3$. We shall prove that $d_3(a(d-1)) \ge n - (a-2)d$. Firstly, we study the sets $A(l)$.

Lemma 4. *Let $l = n + g - a(d - 1) - 1$, $3 \leq a \leq d - 1$. Then*
a) $\#A(l) = n - a(d - 1)$.
b) For $t \geq 0$, $\#A(l + t) = \#A(l) + t = n - a(d - 1) + t$.

We can estimate the strong order bound.

Proposition 7. *For $3 \leq a \leq d - 1$,*

$$D_3^{ORD}(n + g - a(d - 1) - 1) \leq n - (a - 2)(d - 1) - a + 2.$$

Proof. Let $l = n + g - a(d-1) - 1$, $l_1 = l + d - a$, $l_2 = l + d - a + 1$ and $l_3 = l + 2d - a$. Then $l \leq l_1 < l_2 < l_3$ and $n + g - l_i - 1$ is a pole number, $i = 1, 2, 3$. Hence,

$$D_3^{ORD}(n + g - a(d-1) - 1) \leq \#A(l_1, l_2, l_3) = \#A(l_3) = n - (a-2)(d-1) - a + 2.$$

We can rewrite the strong order bound.

Corollary 1. *Let $l = n + g - a(d - 1) - 1$, $3 \leq a \leq d - 1$. Then*

$$D_3^{ORD}(l) = \min\{\#A(l_1, l_2, l_3) \mid l \leq l_1 < l_2 < l_3 \leq l + 2d - a;$$
$$a(d - 1) - l_i + l \text{ is a pole, }\}.$$

For $l = n + g - a(d - 1) - 1$, $3 \leq a \leq d - 1$, we define the set

$$\overline{\mathcal{P}}(l) = \{(l_1, l_2, l_3) \mid l \leq l_1 < l_2 < l_3 \leq l + 2d - a;$$
$$a(d - 1) - l_i + l \text{ is a pole, } i = 1, 2, 3\}.$$

It is easy to prove that if $(l_1, l_2, l_3) \in \overline{\mathcal{P}}(l)$ and $l_3 > l + d - 1$ then $\#A(l_1, l_2, l_3) \geq n - (a - 2)(d - 1) - a + 2$. We suppose that $l_3 \leq l + d - 1$.

Proposition 8. *Let $l = n + g - a(d-1) - 1$, $3 \leq a \leq d - 1$ and $(l_1, l_2, l_3) \in \overline{\mathcal{P}}(l)$.*
(i) If $l + d - a \leq l_1 < l_2 < l_3 \leq l + d - 1$, then $\#A(l_1, l_2, l_3) \geq n - (a - 2)d$.
(ii) If $l_1 = l$, then $\#A(l_1, l_2, l_3) \geq n - (a - 2)d$.

Proof. Since $l_3 - l_1 \neq 1$ and $\#A(l_1, l_2, l_3) \geq \#A(l_1, l_3)$. By computation, we prove (ii)

Theorem 6. *Let $l = n + g - a(d - 1) - 1$, $3 \leq a \leq d - 1$.*

$$D_3^{ORD}(l) = n - (a - 2)d.$$

We consider now $m = n - a(d - 1) + b$, with $1 \leq b \leq d - a - 2$. We computed the second distance. As a corollary, we get the third distance.

Theorem 7. *Let $m = n - a(d - 1) + b$, with $1 \leq b \leq d - a - 2$. Then,*

$$d_3(m) = (a + 1)(d - 1).$$

Proof.

$$d_2(n - a(d-1) + d - a - 2) < d_3(m) \le d_3(n - a(d-1) + 1) = (a+1)(d-1).$$

We consider now $b = d - a - 1$ so that $m = n - a(d-1) + d - a - 1$ for $1 \le a \le d-2$. In the case $a = d-2$, $d_3(m) = (a+1)(d-1)$. Let $2 \le a \le d-3$. We will prove that $d_3(n-a(d-1)+d-a-1) = (a+1)(d-1)$, using the order bound. To do this, we have to prove that $d_3^{ORD}(g+(a-1)(d-1)+a-1) = (a+1)(d-1)$.

Lemma 5. *If $g < l_1 < l_2 < l_3$ and $l_3 - l_1 < d-1$ then $\#A(l_1, l_2, l_3) > \#A(l_2, l_3)$.*

From lemma 3, we distinguish several cases. It is easy to prove that, for $2 \le a \le d-3, g+(a-1)(d-1)+a-1 \le l_1 < l_2 < l_3 \le g + a(d-1) + a - 1$, $\#A(l_1, l_2, l_3) \ge (a+1)(d-1)$. We must also compute $\#A(l_3)$.

Proposition 9. *If $2 \le a \le d-3$, and $l_3 \ge g + a(d-1) + a$, then $\#A(l_3) \ge (a+1)(d-1)$ with equality if $l_3 = g + a(d-1) + a$.*

Proof. We know that $\#A(g + a(d-1) + a) = (a+1)(d-1)$. On the other hand, if $l_3 \ge g + a(d-1) + a$, $\#A(l_3) \ge d_1^{ORD}(g + a(d-1) + a)$. By the property of the sets $A(l)$, $d_1^{ORD}(g + a(d-1) + a) = \min\{\#A(l) \mid g + a(d-1) + a \le l \le g + a(d-1) + a + d - 2\}$. We can write $p_{l+1} = x(d-1) + yd$ with $d - 2 \le x \le a$ and $a \le y \le d - 2$ so we complete the proof ([5], Theorem 4.9).

Theorem 8. *If $m = n - a(d-1) + b$ with $2 \le a \le d-2$ and $b = d - a - 1$ then*

$$d_3(m) = (a+1)(d-1)$$

References

1. A. Barbero and C. Munuera, "The weight hierarchy of Hermitian codes" *SIAM J. Discrete Math.* vol. 13, pp.79–104, 2000.
2. G.D. Forney, Jr., "Dimension/Length Profiles and Trellis Complexity of Linear Block Codes", *IEEE Trans. Inform. Theory,* vol. IT-40, pp. 1741–1752, Nov., 1994.
3. T. Helleseth and P.V. Kumar, "The weight Hierarchy of the Kasami codes", *Discrete Math.*
4. T. Helleseth, T. Kløve, and J. Mykkeleveit, "The weight distribution of irreducible cyclic codes with block lengths $n_1((q^l - 1)/N)$", *Discrete Math.* vol. 18, pp. 179–211, 1977.
5. T. Høholdt, J.H. van Lint, and R. Pellikaan, "Algebraic Geometry Codes", to appear in *Handbook of Coding Theory*, R.A. Brualdi, W.C. Huffman, V. Pless (Eds.), Elsevier.
6. P. Heijnen and R. Pellikaan, "Generalized Hamming weights of q-ary Reed-Muller codes". *IEEE Trans. Inform. Theory*, vol. IT-44. pp. 181–197, January 1998.
7. C. Munuera, "On the Generalized Hamming Weights of Geometric Goppa Codes", *IEEE Trans. Inform. Theory*, vol.IT-40, pp. 2092–2099, November 1994.

8. C. Munuera and D. Ramírez, "The Second and The Third Generalized Hamming Weigths on Hermitian Codes". *IEEE Transactions on Information Theory*, Vol. 45, No. 2, March 1999.
9. V.K. Wei, "Generalized Hamming weights for linear codes", *IEEE Trans. Inform. Theory*, vol. IT-37, pp. 1412–1418, May 1991.
10. K. Yang, P.V. Kumar, and H. Stichtenoth, "On the weight hierarchy of Geometric Goppa Codes", *IEEE Trans. Inform. Theory*, vol. IT-40, pp. 913–920, 1994.

Error Correcting Codes over Algebraic Surfaces

Thanasis Bouganis

Department of Pure Mathematics and Mathematical Statistics[*]
University of Cambridge
ab442@hermes.cam.ac.uk

Abstract. We study error correcting codes over algebraic surfaces. We give a construction of linear error correcting codes over an arbitrary algebraic surface and then we focus on linear codes over ruled surfaces. At the end we discuss another approach to getting codes over algebraic surfaces using sections of rank two bundles. The new codes are not linear but do have a group structure.

1 Introduction

Algebraic-geometric (AG) codes have been proved one of the most important classes of error correcting codes. The main reason for this is that they provide families of codes that beat the Gilbert-Varshamov bound, see [6] or [8].

A natural question to ask is what kind of codes can we get if we consider algebraic varieties of higher dimensions. Or better, can the definition of AG codes be extended to higher dimensional varieties? (But see also the discussion in the last section for another motivation). In this direction the known general constructions are restricted to hypersurfaces where the Lefschetz-Grothendieck trace formula (using étale cohomology) can give reasonable bounds for the minimum distance, see [5] and [8]. Here we consider the case of surfaces.

This paper is organized as follows. We begin by studying a naive extension of the standard definition of AG codes to surfaces. We connect the analysis of this approach with a known, and let us say hard, problem in algebraic geometry. Then we propose another general construction of AG codes over surfaces, analyze it and apply it to a broad family of surfaces, the ruled surfaces. Then we investigate the possibility of getting asymptotically good codes using ruled surfaces.

Also we propose another construction of error correcting codes over surfaces. The main difference from the previous constructions is the use of global sections of rank two bundles instead of functions defined over the surface. We can establish the parameters of this construction assuming the existence of a rank two bundle with a special property. Moreover we explain why this construction seems to be more natural. Finally, as we said, we conclude with our motivation for studying error correcting codes over surfaces.

We would like to note that the paper assumes familiarity with notions of algebraic geometry. We do not prove any fact coming from algebraic geometry.

---

[*] Part of the this work appeared as author's Master Thesis at Boston University, USA

M. Fossorier, T. Hoeholdt, and A. Poli (Eds.): AAECC 2003, LNCS 2643, pp. 169–179, 2003.
© Springer-Verlag Berlin Heidelberg 2003

The reference that we use is mainly [4]. For a self-contained exposition to all of the above the interested reader can see [2].

2 Algebraic Geometric Codes

Let C be a non-singular projective (absolutely irreducible) curve of genus g defined over the field $\mathbf{F}_q$. Let $D := P_1 + \cdots + P_n$ be a divisor on C, where P_i is a rational point of C. Moreover let G be another divisor on C such that $supp(D) \cap supp(G) = \emptyset$ and $n > deg(G) > 2g-2$ (all the divisors that we consider in the paper are rational i.e stable under $Gal(K/\mathbf{F}_q)$ for any Galois extension K). We define $C_L(D, G) := \{(f(P_1), \ldots, f(P_n)) : f \in L(G)\}$. The next theorem establishes the parameters of these codes, for a proof see [6], [7], [8].

Theorem 1. $C_{\mathcal{L}}(D, G)$ is an $[n, k, d]_q$ code where $k = deg(G) - g + 1$ and $d \geq n - deg(G)$.

In order to motivate our later constructions, at this point we define another family of linear codes over curves that have the same parameters as the AG codes previously defined. With notation as before we consider the corresponding (up to isomorphism) line bundle $\mathcal{L}_G$ to the divisor G. We denote by $\mathcal{L}_G(X)$ the vector space of the global sections.

We fix a trivialization $\{(U_i, \phi_i\}_{i \in I}$ of the line bundle $\mathcal{L}_G$. We assign each point P_i to a unique open set U_i. That is, we fix a mapping $\theta : [1, n] \to I$ such that $\theta(j) = i$ implies that the point $P_j \in U_i$. We define the new code as $C_{\mathcal{L}}(D, G) := \{((\phi_{\theta(1)} \circ s)(P_1), \ldots, (\phi_{\theta(n)} \circ s)(P_n)) : s \in \mathcal{L}_G(X)\}$.

Remark 1. Of course the definition of the code $C_{\mathcal{L}}(D, G)$ depends on the choice of the trivialization of the line bundle and the mapping θ. However, since we are only interested in the parameters of these codes, these choices are not important. So we can just write $C_{\mathcal{L}}(D, G)$.

Theorem 2. The code $C_{\mathcal{L}}(D, G)$ is an $[n, k, d]_q$ where $k \geq deg(G) - g + 1$ and $d \geq n - deg(G)$. Moreover if $deg(G) > 2g - 2$ then $k = deg(G) - g + 1$.

Proof. The dimension is an application of Riemann Roch theorem and the designed minimum distance follows from the fact that $(s)_0 \sim G$ where $(s)_0$ the zero divisor of a global section (see [2]) or [4].   □

3 Extending Algebraic Geometric Codes: A Naive Approach

Let X denote a projective nonsingular surface defined over some finite field $\mathbf{F}_q$ with arithmetic genus p_α. Moreover let K be the canonical divisor on X and H be an ample divisor on X. By $N(X)$ we will denote the rational points of X over $\mathbf{F}_q$ ($n := \#N(X)$) and by $D.C$ the intersection number of the divisors D and C.

Let us select an effective divisor $G \geq 0$ on X with the property the $P_i \notin G$ for all i (viewing G as a collection of curves). The code $C_L(X, G)$ is defined as,

$$C_L(X, G) := \{(f(P_1), \ldots, f(P_n)) \ : \ f \in L(G)\} \tag{1}$$

Lemma 1. *The code $C_L(X, G)$ is an $[n, k, d]_q$ code. Assuming that $f(P_i) = 0$ for all i implies $f = 0$ and that $G.H > K.H$ then $k \geq (G^2 - G.K)/2 + p_\alpha + 1$.*

Proof. The code $C_L(X, G)$ is obviously linear. For the dimension of the code, under the given assumptions we can apply the Riemann Roch theorem for surfaces. Indeed by writing $h^i(G)$ for the dimension of the vector space $H^i(X, G)$ we have, $k = h^0(G) = (G^2 - G.K)/2 + p_\alpha + 1 + h^1(G) - h^2(G)$. By Serre duality we have that $H^2(X, G) \cong H^0(X, K - G)^\vee$ and so $h^2(G) = h^0(K - G)$. Since we assume that $G.H > K.H$, $h^2(G) = 0$. $\qquad\square$

Proving a non-trivial lower bound for the minimum distance in the general setting is not as easy as it is for curves. Let $L(G - P_{i_1} - \ldots - P_{i_m})$ denote the vector subspace of $L(G)$ with the property that all $f \in L(G - P_{i_1} - \ldots - P_{i_m})$ have zeros at $P_{i_1}, \ldots, P_{i_m}$. Notice that $G - P_{i_1} - \ldots - P_{i_m} \notin Div\, X$ since P_{i_j} are just points and so we cannot establish the dimension of that space using directly the Riemann Roch theorem. In the context of algebraic geometry the associated linear system $|G - P_{i_1} - \ldots - P_{i_m}|$ is called *linear system with assigned points*.

Let us denote the vector space $L(G - P_{i_1} - \ldots - P_{i_m})$ by δ. Let $\pi : X' \to X$ be the morphism that we obtain by blowing up the points $P_{i_1}, \ldots, P_{i_m}$ and let $E_1, \ldots, E_m$ be the corresponding exceptional divisors on X'. We denote the vector space $L(\pi^*G - E_1 - \ldots - E_m)$ by δ'. The next lemma indicates that we can reduce our study to the space δ' on the surface X'. For a proof see [2] or [4].

Lemma 2. *There is a one-to-one correspondence between the elements of δ and δ'.*

The main theorem of this section is,

Theorem 3. *If we denote by $l(\delta)$ (respectively $l(\delta')$) the dimension of δ (of δ') then, $l(\delta') \leq (G^2 - G.K)/2 + p_\alpha(X) + 1 - (m - h^1(\delta'))$, where $h^1(\delta')$ is the dimension of the vector space $H^1(X', \pi^*G - E_1 - \ldots - E_m)$. Moreover if we assume that $G.H > K.H$ then, $l(\delta') \leq l(\delta) - (m - h^1(\delta'))$.*

Proof. This follows from the theory of monoidal transformation. Due to lack of space we do not include the proof. See [2].

The important conclusion is that in order to reason about the minimum distance of the code $C_L(X, G)$ we need to know the relative "position" of the rational points on the surface X. This is what the quantity $h^1(\delta')$ counts, the failure of the points $P_{i_1}, \ldots, P_{i_m}$ to impose m independent conditions on the linear system δ. This motivates the construction in the next section.

4   Selecting the Rational Points

The main idea of this construction is to obtain some information about the relative position of the points where we evaluate our sections of a selected line bundle. We achieve that by picking curves over the surface X and evaluating the sections on the rational points of X that belong to the selected curves.

With notation as before, consider a divisor $D = \sum_{j=1}^{m} C_j$ where C_j are prime divisors (irreducible curves) on X and an effective divisor G satisfying the following properties,

1. $C_j.H \geq \alpha$ for all j and for some integer $\alpha > 0$.
2. $N(C_j) \geq N$ for some $N > 0$.($N(C_j)$ the rational points of the curve C_j).
3. $G.C_j \leq \beta < N$ for all j and some integer $\beta > 0$.
4. $G.H > K.H$ where K is the canonical divisor of X.

We consider the line bundle $\mathcal{L}_G$ associated to the divisor G. As before we fix an open covering $\mathcal{U} = \cup \mathcal{U}_i$ and trivializations $\phi_i : \pi^{-1}\mathcal{U}_i \to \mathcal{U}_i \times \mathbf{F}_q$ of the line bundle $\mathcal{L}_G$. For simplicity we write $s(P_j)$ and we mean $(\phi_i \circ s)(P_j) \in \mathbf{F}_q$. We define the code $C_{\mathcal{L}}(D, G)$ as follows,

$$C_{\mathcal{L}}(D, G) := \left\{ \left(s(P_1^1), \ldots, s(P_{N(C_1)}^1), \ldots, s(P_i^j) \ldots, s(P_{N(C_m)}^m) \right) : s \in \mathcal{L}_G(S) \right\}$$

where P_i^j denotes a (say the ith) rational point of the curve C_j. Notice that we may evaluate our section at a rational point more than once, namely if that point is a point of intersection of two distinct curves.

Theorem 4. $C_{\mathcal{L}}(D, G)$ *is an* $[n, k, d]_q$ *code with* $n = \sum_{j=1}^{m} N(C_j)$, *the dimension of the code is* $k = (G^2 - G.K)/2 + p_a(X) + 1 + h^1(G)$ *and the minimum distance is* $d \geq (m - G.H/\alpha).(N - \beta)$.

Proof. The code $C_{\mathcal{L}}(D, G)$ is obviously linear. Getting the dimension of the code is an application of the Riemann-Roch theorem and Serre duality (assumption (4)). For the minimum distance, let us define the *component j* of the section $s \in \mathcal{L}_G(X)$ to be the evaluation of s at the rational points of the curve C_j; we denote it by C_j^s. We call C_j^s a *zero component* of s if $s(P_1^j) = \cdots = s(P_{N(C_j)}^j) = 0$. The following lemmas will establish the theorem (for the first one see [4]),

Lemma 3. *If C and D are curves on X having no common irreducible components, then $C.D = \sum_{P \in C \cap D} (C.D)_P$, where $(C.D)_P := dim_k(\mathcal{O}_{X,P}/(f, g)) < \infty$ for f, g local equations of C and D respectively at the point P.*

Lemma 4. *The component C_j^s of a section $s \in \mathcal{L}_G(X)$ is a zero component if and only if s vanishes identically on the irreducible curve C_j. In particular if s vanishes on more than $G.C_j$ points on C_j then C_j^s is a zero component.*

Proof. We prove the interesting direction. Let $(s)_0$ be the zero divisor of the section s. Notice that since C_j is irreducible the only way $(s)_0$ shares an irreducible component with C_j is if $C_j \subset supp((s)_0)$, i.e. the section s vanishes identically on C_j. Let us assume that C_j^s is a zero component but still s is not identically zero on C_j. By the above lemma we have that $(s)_0.C_j = \sum_{P \in (s)_0 \cap C_j}((s)_0.C_j)_P \geq N(C_j)$. But since $(s)_0 \sim G$ we have $(s)_0.C_j = G.C_j < N(C_j)$ by condition (3). Contradiction. The second statement follows similarly. $\qquad\square$

Let us denote by D^s the number of non-zero components of a section $0 \neq s \in \mathcal{L}_G(X)$.

Lemma 5. *For a section $0 \neq s \in \mathcal{L}_G(X)$ we have that $D^s \geq m - G.H/\alpha$.*

Proof. Let us write $(s)_0 = \sum_{j=1}^m n_j C_j + \sum m_i D_i$ with $n_j, m_i \geq 0$ and D_i some other prime divisors of $Pic\,X$ not in the support of D. The fact that $(s)_0 \sim G$ implies that $(s)_0.H = G.H$ or equivalently $\sum_{j=1}^m n_j C_j.H + \sum m_i D_i.H = G.H$. From the Nakai-Moishezon criterion of ampleness (see [4]) we have $C_j.H, D_i.H > 0$. So we obtain the inequality $\sum_{j=1}^m n_j C_j.H \leq G.H$. By condition (1) and the previous lemma we have that $(m - D^s)\alpha \leq G.H$ or that $D^s \geq m - G.H/\alpha$. $\quad\square$

Now we can conclude the theorem for the minimum distance d. Let $0 \neq s \in \mathcal{L}_G(X)$. If a component C_j^s is not a zero component then the number of rational points of C_j that s vanishes on cannot exceed the number $G.C_j \leq \beta$ since otherwise by Lemma 8 C_j^s would be a zero component. So for the minimum distance we have, $d \geq D^s \cdot (N - \beta)$ or $d \geq (m - G.H/\alpha)(N - \beta)$. $\qquad\square$

Example: Let us apply the above construction when $X = \mathbf{P}^2$. Let us denote by L a divisor corresponding to a line on $\mathbf{P}^2$. We pick $D = \sum_{j=1}^{q+1} L_j$ for any distinct $q + 1$ lines of the projective plane, and $G = L'$ for some line. Notice that any line is an ample divisor on $\mathbf{P}^2$ and since $Pic\,\mathbf{P}^2 = \mathbf{Z}$ we have for any two curves $D_1 = d_1 L$ and $D_2 = d_2 L$, $D_1.D_2 = d_1 d_2$. The code $C_\mathcal{L}(\mathbf{P}^2, G)$ has $n = (q+1)^2$, $k = \frac{1}{2}(1+3) + 1 = 2$ since $K = -3L$, $p_\alpha(\mathbf{P}^2) = 0$ and $H^1(\mathbf{P}^2, L) = 0$. For the minimum distance, $d \geq (q + 1 - 1)(q + 1 - 1) = q^2$.

5   Linear Codes over Ruled Surfaces

In this section we apply the previous construction to ruled surfaces. The knowledge of their intersection theory will help us to establish explicitly their parameters. We briefly recall the definition and the main properties of ruled surfaces. For all of them see [4]. A notation clarification: we write $D \equiv C$ to denote that the divisors D and C are numerically equivalent.

Definition 1. *A ruled surface, is a surface X together with a surjective morphism $\pi : X \to C$ to a nonsingular curve C, such that the fibre X_y is isomorphic to $\mathbf{P}^1$ for every point $y \in C$ and π admits a section.*

The main properties of ruled surfaces that we will use are (for the last two we consider a *normalized* ruled surface),

1. If $\pi : X \to C$ is a ruled surface, then there exists a locally free sheaf $\mathcal{E}$ of rank 2 on C such that $X \cong \mathbf{P}(\mathcal{E})$ over C. Conversely, every such $\mathbf{P}(\mathcal{E})$ is a ruled surface over C.
2. If we define $NumX := DivX/(\equiv)$ then, $NumX \cong \mathbf{Z} \oplus \mathbf{Z}$, generated by C_0, f and satisfying $C_0.f = 1$ and $f^2 = 0$, where f a general fibre and C_0 a hyperplane section.
3. If the genus of C is g, then $p_\alpha(X) = -g$.
4. There is a one-to-one correspondence between sections $\sigma : C \to X$ and surjections $\mathcal{E} \to \mathcal{L} \to 0$, where $\mathcal{L}$ is an invertible sheaf on C, given by $\mathcal{L} = \sigma^* \mathcal{O}_X(1)$.
5. If D is any section of X, corresponding to a surjection $\mathcal{E} \to \mathcal{L} \to 0$, and if $\mathcal{L} = \mathcal{L}(\mathcal{D})$ for some divisor $\mathcal{D}$ on C then $\delta := deg(\mathcal{D}) = C_0.D$ and, $D \equiv C_0 + (\delta + e)f$, where $e := -deg(\mathcal{E})$. In particular, we have $C_0^2 = -e$.
6. For the canonical divisor K we have, $K \equiv -2C_0 + (2g - 2 - e)f$.

Linear Codes over Ruled Surfaces with $e \geq 0$: Let us fix a nonsingular irreducible curve C defined over $\mathbf{F}_q$ and consider a locally free sheaf $\mathcal{E}$ of rank 2 over C that is normalized with $e \geq 0$. Let us consider a collection $\{D_i\}$ with $1 \leq i \leq m$ of divisors on C such that there are surjections $\mathcal{E} \to \mathcal{L}(D_i) \to 0$. Let δ_i be the degree of D_i and we set $\delta := \min_i \delta_i$ and $\Delta := \max_i \delta_i$.

Let us denote by C_i the corresponding divisor on $X = \mathbf{P}(\mathcal{E})$ of the surjection $\mathcal{E} \to \mathcal{L}(D_i) \to 0$. Let moreover N denote the number of rational points on C. We consider the divisor $G := aC_0 + bf$ with $a, b \in \mathbf{Z}$. In order to pick an ample divisor we need the following lemma, for a proof see [4].

Lemma 6. *Let X be a ruled surface over a curve C with invariant $e \geq 0$. Then a divisor $H \equiv aC_0 + bf$ is ample if and only if $a > 0$ and $b > ae$. In particular the divisor $H := C_0 + (e + 1)f$ is an ample divisor.*

We define the code $C_{\mathcal{L}}(D, G)$ over the ruled surface X with $D = \sum_i^m C_i$. We check the four conditions stated in the general construction.

1. *The intersection product $C_i.H$:* $C_i.H = (C_0 + (\delta_i + e)f).(C_0 + (e + 1)f) = \delta_i + e + 1$. In particular we have, $\delta + e + 1 \leq D_i.H \leq \Delta + e + 1$.
2. *The rational points on C_i:* Since C_i are isomorphic to C, they have $N(C_i) = N(C) = N$ rational points.
3. *The intersection product $G.C_i$:* $G.C_i = (aC_0 + bf).(C_0 + (\delta_i + e)f) = a\delta_i + b$.
4. *The inequality $G.H > K.H$:* $a + b > 2g - e - 4$

We conclude,

Theorem 5. *The code $C_{\mathcal{L}}(D, G)$ defined over the ruled surface X is an $[n, k, d]_q$ code with $n = mN$, $k = ab + a + b + 1 - \frac{1}{2}e(a^2 + a) - g(a + 1) + h^1(G)$ and $d \geq (m - \frac{a+b}{\delta+e+1})(N - (a\Delta + b))$*

Proof. The proof follows using the above computations and Theorem 6.   □

Example: Let $X = \mathbf{P}(\mathcal{E})$ be a rational ruled surface X (i.e. $C = \mathbf{P}^1$) with $\mathcal{E} = \mathcal{O}(\mathbf{P}^1) \oplus \mathcal{O}(\mathbf{P}^1)$, (this is the quadric $\mathbf{P}^1 \times \mathbf{P}^1$). It is known that to every point P_i of $\mathbf{P}^1$ there corresponds a section $D_i \sim C_0 + f$. We consider the collection of the rational divisors $\{P_i\}$ of $\mathbf{P}^1$. Let $\{C_i\}$ be the corresponding sections on X and let us pick $D = \sum_i^{q+1} C_i$ and $G = C_0 + f$. Then we get the linear code $C_{\mathcal{L}}(D, G)$ with $[(q+1)^2, 4, q(q-1)]_q$.

Linear Codes over Ruled Surfaces with $e < 0$: Considering the same setting as before we need to identify an ample divisor, for a proof see [4] ,

Lemma 7. *Let X be a ruled surface over a curve C of genus $g \geq 2$, with invariant $e < 0$ and assume that the characteristic of the base field is $p > 0$. If $a > 0$ and $b > a(\frac{1}{2}e + \frac{1}{p}(g-1))$, then any divisor $H \equiv aC_0 + bf$ is ample. In particular the divisor $H := C_0 + \frac{1}{p}(g-1)f$ is ample (wlog we assume p divides $(g-1)$).*

Doing the same calculations as for $e \geq 0$ we conclude,

Theorem 6. *The code $C_{\mathcal{L}}(D, G)$ defined over the ruled surface X is an $[n, k, d]_q$ code with $n = mN$, $k = ab + a + b + 1 - \frac{1}{2}e(a^2 + a) - g(a+1) + h^1(G)$ and $d \geq (m - \frac{a\frac{1}{p}(g-1)+b-ae}{\delta+\frac{1}{p}(g-1)})(N - (a\Delta + b))$.*

6   Asymptotically Good Codes over Ruled Surfaces

We investigate the conditions needed to get asymptotically good codes over ruled surfaces. We give a sufficient condition on properties of the base curve. We can prove that there are curves satisfying this condition but we defer this to an extended version of this paper. We consider the case $e < 0$, (due to lack of space we will not explain why we are not considering the case $e \geq 0$). We need two lemmas, proofs of which can be found in [4],

Lemma 8. *Let C be a curve of genus $g \geq 1$, and D be a divisor of degree $d = -e$ for some $e < 0$. Consider the non splitting short exact sequence, $0 \rightarrow \mathcal{O}_C \rightarrow \mathcal{E} \rightarrow \mathcal{L}(D) \rightarrow 0$, corresponding to the nonzero element $\xi \in H^1(C, \mathcal{L}(-D))$. For $-g \leq e < 0$, and any given D of degree $d = -e$ there always exists a $\xi \in Ext^1(D, \mathcal{O}_C) \cong H^1(C, \mathcal{L}(-D))$ such that $\mathcal{E}$ is normalized. Moreover the bounds on e are tight.*

Lemma 9. *Let us call a locally free sheaf of rank two $\mathcal{E}$ stable (respectively semistable) if for every quotient locally free $\mathcal{E} \rightarrow \mathcal{F} \rightarrow 0$ we have that, $deg(\mathcal{F}) > (deg(\mathcal{E}))/2$ ($\geq$ for semistable). Let us assume that $\mathcal{E}$ is normalized. Then $\mathcal{E}$ is stable (respectively semistable) if and only if $deg(\mathcal{E}) > 0$ (respectively $deg(\mathcal{E}) \geq 0$).*

Let us assume that for the curve C we have $\frac{N}{g} = 1 + \epsilon$ with $\epsilon > 0$. We set $a = 1$, $b = c_b g$, $e = -c_e g$ with $0 < c_e \leq 1$ and $\delta = c_\delta g$, $\Delta = c_\Delta g$. For the rate of the code we have, $k/mN \geq \frac{2c_b g - e - 2g}{mN} = \frac{2c_b + c_e - 2}{m} \cdot \frac{g}{N}$ and for the relative distance $d/mN \geq (1 - \frac{c_b + c_e + 1/p}{c_\delta + 1/p} \cdot \frac{1}{m})(1 - \frac{(c_\Delta + c_b)g}{N})$. So we have the conditions,

1. $\frac{c_b + c_e + 1/p}{c_\delta + 1/p} < m < c$ where $c > 0$ some constant.
2. $c_\Delta + c_b < 1 + \epsilon$
3. $2c_b + c_e > 2$
4. $c_\delta > \frac{c_e}{2}$, (from Lemma 16)
5. $c_b + 2c_e > 2 - \frac{3}{p}$ (from the condition $G.H > K.H$)

Let us write $c_b = 1 - \frac{c_e}{2} + \delta_1$ and $c_\delta = \frac{c_e}{2} + \delta_2$ with $\delta_1, \delta_2 > 0$, as we are forced from (3) and (4). From (2) we have that $\delta_1 + \delta_2 < \epsilon$ and from (5) we have $2 - \frac{6}{p} < 3c_e \leq 3$. Condition (1) implies, $m > (2 + c_e + 2\delta_1)/(c_e + 2\delta_2)$. We conclude,

Theorem 7. *Let $\{C_i\}$ be a family of irreducible curves with $\frac{N_i}{g_i} = 1 + \epsilon_i$ where $\epsilon_i > 0$. There exists a family of asymptotically good codes over surfaces $\{X_i\}$ if for every curve C_i there is a normalized locally free sheaf $\mathcal{E}_i$ of rank two with at least $m := (2 + c_{e_i} + 2(\epsilon_i - \delta_i))/(c_{e_i} + 2\delta_i)$ surjections $\mathcal{E}_i \to \mathcal{F}_{i_j} \to 0$, where $c_{e_i} := (deg(\mathcal{E}_i))/(g_i)$ such that 1) $0 < \delta_i < \epsilon_i$ and 2) $deg(\mathcal{F}_{i_j}) = c_{e_i} g_i/2 + \delta_i \cdot g_i$, for all $1 \leq j \leq m$*

Proof. Follows from above by taking $X_i := \mathbf{P}(\mathcal{E}_i)$.

Notice that if $\epsilon > 1$ then we can consider a single component in the surface X. For $0 < \epsilon < 1$ we need always at least 2 component. As $\epsilon \to 0$ the number of components increases. An interesting question is what if $\epsilon = 0$? Can we get asymptotically good codes over ruled surfaces coming from such curves. Notice that C cannot give asymptotically good AG codes and it is known that there is no operation on codes (direct sum or tensor) that can give an asymptotically good code from codes that are not asymptotically good.

7   Group Codes over Algebraic Surfaces

In this section we give another construction of error correcting codes over surfaces using sections of rank two bundles. The codes that we will define are not linear over the base field. However the codes maintain their group structure, namely the code C that we define is a subgroup of $\mathbf{F}_{q^2} \times \cdots \times \mathbf{F}_{q^2}$.

We are still investigating the parameters of these codes in the general setting but under some assumption we can fully determine them. It will become obvious why we believe that these codes are more naturally defined over surfaces. We note also that this construction generalizes to higher dimensions. For the needed background used here see [3].

With notation as before, we let $\mathcal{E}$ denote a locally free sheaf of rank 2. By $c_1(\mathcal{E})$ and $c_2(\mathcal{E})$ we denote the first and the second Chern classes of the sheaf $\mathcal{E}$.

For what follows we will assume that $\mathcal{E}$ is tensored with an invertible sheaf $\mathcal{L}^{\otimes n}$ with $n \gg 0$ and $\mathcal{L}$ ample so we have $H^i(X, \mathcal{E} \otimes \mathcal{L}^{\otimes n}) = 0$ for $i > 0$ (see [4]). By abuse of notation we will keep writting $\mathcal{E}$ for the sheaf $\mathcal{E} \otimes \mathcal{L}^{\otimes n}$.

Let us fix a covering $\mathcal{U} = \cup \mathcal{U}_i$ of the surface X and trivializations $\phi_i :$ $\pi^{-1}\mathcal{U}_i \to \mathcal{U}_i \times \mathbf{F}_q^2$ of the vector bundle (E, π) corresponding to the sheaf $\mathcal{E}$. For simplicity we will write just $s(P) \in \mathbf{F}_q^2$ for the vector $(\phi_i \circ s)(P)$ for $P \in \mathcal{U}_i$ and $s \in \mathcal{E}(X)$. We define, $C_{\mathcal{E}}(X) := \{(s(P_1), \ldots, s(P_n)) : s \in \mathcal{E}(X)\}$.

Theorem 8 (Riemann Roch). *With notation as before,* $\chi(\mathcal{E}) = (c_1(\mathcal{E})^2 - K.c_1(\mathcal{E}))/2 + 2(p_\alpha(X) + 1) - c_2(\mathcal{E})$.

Since $\chi(\mathcal{E}) = h^0(\mathcal{E}) - h^1(\mathcal{E}) + h^2(\mathcal{E})$ and by our assumption on $\mathcal{E}$ we have,

Lemma 10. *The code* $C_{\mathcal{E}}$ *is an* $(n, M, d)_{q^2}$ *code with* $\log_q M = (c_1(\mathcal{E})^2 - K.c_1(\mathcal{E}))/2 + 2(p_\alpha(X) + 1) - c_2(\mathcal{E})$.

Let us try to give some intuition about this construction. Consider a global section $0 \neq s \in \mathcal{E}(X)$ over some open neighborhood $U \subset X$ where $\mathcal{E}$ is trivial. Over U we can write s as $s = (s_1, s_2)$ where $s_1, s_2 \in \mathcal{O}_X(U)$. So the locus of s in the neighborhood U is the intersection of the curves defined by $s_1 = 0$ and $s_2 = 0$. If we assume that s_1 and s_2 do not have common components then the locus of s will be collection of points. Recall that a section s is called *regular* if $(s)_0$ has codimension 2, i.e collection of points (scheme theoretically).

The following lemma indicates that $c_2(\mathcal{E})$ is for $\mathcal{E}$ the analogue of the divisor for the invertible sheaves. If we use the notation $D_1 \sim D_2$ to indicate that the zero cycles D_1 and D_2 are *rational equivalent* then (for a proof see [3]),

Lemma 11. *If s is a regular global section of $\mathcal{E}$, then* $(s)_0 \sim c_2(\mathcal{E})$. *In particular* $deg((s)_0) = deg(c_2(\mathcal{E}))$.

Now let us assume that there exists a locally free sheaf of rank two with the property that *All nonzero global sections of $\mathcal{E}(X)$ are regular.* (Or a subspace of $\mathcal{E}(X)$ of high enough dimension). Then from the last lemma we have,

Lemma 12. *Let $\mathcal{E}$ be a locally free sheaf of rank two with the above property. Then for the minimum distance d of the code $C_{\mathcal{E}}(X)$ is* $d \geq n - c_2(\mathcal{E})$.

We do not know yet how to construct locally free sheaves $\mathcal{E}$ having this property. However in this direction we have constructed one which is the direct sum of two invertible sheaves i.e $\mathcal{E} = \mathcal{L}_1 \oplus \mathcal{L}_2$ such that for every nonzero section $s \in \mathcal{E}(X)$, written in the form $s = (s_1, s_2)$ with $0 \neq s_1 \in \mathcal{L}_1(X)$ and $0 \neq s_2 \in \mathcal{L}_2(X)$, is regular. Notice that this is not a trivial task. For example over $\mathbf{P}^2$ since $Pic\,\mathbf{P}^2 = \mathbf{Z}$ this is impossible. The construction is over ruled surfaces with $e \geq 0$. This indicates that the property that we demand is not unnatural; and as we said even a large enough subspace will do. For more details on that see [2].

We believe that these codes are more natural over surfaces. The similar way to compute the designed distance with the AG codes is one indication of that.

Since we collect information about our objects (functions or sections) on points and we are interested in their zero locus it is more natural to consider sections of rank two bundles. The zero locus of a typical section (of a rank two bundle) has codimension 2 (points) as opposed to the codimension 1 (curves) of functions.

8   Discussion

Another reason for studying error correcting codes over algebraic surfaces is to construct asymptotically good *locally decodable codes*.

A locally decodable code C is an error correcting code $[n, k, d]_q$ that has the following property: for any word x with $d(x, c) < \alpha d/2$ where $c \in C$ and $0 < \alpha < 1$, and index $i \in \{1, \ldots, n\}$, we can determine c_i with high probability by querying $\mathcal{O}(\log^t(n))$ many positions of x for some constant $t > 0$.

In [1], they give a construction of a locally decodable code. Their construction can be viewed as linear codes defined over $\mathbf{P}^n$ by evaluating homogeneous polynomials. The main idea is that any given position can be locally decoded by the following steps,

- consider "enough" lines (or more general curves) that pass through the corresponding point.
- consider the Reed Solomon (or more general AG) codes that are defined over the selected lines (curves).
- decode them and let the majority be the symbol under question.

However this construction cannot give a family of asymptotically good codes. It is an open question if there exist locally decodable codes that are asymptotically good. Our goal is to extend this construction from the projective space to other higher dimension varieties. In this direction we have obtained families of asymptotically good codes defined over surfaces.

Acknowledgements. The author is grateful to Emma Previato, Peter Gács and Drue Coles for many discussions.

References

1. L. Babai, L. Fortnow, L. Levin, and M. Szegedy. *Checking Computations in Poly-logarithmic Time.* In 23rd STOC, pages 21–31, 1991
2. Thanasis Bouganis, *Error Correcting Codes over Algebraic Surfaces.* MA Thesis, Department of Computer Science, Boston University 2002
3. William Fulton, *Intersection Theory*, Springer-Verlag, Berlin Heidelberg New York Tokyo, 1984
4. Robin Hartshorne, *Algebraic Geometry*, Graduate Texts in Math., Springer-Verlag, Berlin Heidelberg New York 1977
5. Gilles Lachaud, *Plane sections and codes on algebraic varieties*, Fourth International Conference on Arithmetic, Geometry and Coding Theory, CIRM, France 1996

6. J.H. van Lint, *Introduction to Coding Theory*, Graduate Texts in Math., Springer-Verlag, Berlin Heidelberg New York 1999
7. Carlos Moreno, *Algebraic Curves over Finite Fields*, Cambridge Tracts in Math., vol. 97 Cambridge University Press, Cambridge, 1991
8. M.A. Tsfasman and S.G. Vladut, *Algebraic-Geometric Codes*, Amsterdam, The Netherlands: Kluwer, 1991

A Geometric View of Decoding AG Codes

Thanasis Bouganis[1] and Drue Coles[2]

[1] Department of Pure Mathematics and Mathematical Statistics
University of Cambridge
ab442@hermes.cam.ac.uk
[2] Computer Science Department, Boston University
dcoles@cs.bu.edu

Abstract. We investigate the use of vector bundles over finite fields to obtain a geometric view of decoding algebraic-geometric codes. Building on ideas of Trygve Johnsen, who revealed a connection between the errors in a received word and certain vector bundles on the underlying curve, we give explicit constructions of the relevant geometric objects and efficient algorithms for some general computations needed in the constructions. The use of vector bundles to understand decoding as a geometric process is the first application of these objects to coding theory.

1 Introduction

We investigate the use of vector bundles over finite fields to obtain a geometric view of decoding algebraic-geometric (AG) codes. Such codes form a large and important class of linear codes obtained by evaluating rational functions with prescribed poles and zeros on an algebraic curve. Various general and efficient decoding algorithms have been developed since the discovery of these codes by Goppa [3] in the early 1980s. These algorithms (surveyed in [7]) work by solving systems of linear equations or performing other distinctly algebraic operations. The use of vector bundles to understand decoding of AG codes as a geometric process is the first application of these objects to coding theory.

We build on ideas of Trygve Johnsen, who in [8] observed that the syndrome of a received word can be identified with a rank two extension of two fixed line bundles on the curve, and the error locations can be obtained from the minimal quotient bundle of a rank two bundle corresponding to that extension.

We give explicit constructions of these extensions and efficient algorithms for some general computations needed in the constructions. In particular, we give an efficient (under general conditions) algorithm for computing an open cover of the curve that trivializes certain line bundles of interest. We also discuss how to determine functions that correspond via Serre duality to given points in projective space. Our aim is to specify in a natural way the geometric objects that a decoding algorithm based on Johnsen's insight would take as input.

Some simple and efficient decoding algorithms for AG codes correct up to $(d^* - 1 - g)/2$ errors, where d^* is the designed minimum distance and g is the genus of the curve (see Sect. 2 for definitions and a review of AG codes). More

M. Fossorier, T. Hoeholdt, and A. Poli (Eds.): AAECC 2003, LNCS 2643, pp. 180–190, 2003.
© Springer-Verlag Berlin Heidelberg 2003

complicated but still efficient decoders correct up to the designed error capacity $(d^* - 1)/2$, and sometimes beyond. The geometric approach described in this paper can correct up to the designed error capacity.

Outline of the Paper. We review the construction and basic facts about AG codes in Sect. 2. Vector bundles enter the picture in Sect. 3. A concrete geometric picture of rank two bundles on a curve was given in [9], and we state what is needed for our purposes and then describe the connection to decoding. This material was presented in [8], but we elaborate on a few points and identify more precisely the error divisor corresponding to the rank two bundles of Johnsen's theorem. Section 4 contains explicit constructions of the rank two extensions, and Sect. 5 concludes with an example of how general problems of coding theory may be translated into vector bundle language via the connection between syndromes of received words and rank two vector bundles.

2 Review of AG Codes and Preliminaries

If C is a smooth projective (absolutely irreducible) curve of genus g defined over a finite field $\mathbb{F}_q$, and $D = P_1 + \cdots + P_n$ a divisor consisting of $\mathbb{F}_q$-rational points P_i, then a divisor G with support disjoint from that of D determines the AG code

$$C(D, G) = \left\{ (f(P_1), \ldots, f(P_n)) : f \in \mathcal{L}(G) \right\}$$

Such codes are also called geometric Goppa codes.

We assume $n > \deg G > 2g - 2$, so the dimension $C(D, G)$ is $\deg G - g + 1$ as a consequence of the Riemann-Roch theorem. Any non-zero function $f \in \mathcal{L}(G)$ has at most $\deg G$ zeros, so the minimum distance is at least $d^* = n - \deg G$. The number d^* is called the *designed* distance.

It is a standard result that $C(D, G)^{\perp} = C(D, K + D - G)$ for a suitably chosen canonical divisor K. See [10] for a proof of this fact as well as a general introduction to AG codes, or [15] for a full account of AG codes.

We fix the following curve, divisors and parameters for the rest of this paper.

Notation. *Fix a smooth projective (absolutely irreducible) curve C of genus g defined over a finite field $\mathbb{F}_q$. Let $C(\mathbb{F}_q)$ denote the set of $\mathbb{F}_q$-rational points on C. Fix a divisor $D = P_1 + \cdots + P_n$ with $P_i \in C(\mathbb{F}_q)$ and a divisor G with support disjoint from that of D and with $\deg(D) > \deg(G) + 2 > 2g$. Let $H = D - G$. Let K be a canonical divisor such that $C(D, G)^{\perp} = C(D, K + H)$ and let k be the integer such that $\dim C(D, K + H) = l(K + H) = k + 1$.*

The rational map $\varphi : C \to \mathbb{P}^k$ determined by the linear system $K + H$ is an embedding since $\deg(K + H) > 2g$ (Corollary 4 in Chap. III, Sect. 5.6 of [12]). Note that $\varphi(P_i)$, viewed as an element of $\mathbb{F}_q^{k+1}$, is the i-th column of a generator matrix for $C(D, K + H)$ and thus the i-th column of a parity check matrix for $C(D, G)$. Therefore if a codeword in $C(D, G)$ is transmitted, we may view the syndrome of the received word as a linear combination of the points $\varphi(P_i)$. This observation motivates the following definition.

Definition 1. *An effective divisor with no repeated points will be called* simple. *We define the* span *of a simple divisor $A = Q_1 + \cdots + Q_\alpha$ to be the projective linear subspace of $\mathbb{P}^k$ spanned by $\{\varphi(Q_1), \ldots, \varphi(Q_\alpha)\}$. We will refer to a simple divisor with support in $C(\mathbb{F}_q)$ as an $\mathbb{F}_q$-simple divisor.*

Our definition of *Span A* is a special case of the definition in [9].

Lemma 1. *Let $x \in C(D, G)$ and let $y = x + e$ denote the received word. Suppose that e is non-zero precisely in positions $i_1, \ldots, i_\alpha$, and let $S(y)$ denote the syndrome of y. If $2\alpha < \deg H$ then $A = P_{i_1} + \cdots + P_{i_\alpha}$ is the unique divisor of degree at most α containing $S(y)$ in its span.*

Proof. We use the fact that if B is a simple divisor of degree $b < \deg H$ then $\dim Span(B) = b - 1$. This is an easy application of Riemann-Roch.

Clearly $S(y) \in Span(A)$. But suppose that A' is another divisor of degree $\alpha' \leq \alpha$ such that $S(y) \in Span(A')$. This implies a linear dependency among the points in the support of $B = A + A'$, which as noted in the previous paragraph is impossible since $\deg B < \deg H$. $\qquad\square$

Lemma 2. *For a simple divisor A of degree $\alpha < \deg H$, consider the projection $\pi : H^0(C, K + H)^* \to H^0(C, K + H - A)^*$ that maps $(c_0 : \cdots : c_j : \cdots : c_k)$ to $(c_0 : \cdots : c_j)$. We have $ker(\pi) = Span(A)$.*

Proof. Write $A = Q_1 + \cdots + Q_\alpha$. Each point $\varphi(Q_i)$ is zero in the first j coordinates, so clearly the span of such points is contained in the kernel of π. On the other hand, $\alpha = k - j$ by Riemann-Roch. This fact combined with Lemma 1 tells us that the points $\varphi(Q_i)$ form a basis for the particular $\mathbb{P}^{\alpha-1}$ consisting of points with zero in the first j coordinates. It follows that the kernel of π is contained in the Span of A. $\qquad\square$

3   Vector Bundles

We now explain how to identify the syndrome of a received word (actually *any* point of $\mathbb{P}^k$) with certain rank two vector bundles. We will use the same notation for a divisor B and the associated invertible sheaf or line bundle.

Let $Ext(H, \mathcal{O}_C)$ denote the extensions (considered up to isomorphism) of $\mathcal{O}_C$ by H; that is, short exact sequences of the form $0 \to \mathcal{O}_C \to E \to H \to 0$. We indicate such extensions by their middle term E, a rank two vector bundle. These extensions are identified with $H^0(C, K + H)^*$ as follows:

$$Ext(H, \mathcal{O}_C) \cong Ext(\mathcal{O}_C, -H) \qquad \text{(tensoring by -H)} \qquad (1)$$
$$\cong Ext^1(\mathcal{O}_C, -H) \qquad \text{(by Exercise III.6.1 in [6])} \qquad (2)$$
$$\cong H^0(C, K + H)^* \qquad \text{(by Serre duality)}. \qquad (3)$$

The projectivization of $H^0(C, K + H)^*$ is identified with $\mathbb{P}^k$ in a natural way: $\psi \mapsto (\psi(s_0) : \cdots : \psi(s_k))$, where $\{s_i\}$ is a basis for $H^0(C, K + H)$. Thus we identify $\mathbb{P}^k$ with $\mathbf{P}(Ext(H, \mathcal{O}_C))$, and this plays a crucial role in what follows.

Johnsen's idea, formalized in the next theorem, is to identify the syndrome of a received word corrupted in fewer than $(\deg H)/2$ places with the corresponding rank two extension $E \in Ext(H, \mathcal{O}_C)$. If A is the minimal quotient bundle of E, then we get the error locations from the support of A (with a caveat; see the remarks after the proof). The error values can then be computed by solving the linear system determined by the "error columns" of the parity check matrix, since the syndrome of the received word is a linear combination of these columns.

Theorem 1. *If $E \in Ext(H, \mathcal{O}_C)$ is a non-split extension corresponding to the syndrome of a received word with errors in positions $\{i_1, \ldots, i_\alpha\}$ and $2\alpha < \deg H$, then the unique minimal quotient bundle of E corresponds to the "error divisor" $A = P_{i_1} + \cdots + P_{i_\alpha}$.*

Proof. We assemble ideas already present in Johnsen's argument. Consider the map $\pi' : Ext(H, \mathcal{O}_C) \to Ext(H - A, \mathcal{O}_C)$ that sends the bottom row of the following diagram to the top row:

$$
\begin{array}{ccccccccc}
0 & \longrightarrow & \mathcal{O}_C & \longrightarrow & E \times_H (H - A) & \longrightarrow & H - A & \longrightarrow & 0 \\
& & \| & & \downarrow & & \downarrow & & \\
0 & \longrightarrow & \mathcal{O}_C & \longrightarrow & E & \longrightarrow & H & \longrightarrow & 0
\end{array}
\tag{4}
$$

It is a fact (Lemma 2.3 in [9]) that the following diagram commutes:

$$
\begin{array}{ccc}
Ext(H, \mathcal{O}_C) & \xrightarrow{\ \pi'\ } & Ext(H - A, \mathcal{O}_C) \\
\downarrow & & \downarrow \\
H^0(C, K + H)^* & \xrightarrow{\ \pi\ } & H^0(C, K + H - A)^*
\end{array}
\tag{5}
$$

Using this fact and Lemma 2 we have

$$
E \in Span\, A \leftrightarrow E \in ker(\pi) \tag{6}
$$
$$
\leftrightarrow E \in ker(\pi') \ . \tag{7}
$$

The zero element of a group of extensions is the split exact sequence, and it is easily verified that $\pi'(E)$ splits if and only if $H - A$ is a subbundle of E, which is to say that A is a quotient bundle of E. These observations combined with Lemma 1 prove that A is the unique minimal quotient bundle of E. □

The theorem as originally formulated in [8] claims that the process of error location can be described as finding the unique effective divisor A of degree at most α such that the corresponding line bundle is a quotient bundle of E. We observe here, however, that the bundle (= divisor class) is unique, not the divisor. We show concretely in the next section how different equivalent divisors can be extracted from a representation of the associated line bundle.

Of course, one could easily check which of two equivalent divisors is in fact an error divisor for the received word: just "correct" the received word in both cases and see which leads to a true codeword. The practicality of this solution depends on the number of equivalent $\mathbb{F}_q$-simple divisors. This number may be large. Families of curves attaining the Drinfeld-Vlăduţ bound, for example, have divisor classes for which the number of $\mathbb{F}_q$-simple representatives is exponential in the number of rational points (implicit in the proof of Theorem 2 in [1]).

The original version of Johnsen's theorem also states that if E corresponds to the syndrome of a received word with fewer than $(\deg H)/2$ errors, then E is an unstable extension. We omitted this fact from Theorem 1 because we do not discuss the issue of stability and its role in Geometric Invariant Theory. We refer instead to Remark 3.3 of Johnsen's paper [8] for a discussion of these ideas.

4   Explicit Constructions

The syndrome of a received word y corrupted in at most $(\deg H)/2$ positions has been identified with a rank two vector bundle $E \to C$ in such a way that the error divisor can be obtained from the minimal quotient bundle of E. We now show how to construct a representation of E given y.

4.1   Representing Rank Two Bundles

The local trivialization of a vector bundle is not unique, but in order to perform computations a suitable open cover U_i must somehow be chosen. See the discussion below on this point. Once a choice has been made, the rank two extension $E \in Ext(H, \mathcal{O}_C)$ can be represented as a system $E_{i,j}$ of 2×2 transition matrices of the form

$$E_{ij} = \begin{pmatrix} 1 & 0 \\ f_{ij} & h_{ij} \end{pmatrix} \tag{8}$$

with entries in $\mathcal{O}_C(U_i \cap U_j)$ and the usual pasting conditions satisfied (see Chap. VI, Sect. 1.2 of [12]). The functions $h_{i,j}$ are transition functions for H with respect to the chosen cover. The particular element of $Ext(H, \mathcal{O}_C)$ is determined by the $f_{i,j}$ entries. Actually, we have identified syndromes (as points in projective space) with the *projectivization* of $Ext(H, \mathcal{O}_C)$, so we should really say that the particular element of $Ext(H, \mathcal{O}_C)$ is determined by a *class* of functions $f_{i,j}$. We define the equivalence relation concretely and discuss the issue of computing the $f_{i,j}$ entries in Sect. 4.3.

Let π denote the projection $E \to C$. We can change the isomorphisms $\pi^{-1}(U_i) \cong U_i \times \overline{\mathbb{F}}_q$ to obtain a different representation of the same bundle by setting $E'_{i,j} = M_i E_{i,j} M_j^{-1}$, where the 2×2 matrices M_i have entries in $\mathcal{O}_C(U_i)$ and an inverse of the same form. Therefore, saying that A is a quotient bundle of E means that it can be expressed by transition functions $g_{i,j}$ with respect to the chosen cover and that for each U_i there is a 2×2 matrix M_i with entries in $\mathcal{O}_C(U_i)$ and M_i^{-1} of the same form, such that

$$M_i \begin{pmatrix} 1 & 0 \\ f_{i,j} & h_{i,j} \end{pmatrix} M_j^{-1} = \begin{pmatrix} * & 0 \\ * & g_{i,j} \end{pmatrix} . \tag{9}$$

Notice that we can extract an effective divisor from the transition functions $g_{i,j}$, but there is no unique or canonical way to do this. For example, suppose that C is covered by two open sets U_α and U_β. The transition function $g_{\alpha,\beta}$ defines a line bundle and we want to find local equations $g_\alpha \in \mathcal{O}_C(U_\alpha)$ and $g_\beta \in \mathcal{O}_C(U_\beta)$ for the corresponding divisor. One easy way to do this is to take $g_\alpha = 1$ and $g_\beta = 1/g_{\alpha,\beta}$, in which case we get the divisor by finding the zeros of $g_\beta|_{U_\beta}$. But we could just as well have taken $g_\beta = 1$ and $g_\alpha = g_{\alpha,\beta}$, and in this case we get a different (but equivalent) divisor. Both of these divisors may happen to be $\mathbb{F}_q$-simple, as we remarked at the end of the previous section, even though at most one of them is the actual error divisor.

4.2   Constructing a Suitable Cover

We call a cover *good* for a divisor if it trivializes the corresponding line bundle. This property is described more concretely by saying that there exists an open cover U_i of the curve and local equations f_i for the divisor restricted to U_i. The system (U_i, f_i) is called a *locally principle* divisor and is essentially a line bundle. This connection is described in Chap. VI, Sect. 1.4 of [12].

We need to determine a cover that is good for $H = D - G$ as well as for an unknown error divisor A. The next example shows that not every cover is good for a given divisor. (We remark, however, that any line bundle on a non-compact Riemann surface has an analytic trivialization; see, for example, Theorem 30.3 in [2].)

Example 1. Let $U = C \setminus Q$ for some point $Q \in C$. The line bundle P for any point $P \in U$ is not trivial over U unless $C \cong \mathbb{P}^1$. Indeed, if P is trivial over U then there is a local equation $f \in \mathcal{O}_C(U)$ for P in U. The morphism $f : U \to \mathbb{P}^1$ can be uniquely extended to a morphism $\tilde{f} : C \to \mathbb{P}^1$ by Proposition 6.8 in Chap. 1 of [6]. Now viewing $\tilde{f}$ as a function on C, we have $div(\tilde{f}) = P - Q$, which implies $P \sim Q$, and this is possible only when $C \cong \mathbb{P}^1$.

We assume that G has $\mathbb{F}_q$-rational support. Actually, G is usually taken to be a multiple of a single $\mathbb{F}_q$-rational point because the generally hard problem of computing a basis for $\mathcal{L}(G)$ may be easy in this case, and as far as the dimension and designed distance of the code are concerned, all that matters is the *degree* of G.

We will prove in Theorem 2 that for any set S of points on C there exists a cover by two open sets that is good for any divisor with support in S, and that if $S \subset C(\mathbb{F}_q)$ then a representation of this cover can be efficiently computed in many cases of interest. (We define precisely what we mean by "representation" immediately prior to stating the theorm.) In particular, if we take $S = C(\mathbb{F}_q)$ then we get a cover good for $H = D - G$ as well as for any error divisor A. The computational details of the proof are worked out in Lemmas 3 and 4.

We assume that our curve $C \subset \mathbb{P}^m$ is given explicitly as the zero locus of an ideal $F = (F_1, \dots F_l)$, where $F_i \in \mathbb{F}_q[x_0, \dots, x_m]$. Let $wt(F_i)$ denote the number of non-zero coefficients in F_i. We define $size(F) = \sum_{i=1}^{l} \deg(F_i) \cdot wt(F_i)$. We assume unit cost to perform an arithmetic operation in $\mathbb{F}_q$.

Lemma 3. *Let $P \in C(\mathbb{F}_q)$. A local parameter at P can be computed in time $\mathcal{O}(l^2 m + size(F))$.*

Proof. Write $P = (p_0 : \cdots : p_m)$ with $p_0 = 1$, and consider the affine algebraic curve defined by the polynomials $\hat{F}_i = F_i(1, x_1, \ldots, x_m)$. The tangent line $T_{\hat{P}}$ at $\hat{P} = (p_1 : \cdots : p_m)$ is the set of points $(x_1 : \cdots : x_m)$ such that

$$
\begin{bmatrix}
\frac{\partial \hat{F}_1}{\partial x_1}\big|_{\hat{P}} & \cdots & \frac{\partial \hat{F}_1}{\partial x_m}\big|_{\hat{P}} \\
\vdots & & \vdots \\
\frac{\partial \hat{F}_l}{\partial x_1}\big|_{\hat{P}} & \cdots & \frac{\partial \hat{F}_l}{\partial x_m}\big|_{\hat{P}}
\end{bmatrix}
\begin{bmatrix}
x_1 - p_1 \\
\vdots \\
x_m - p_m
\end{bmatrix}
= 0 .
\tag{10}
$$

The left matrix can clearly be computed in time $\mathcal{O}(size(F))$. An $\mathbb{F}_q$-rational point $(q_1 : \cdots : q_m) \neq \hat{P}$ on $T_{\hat{P}}$ can be found in $\mathcal{O}(l^2 m)$ steps by Gaussian elimination; choose a coordinate j for which $p_j \neq q_j$. Then $x_j - p_j$ vanishes at $\hat{P}$ but is not zero on $T_{\hat{P}}$, and thus is a local parameter at $\hat{P}$. It follows that the function $f(x_0, \ldots, x_m) = \frac{x_j - p_j x_0}{x_0}$ as a rational function on C is a local parameter at P. $\square$

Lemma 4. *There exists for any set $S = \{P_0, P_1, \ldots, P_n\}$ of points on C a local parameter π at P_0 that has no poles or zeros in $S \setminus P_0$. If $S \subseteq C(\mathbb{F}_q)$ and $S \subset C \setminus Y$ for a hyperplane Y given by a linear equation with coefficients in $\mathbb{F}_q$, then an explicit representation of such a function π can be computed in time $\mathcal{O}(mn + l^2 m + size(F))$.*

Proof. The first claim follows from Theorem 1 in Chap. III, Sect. 1.3 of [12], which shows that for every divisor on a smooth variety X and every finite set S of points on X, there exists an equivalent divisor with support disjoint from S. We construct π as in the proof of that theorem and use the same argument to show that it is a local parameter at P_0, but we specify some details more concretely in order to reason about computational complexity.

We may assume after a linear change of coordinates that Y is given by $x_0 = 0$. Let $U = C \setminus Y$. We will specify functions $h_i \in \mathcal{O}_C(U)$ for $1 \leq i \leq n$ such that $h_i(P_j) = 0$ if and only if $i \neq j$, but first we show how to use the functions h_i to determine π.

Compute a local parameter $\pi' \in \mathcal{O}_C(U)$ at P_0 and constants $\alpha_i = \frac{1 - \pi'(P_i)}{h_i(P_i)^2}$, and define

$$
\pi = \pi' + \sum_{i=1}^{n} \alpha_i h_i^2 .
\tag{11}
$$

Now $\pi \in \mathcal{O}_C(U)$ and $\pi(P_i) = 1$ for $1 \leq i \leq n$, so it remains only to show that π is a local parameter at P_0. Indeed, $h_i(P_0) = 0$ for $1 \leq i \leq n$, so π' divides h_i in the local ring $\mathcal{O}_{P_0}$ and therefore $\sum_{i=1}^{n} \alpha_i h_i^2 = (\pi')^2 h$ for some $h \in \mathcal{O}_{P_0}$. Write $\pi = \pi'(1 + \pi' h)$ and note that $(1 + \pi' h)(P_0) = 1$, from which it follows that π is a local parameter at P_0.

We now specify the functions $h_i \in \mathcal{O}(U)$. Express the points $P_i \in U$ in affine coordinates by writing $P_i = (p_{i,1} : \cdots : p_{i,m})$. Now for each point P_i and each affine coordinate x_k, define the set $Q_{i,k} = \mathbb{F}_q \setminus p_{i,k}$. Finally, define $g_i \in \mathbb{F}_q[C]$ for $1 \leq i \leq n$ by

$$g_i(x_1, \ldots, x_m) = \prod_{k=1}^{m} \left(\prod_{c \in Q_{i,k}} (x_k - c) \right). \tag{12}$$

By this construction we have $g_i(P_j) = 0$ if and only if $j \neq i$. Let $\tilde{g}_i$ denote the homogenization of g_i and take $h_i = \tilde{g}_i / x_0^a$, where $a = m(q-1) = \deg g_i$.

We view q as a constant, so each $h_i(P_i)$ can be computed in $\mathcal{O}(m)$ steps. Note that π' as specified in Lemma 3 can be evaluated at any point in constant time. The total time to compute all constants α_i is therefore $\mathcal{O}(mn)$, and to this quantity we add the time to compute π' in Lemma 3 to complete the proof. We have specified π as a function f / x_0^a, where f is a sum of $n + 1$ products having $m(q-1)$ terms of the form $(x_k - c)$. □

Let $V(I)$ denote the vanishing set, or zero locus, of the homogeneous ideal I of polynomials in $m + 1$ variables. An open set $U \subseteq \mathbb{P}^m$ is given concretely by specifying generators for ideals I_0 and I_1 satisfying $U = V(I_0) \setminus V(I_1)$. We define a representation of an open cover of a projective variety to be a collection of ideals representing the sets comprising the cover.

Theorem 2. *There exists for any set $S = \{P_0, \ldots, P_n\}$ of points on C a cover by two open sets that is good for any divisor with support in S. If $S \subseteq C(\mathbb{F}_q)$ and $S \subset C \setminus Y$ for a hyperplane Y given by a linear equation with coefficients in $\mathbb{F}_q$, then a representation of this cover can be computed in time $\mathcal{O}(mn^2 + l^2mn + n \cdot \text{size}(F))$.*

Proof. Let π_i be a local parameter at P_i for $0 \leq i \leq n$, and let J_i be the divisor such that $div(\pi_i) = P_i + J_i$. If each π_i is constructed as in Lemma 4, then the support of J_i is disjoint from S. Cover C with the following two open sets: $U_0 = C \setminus Supp(J_0 + \cdots + J_n)$ and $U_1 = C \setminus S$. This cover is good for $c_0 P_0 + \cdots + c_n P_n$ since it has a local equation $\pi_0^{c_0} \cdots \pi_n^{c_n}$ in U_0, and 1 in U_1.

Let I denote the ideal generating S and recall that $C = V(F_1, \ldots, F_l)$. The function $\pi_0 \cdots \pi_n$ is obtained as a ratio f / x_0^r of polynomials by $n + 1$ applications of Lemma 4 (establishing the computational complexity claim), and we have:

$$U_0 = V(F_1, \ldots, F_l, x_0 \cdot f) \setminus V(I) \tag{13}$$
$$U_1 = V(F_1, \ldots, F_l) \setminus V(I). \tag{14}$$

The ideal I is generated by linear polynomials depending on the coordinates of the points P_i. □

We apply Theorem 2 by taking $S = Supp(H) \subseteq C(\mathbb{F}_q)$. The requirement that the $\mathbb{F}_q$-rational points be contained in the complement of a single hyperplane can be relaxed in a number of ways (with easy adjustments to the proof). For example, if C is a plane curve then it suffices to require that it not contain the point $[0 : 0 : 1]$, or at least that this point is not used to build the code.

4.3   Computing Serre Duality

We now discuss the issue of computing the lower-left entries of the 2×2 transition matrices $E_{i,j}$ representing $E \in Ext(H, \mathcal{O}_C)$. The Čech cohomology agrees with the cohomology of coherent sheaves in our setting, so it suffices for us to compute the perfect pairing $\check{H}^0(\mathcal{U}, K+H) \times \check{H}^1(\mathcal{U}, -H) \to \mathbb{F}_q$ using the open two-covering $\mathcal{U}$ of C computed in the previous section.

  If we are able to compute this pairing then we are done, since the identification $\check{H}^1(\mathcal{U}, -H) \to Ext^1(\mathcal{O}_C, -H)$ is essentially the identity map, and we plug a function from the latter group into a 2×2 transition matrix to get an element of $Ext(\mathcal{O}_C, -H)$. Then multiplying each entry by a transition function for H gives an element of $Ext(H, \mathcal{O}_C)$. The task of computing a transition function for H was handled in Theorem 2.

  One way to determine an explicit form of the pairing is to identify $\check{H}^1(\mathcal{U}, -H)$ with a class of repartitions (adeles) and use the perfect pairing between differentials ω and repartitions r given by Serre [13] in the proof of the Duality theorem, namely $(\omega, r) \mapsto \sum_{P \in C} Res_P(r_P \omega)$. Adeles over a non-algebraically closed field are considered in [14].

  Another approach is to take the mapping used by Griffiths [4] in his proof of the Riemann-Roch theorem. Let $K = div(\omega)$ and recall that we write $\{P_0, \ldots, P_n\}$ for the support of H. We define the pairing $\check{H}^0(\mathcal{U}, K + H) \times \check{H}^1(\mathcal{U}, -H) \to \mathbb{F}_q$ as follows:

$$(f, g) \mapsto \sum_{i=0}^{n} Res_{P_i}(fg\omega) \ . \tag{15}$$

  This gives us a concrete way to determine the lower-left entry of the transition matrix for the extension corresponding to the syndrome $S(y)$ of a received word. Let $S(y)_i$ denote the i-th coordinate of the syndrome and fix a basis $\{f_i\}_{i=0}^k$ for $\check{H}^0(\mathcal{U}, K + H)$. We are looking for a function $g \in \check{H}^1(\mathcal{U}, -H)$ that satisfies $(f_i, g) \mapsto S(y)_i$ for $0 \le i \le k$. Given a basis for the two cohomology groups, this task reduces to solving a linear system of equations.

  As a by-product of these observations, let us state what it means to projectivize $Ext(H, \mathcal{O}_C)$ in terms of the transition matrices for the corresponding rank two vector bundles. We want to identify non-isomorphic extensions of $\mathcal{O}_C$ by H determined by the transition matrices

$$\begin{pmatrix} 1 & 0 \\ g & h \end{pmatrix} \quad \text{and} \quad \begin{pmatrix} 1 & 0 \\ \tilde{g} & h \end{pmatrix} \tag{16}$$

if they correspond to the same point in projective space. Thus in view of (15) and the subsequent paragraph, projectivizing $Ext(H, \mathcal{O}_C)$ amounts to identifying g and $\tilde{g}$ whenever we have $(f_i \times g) = \gamma(f_i \times \tilde{g})$ for $0 \le i \le k$ and some non-zero constant $\gamma \in \overline{\mathbb{F}}_q$.

5 An Application

We conclude with an example of how general problems of coding theory may be translated into vector bundle language via the machinery of Sect. 3. Consider the following open question:

> *For $0 < \delta < 1/2$, is there a family of linear codes of length n and relative distance δ with an exponential (in terms of n) number of codewords contained in some Hamming ball of relative radius less than δ?*

The question has obvious implications for bounds on the list decoding radius of a linear code. Recall that the list decoding problem is to generate in polynomial time a list of all codewords within a certain distance of a given word. The question has been answered affirmatively for random non-linear codes, but the linear case is evidently quite difficult. See [5] for related questions and results.

We are going to give a sufficient condition in vector bundle language for an AG code to have exponentially many codewords on the boundary of some ball of relative radius less than δ. Note that AG codes were constructed in [1] having an exponential number of minimum weight words, which is to say that the ball of relative radius exactly δ centered at zero (or by linearity at any other codeword) contains exponentially many codewords on its boundary.

The assumption in Theorem 1 that $2\alpha < \deg H$ guaranteed that there is a unique closest codeword and a unique minimal quotient bundle corresponding to the error divisor. If we demand only that $\alpha < \deg H$ then there may be *many* error vectors of minimum weight α, each corresponding to an error divisor. The open question is answered affirmatively if there is a family of AG codes $C(D, G)$ and rank two extensions $E \in Ext(D - G, \mathcal{O}_C)$ with exponentially many minimal quotient bundles corresponding to $\mathbb{F}_q$-simple divisors.

Let $M(E)$ denote the set of maximal subbundles (or equivalently, minimal quotient bundles) of E. It is known that $M(E)$ is a closed subset of $Pic^m(C)$, where m is the degree of E, and that $\dim M(E) = 1$ when the s-invariant of E equals the genus of the curve. Moreover, two distinct minimal quotient bundles cannot be isomorphic except in special cases (Lemma 1.5 in [11]). We want to count the points of $M(E)$ that correspond to $\mathbb{F}_q$-simple divisors. One question to investigate is whether these divisors correspond to the rational points of $M(E)$, in which case we are interested in curves C with many rational points (so that they are good for coding) and rank two bundles $E \in Ext(D - G, \mathcal{O}_C)$ such that $M(E)$ also has many rational points.

Acknowledgements. We are grateful to Emma Previato for encouragement and expert advice about the content and presentation of this material. We also thank Trygve Johnsen and Angela Vierling for helpful remarks, and the anonymous referees for their suggestions.

References

1. Alexei Ashikhmin, Alexander Barg, and Serge Vlăduţ. *Linear codes with many light vectors*, Journ. Combin. Theory Ser. A 96, no. 2, 396–399, 2001.
2. Otto Forster. *Lectures on Riemann surfaces*, Springer-Verlag, New York, 1981. Translated by Bruce Gilligan.
3. V.D. Goppa. *Codes on algebraic curves*, Soviet Math. Dokl., vol. 24, 170–172, 1981.
4. P.A. Griffiths. *Introduction to algebraic curves*, Translations of Math. Monographs 76, Amer. Math. Soc., 1989.
5. Venkatesan Guruswami. *Limits to List Decodability of Linear Codes*, 34th Annual ACM Symposium on Theory of Computing, 802–811, 2002.
6. Robin Hartshorne. *Algebraic Geometry*, Springer-Verlag, New York, 1977.
7. Tom Høholdt and Ruud Pellikaan. *On the decoding of algebraic-geometric codes*, IEEE Trans. Inform. Theory, vol. 41, 1589–1614, 1995.
8. Trygve Johnsen. *Rank two bundles on Algebraic Curves and decoding of Goppa Codes*, electronic preprint alg-geom/9608018, downloadable at http://arxiv.org/list/alg-geom/9608. To appear in International Journal of Pure and Applied Mathematics.
9. H. Lange and M.S. Narasimhan. *Maximal Subbundles of Rank Two Vector Bundles on Curves*, Math. Ann., vol. 266, 55–72, 1983.
10. J.H. van Lint. *Introduction to Coding Theory*, 3rd edition, Springer-Verlag, 1999.
11. Masaki Maruyama. *On Classification of Ruled Surfaces*, Lectures in Mathematics, Kinokuniya Book Store Co., Ltd. 1970.
12. Igor R. Shararevich. *Basic Algebraic Geometry 1-2*, 2nd edition, Springer-Verlag, New York, 1994. Translated by Miles Reid.
13. Jean-Pierre Serre. *Algebraic Groups and Class Fields*, Springer-Verlag, New York, 1988.
14. Henning Stichtenoth. *Algebraic Function Fields and Codes*, Universitext, Springer-Verlag, Berlin, 1993.
15. M.A. Tsfasman and S.G. Vlăduţ. *Algebraic-geometric codes*, Mathematics and its Applications, vol. 58, Kluwer Acad. Publ., Dordrecht, 1991.

Performance Analysis of M-PSK Signal Constellations in Riemannian Varieties

Rodrigo Gusmão Cavalcante and Reginaldo Palazzo Jr.

Departamento de Telematica, FEEC-UNICAMP, Campinas, Brazil
{rgc,palazzo}@dt.fee.unicamp.br

Abstract. In this paper we consider the performance analysis of a digital communication system under the hypothesis that the signal constellations are on Riemannian manifolds. As a consequence of this new approach, it was necessary to extend the concepts related to signal constellations, symbol error probability and average energy of the signal constellation to the theory of differentiable manifolds. The important result coming out of this formalism is that the sectional curvature of the variety is a relevant parameter in the design and in the performance analysis of signal constellations.

1 Introduction

The main objectives to achieve when designing a new communication system are that the new system be less complex and have a better performance under the error probability criterion. Equivalently, for a fixed error rate, the signal-to-noise ratio be less than that of the known systems.

In this direction, we consider each one of the block diagrams, Fig. 1, as a set of points E_i together with a metric d_i. This allows an interpretation of each one of the blocks as a metric space (E_i, d_i). As an example, the source encoder consists of a set of codewords E_1 with an associated distance d_1, where this distance may be the chi-squared distance. The channel encoder has associated a set of codewords E_2, in general, with the Hamming distance d_2. The channel consists of a set of random points with the Euclidean distance (when the additive white gaussian noise is taken into consideration). In general, each one of the decoders make use of the same metric as the corresponding encoders with the purpose of matching the corresponding metric spaces.

Therefore, the objective of the metric space point of view is to determine the geometric and algebraic characteristics associated to each one of the set of points E_i as well as the properties and conditions that must be satisfied by the transformations that will connect the distinct metric spaces such that a better performance be achieved by the new communication system.

In general, the metric space (E_3, d_3), consisting of the blocks modulator, channel, and demodulator, is associated to the Euclidean space with its usual metric. This is the reason the Euclidean geometry is used in the design of the signal constellations in the modulator in order to minimize the noise action which in general is the gaussian noise.

M. Fossorier, T. Hoeholdt, and A. Poli (Eds.): AAECC 2003, LNCS 2643, pp. 191–203, 2003.

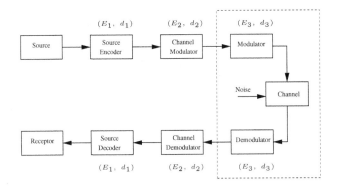

Fig. 1. Communication system model

On the other hand, the dashed block associated with (E_3, d_3) may be characterized as a discrete memoryless channel. Under this characterization, the main motivation of this paper comes from the results shown in [6], where it is identified and analyzed a substantial number of discrete memoryless channels and their embedding on compact surfaces with and without borders. Still in [6] it is considered some channels, for instance the $C_8[3]$ channel with 8 input, 8 output and 3 transitions per input, having the property that its embedding occurs in a minimal surface (catenoid). To the best of our knowledge, minimal surfaces were not considered previously in the literature. Since there are real channels meeting the previous condition, this motivates us to design signal constellations and to realize the performance analysis of the corresponding communication system.

Based on these facts, naturally comes the following question: What is the performance of a communication system when the geometric structure associated with a signal constellation is not the Euclidean one? A partial answer to this question can be found in [1] and [2], for the particular case of signal constellations in hyperbolic spaces. In [8] it was considered the case when the signal constellations are Riemannian varieties, however focusing on signal constellations on minimal surfaces.

In this paper, we focus on the analysis of the metric space consisting of the three blocks, namely: the modulator, channel and demodulator as shown in Fig. 1, aiming at the construction of signal constellations in Riemannian varieties and the corresponding performance analysis. In order to achieve this goal, we have to extend the concepts of error probability, average energy of the signal constellation and of the noise to the context of Riemannian varieties.

2   Review of Riemannian Geometry

To the best of our knowledge, the problem of designing signal constellations in Riemannian variety has not been considered previously in the literature. Thus, in order to do that we review some important concepts needed to what follows.

Definition 1. *A differentiable variety of dimension n consists of a set M and a family of bijective mappings $\mathbf{x}_\alpha : U_\alpha \subset \mathbb{R}^n \to M$ of open sets U_α of $\mathbb{R}^n$ in M such that*

1. *$\cup_\alpha \mathbf{x}_\alpha(U_\alpha) = M$.*
2. *For every pair α and β, with $\mathbf{x}_\alpha(U_\alpha) \cap \mathbf{x}_\beta(U_\beta) = W \neq \phi$, the sets $\mathbf{x}_\alpha^{-1}(W)$ and $\mathbf{x}_\beta^{-1}(W)$ are open sets in $\mathbb{R}^n$ and the mappings $\mathbf{x}_\beta^{-1} \circ \mathbf{x}_\alpha$ are differentiable.*
3. *The family $\{(U_\alpha, \mathbf{x}_\alpha)\}$ is maximal with respect to (1) e (2).*

In order to deal with certain quantities such as length, area, angles, and so on, with respect to differentiable varieties we have to establish the concept of a Riemannian metric.

Definition 2. *A Riemannian metric in a differentiable variety M is a correspondence associating to each point p of M a dot product $\langle \, , \, \rangle_p$ (that is, a positive definite, symmetric, bilinear form) in the tangent space T_pM, varying differentiably in the following sense: If $\mathbf{x} : U \subset \mathbb{R}^n \to M$ is a system of local coordinates at p, with $\mathbf{x}(\boldsymbol{x}_1, \boldsymbol{x}_2, \ldots, \boldsymbol{x}_n) = q \in \mathbf{x}(U)$, then $\langle \frac{\partial}{\partial \boldsymbol{x}_i}(q), \frac{\partial}{\partial \boldsymbol{x}_j}(q) \rangle_q = g_{ij}(\boldsymbol{x}_1, \ldots, \boldsymbol{x}_n)$ is a differentiable function in U.*

A differentiable variety with a given Riemannian metric is called a Riemannian variety. We may use the Riemannian metric to determine the length of a curve $c : I \subset \mathbb{R} \to M$ constrained to the closed interval $[a, b] \subset I$ as follows

$$l_a^b(c(t)) = \int_a^b \left\langle \frac{dc}{dt}, \frac{dc}{dt} \right\rangle^{1/2} dt \, , \tag{1}$$

where $\frac{dc(t)}{dt}$ denotes a vector field originating from $c(t)$, and called the tangent field of $c(t)$.

At this point it is worthwhile introducing a class of curves in M, called geodesics, having the property of minimizing the length of a segment joining two closest points in M.

A parameterized curve $\gamma : I \to M$ is a geodesic $\gamma(t) = (\boldsymbol{x}_1(t), \ldots, \boldsymbol{x}_n(t))$ in a system of coordinates $(U, \mathbf{x})$ if and only if it satisfies the following second order differential equations

$$\frac{d^2 \boldsymbol{x}_k}{dt^2} + \sum_{i,j} \Gamma_{ij}^k \frac{d\boldsymbol{x}_i}{dt} \frac{d\boldsymbol{x}_j}{dt} = 0, \quad k = 1, \ldots, n \, , \tag{2}$$

where Γ_{ij}^m are the Christoffel symbols of a Riemannian connection M given by

$$\Gamma_{ij}^m = \frac{1}{2} \sum_k \left\{ \frac{\partial}{\partial \boldsymbol{x}_i} g_{jk} + \frac{\partial}{\partial \boldsymbol{x}_j} g_{ki} - \frac{\partial}{\partial \boldsymbol{x}_k} g_{ij} \right\} g^{km} \, ,$$

where g^{km} is an element of the matrix G^{km}, whose inverse is G_{km}.

Next, we present the definition of sectional curvature, a parameter which has been shown to be relevant in the design of signal constellations in Riemannian varieties.

Definition 3. *[4] Given a point $p \in M$ and a bi-dimensional subspace $\Sigma \subset T_pM$ the real number $K(x,y) = K(\Sigma)$, where $\{x,y\}$ is any basis for Σ, is called sectional curvature of Σ in M, defined by*

$$K(x,y) = \frac{\langle R(x,y)x, y \rangle}{|x \wedge y|^2} ,$$

where $R(x,y)$ denotes the curvature of a Riemannian variety M, which intuitively measures how close is the Riemannian variety to the Euclidean one, and $|x \wedge y| = \sqrt{|x|^2 + |y|^2 - \langle x,y \rangle^2}$ denotes the area of a bi-dimensional parallelogram determined by the pair of vectors x, y.

3   Signal Constellations in Riemannian Varieties

In designing signal constellations in a Riemannian variety we have to consider a Riemannian metric in each one of the blocks, namely: modulator, channel and demodulator, as shown in Fig. 1, such that the combination of them leads to a metric space (E_3, d_3). These metrics do not need to be the same, however for the communication system to achieve a better performance it is necessary to satisfy the metric matching condition.

Definition 4. *A signal constellation $\mathcal{X} = \{x_1, \ldots, x_m\}$ in an n-dimensional Riemannian variety M with a coordinate system $(U, \mathbf{x})$ is a set of n-dimensional points.*

$$\{x_1 = (\boldsymbol{x}_{11}, \ldots, \boldsymbol{x}_{n1}), \ldots, x_m = (\boldsymbol{x}_{1m}, \ldots, \boldsymbol{x}_{nm})\} \subset U.$$

Since the signal constellations are defined in a Riemannian variety M, then we have to use the Riemannian metric G_{ij}, in order to construct and to realize the performance analysis of such signal constellations. Therefore, the distance to be used between any two given points in M will be the least geodesic distance between the points in M.

We may think of the noise action in the channel as a transformation that takes the transmitted signal $x_m \in M$ to the received signal $y \in M$. We consider that this transformation is given by

$$y = \exp_{x_m}(v) , \quad v \in T_{x_m}M , \tag{3}$$

where $T_{x_m}M$ is the tangent plane of M at x_m, and the mapping $\exp_{x_m} : T_{x_m}M \to M$ is called an exponential mapping. Geometrically, $\exp_{x_m}(v)$ denotes a point in M when it moves a distance equal to $|v|$ from x_m on a geodesic passing by x_m with velocity $\frac{v}{|v|}$.

Notice that the way we have defined the noise in (3) requires that the exponential mapping be defined for every $v \in \mathcal{V} \subset T_{x_m}M$, where $\mathcal{V}$ is an open set of $T_{x_m}M$. Therefore, if we assume that $\mathcal{V}$ is the sample space of the random vector $v = (v_1, \ldots, v_n)$, then the noise is characterized by the probability density function of v. Hence, the probability density function of y, given that x_m was transmitted, is given by

$$p_Y(y/x_m) = p_V(v = (\exp_1^{-1}(y), \ldots, \exp_n^{-1}(y))|J| , \tag{4}$$

where $|J|$ denotes the absolute value of the Jacobian of J, given by

$$J = \begin{bmatrix} \dfrac{d\exp_1^{-1}(y)}{dy_1} & \dfrac{d\exp_1^{-1}(y)}{dy_2} & \cdots & \dfrac{d\exp_1^{-1}(y)}{dy_n} \\ \vdots & \vdots & \ddots & \vdots \\ \dfrac{d\exp_n^{-1}(y)}{dy_1} & \dfrac{d\exp_n^{-1}(y)}{dy_2} & \cdots & \dfrac{d\exp_n^{-1}(y)}{dy_n} \end{bmatrix} .$$

Let R_m be the decision region of x_m, that is, the set representing all the points which are decided as x_m. Thus, if x_m is the transmitted signal, then the probability that the decision will be in favor of a signal which is not in R_m, is given by

$$P_{e,m} = 1 - \int_{R_m} p_Y(y/x_m)\, d\boldsymbol{x}_1 \ldots d\boldsymbol{x}_n . \tag{5}$$

It is important to realize that the error probability associated to x_m, (5), does not depend on a particular coordinate system $(U, \mathbf{x})$ and that it is invariant by isometry. The average symbol error probability, P_e, of a signal constellation in a Riemannian variety may be written as

$$P_e = \sum_{\mathcal{X}} P(x_m) P_{e,m} , \tag{6}$$

where $P(x_m)$ denotes the a priori probability of the signal x_m.

The average energy of a signal constellation $\mathcal{X}$ is given by

$$E_t = \sum_{\mathcal{X}} P(x_m) d^2(x_m, \bar{x}) , \tag{7}$$

where $d^2(x_m, \bar{x})$ is the squared geodesic distance between the signal x_m and the center of mass of the signal constellation $\bar{x}$. It is known that the center of mass of a signal constellation is the one that minimizes the average energy. Therefore, to determine $\bar{x}$ we have to take the derivative of the average energy with respect to x, and assume that it is zero at $x = \bar{x}$. Hence, $\bar{x}$ is the unique solution to

$$\left.\frac{\partial E_t}{\partial x_j}\right|_{x=\bar{x}} = \sum_{\mathcal{X}} p(x_m) d(x_m, \bar{x}) \left.\frac{\partial d(x_m, x)}{\partial x_j}\right|_{x=\bar{x}} = 0 , \quad j = 1, \ldots, n . \tag{8}$$

In order to have an explicit equation for the signal-to-noise ratio, we have to determine the noise power in a Riemannian variety M given that x_m was transmitted. This is accomplished by defining the noise power as

$$\sigma^2 = \int_M d^2(y, x_m) p_Y(y/x_m)\, d\boldsymbol{x}_1 \ldots d\boldsymbol{x}_n . \tag{9}$$

As previously mentioned, the noise probability density function in a Riemannian variety M is associated to the probability density function of the random vector $v \in T_{x_m}M$. For instance, if $v = (v_1, \ldots, v_n)$ is a random vector such that its components v_j, $j = 1, \ldots, n$, are gaussian random variables, then v is also gaussian, since $T_{x_m}M$ besides being an n-dimensional vector space the transformation is linear. In order for the probability density function of y be well defined we have to assume that the Riemannian variety M is complete, that is, for every $x_m \in M$ the exponential mapping $\exp_{x_m}(v)$, is defined for every $v \in T_{x_m}M$.

In the particular case when the random variables v_j, $j = 1, \ldots, n$, are gaussian with zero mean and equal variances, we define the probability density function of y given that x_m was transmitted, as

$$p(y/x_m) = k_1 e^{-k_2 d^2(y,x_m)} \sqrt{det(G_{ij})} , \tag{10}$$

where $d^2(y, x_m)$ is the squared geodesic distance between the received signal y and the transmitted signal x_m, $\sqrt{det(G_{ij})}$ is the volume element of the variety M, and k_1, k_2 are constants satisfying the condition

$$\int_M k_1 e^{-k_2 d^2(y,x_m)} \sqrt{det(G_{ij})} \, d\boldsymbol{x}_1 \ldots d\boldsymbol{x}_n = 1 . \tag{11}$$

As in the case of finding the noise power given by (9), the constants k_1 and k_2 given by (11) depend on the transmitted signal x_m. However, if there exists an isometry taking one signal to the other, then σ^2, k_1 and k_2 assume the same values independently of the signal in consideration. This fact was used in [8] in the performance analysis of two 4-PSK signal constellations on the Enneper minimal surface.

Among the Riemannian varieties, the ones with constant sectional curvature are the simplest to deal with when considering the construction and the performance analysis of the signal constellations. This is a consequence of the fact that these spaces are homogeneous, that is, every point is on an equal footing from every other point. As a consequence, the noise always acts in the same way independent of the point in consideration. Hence, σ^2, k_1 and k_2 assume the same values for every point in M. Since the spaces with constant curvature have a great number of local isometries, it is clear that we may naturally have geometrically uniform signal constellations in such spaces.

On the other hand, determining all the geometrically uniform signal constellations in the space of constant curvature implies that we have to find all the subgroups that act transitively in these spaces. However, this is not an easy problem to solve.

4   Establishing the PDF in Spaces with Constant Curvature K

In this section, we consider an example of a signal constellation of the type M-PSK in a bi-dimensional Riemannian variety. Due to the geometric structure of this type of signal constellation, we employ the geodesic polar coordinate system, (ρ, θ), [3]. In this system, the coefficients $g_{11}(\rho, \theta)$, $g_{21}(\rho, \theta) = g_{12}(\rho, \theta)$ and $g_{22}(\rho, \theta)$ of the Riemannian metric must satisfy the following conditions

$$g_{11} = 1, \quad g_{12} = 0, \quad \lim_{\rho \to 0} g_{22} = 0, \quad \lim_{\rho \to 0} (\sqrt{g_{22}})_\rho = 1. \tag{12}$$

According to (2), a geodesic $\gamma : I \subset R \to M$ in polar coordinates, $\gamma(t) = (\rho(t), \theta(t))$ must satisfy the following system of second order partial differential equations

$$\begin{cases} \rho'' - \frac{1}{2}(g_{22})_\rho(\theta')^2 = 0 \,, \\ \theta'' + \frac{(g_{22})_\rho}{g_{22}}\rho'\theta' + \frac{1}{2}\frac{(g_{22})_\theta}{g_{22}}(\theta')^2 = 0 \,. \end{cases}$$

From (1) the arc length of the geodesic $\gamma(t)$, from t_1 to t_2, is given by

$$l_{t_1}^{t_2}(\gamma(t)) = \int_{t_1}^{t_2} \sqrt{(\rho')^2 + g_{22}(\theta')^2} \, dt \,.$$

If the coordinate system is the polar one, then g_{22} may be found as the solution to the following second order differential equation, [3],

$$(g_{22})_{\rho\rho} + K g_{22} = 0 \,, \tag{13}$$

where K denotes the sectional curvature of the variety M. In the particular case when M is bi-dimensional, K is known as the gaussian curvature. For simplicity, we assume that K is constant. Next we consider the solutions to (13) for $K = 0$, $K > 0$ and $K < 0$. Since g_{22} must also satisfy (12), we finally have

$$g_{22}(\rho, \theta) = \begin{cases} \rho^2 \,, & \text{if } K = 0 \,, \\ \frac{1}{K}\sin^2(\sqrt{K}\rho) \,, & \text{if } K > 0 \,, \\ \frac{1}{-K}\sinh^2(\sqrt{-K}\rho) \,, & \text{if } K < 0 \,. \end{cases}$$

We associate to the Euclidean space E^2 the curvature $K = 0$, to the spherical space S^2 the curvature $K > 0$ and constant, and to the hyperbolic plane H^2 the curvature $K < 0$ and constant.

When considering the Euclidean space E^2, the geodesic distance between any two given points $z_1, z_2 \in E^2$ is given by

$$d_E = |z_1 - z_2| \,, \tag{14}$$

where $|\ |$ denotes the absolute value of $z_i = r_i e^{j\theta_i}$, $i = 1, 2$.

When considering the spherical space S^2, we can show that the geodesic distance between any two given points $z_1, z_2 \in S^2$ is given by

$$d_S = \frac{2\pi l}{\sqrt{K}} \pm \frac{1}{j\sqrt{K}} \log \frac{|1 + z_1\bar{z}_2| + j|z_1 - z_2|}{|1 + z_1\bar{z}_2| - j|z_1 - z_2|} \,, \tag{15}$$

where l is the number of times that a geodesic passes by the point z_1 or its antipodal, until arriving at z_2, $z_i = r_i e^{j\theta_i}$ and $r_i = -j(e^{j\rho_i \sqrt{K}} - 1)/(e^{j\rho_i \sqrt{K}} + 1)$, $i = 1, 2$.

When considering the hyperbolic space H^2, we can show that the geodesic distance between any two given points $z_1, z_2 \in H^2$ is given by

$$d_H = \frac{1}{\sqrt{-K}} \log \frac{|1 - z_1 \bar{z}_2| + |z_1 - z_2|}{|1 - z_1 \bar{z}_2| - |z_1 - z_2|}, \tag{16}$$

where $z_i = r_i e^{j\theta_i}$ and $r_i = (e^{\rho_i \sqrt{-K}} - 1)/(e^{\rho_i \sqrt{-K}} + 1)$, $i = 1, 2$.

By use of (10), (11) and (9) we can find the probability density function, $p_Y(y/x_m)$, as well as the noise average energy, σ^2, for each one of the spaces E^2, S^2 and H^2. As previously mentioned, these spaces are homogeneous and, therefore, the values of k_1, k_2 and σ^2 are the same independent of the transmitted signal x_m. Thus, in order to simplify the calculations, we assume that the transmitted signal is $x_m = (0, \theta)$. Hence, for $y = (\rho, \theta)$, we end up with the following cases:

1. For $K = 0$, the probability density function is given by

$$p_E(\rho, \theta) = k_1 e^{-k_2 \rho^2} \rho \,,$$

   where $k_1 = k_2/\pi$ and $\sigma^2 = 1/k_2$. Since the coordinate system is polar, the resulting probability density function $p_E(\rho, \theta)$ is Rayleigh.

2. For $K < 0$, the probability density function is given by

$$p_H(\rho, \theta) = \frac{k_1}{\sqrt{-K}} e^{-k_2 \rho^2} \sinh\left(\sqrt{-K}\rho\right),$$

   with

$$k_1 = \frac{\pi^{-3/2} e^{K/4k_2} \sqrt{-K k_2}}{\mathrm{erf}(\sqrt{-K}/2\sqrt{k_2})},$$

   where $\mathrm{erf}(x)$ denotes the error function defined for every $w \in \mathbb{C}$ as

$$\mathrm{erf}(w) = \frac{2}{\sqrt{\pi}} \int_0^w e^{-z^2} dz \,.$$

   The noise average energy is given by

$$\sigma^2 = \frac{2\sqrt{-K k_2} e^{K/4k_2} + \sqrt{\pi}\mathrm{erf}(\sqrt{-K/4k_2})(2k_2 - K)}{4k_2^2 \sqrt{\pi}\mathrm{erf}(\sqrt{-K/4k_2})} \,.$$

3. For $K > 0$, the probability density function is given by

$$p_S(\rho, \theta) = \frac{k_1}{\sqrt{K}} e^{-k_2 \rho^2} |\sin\left(\sqrt{K}\rho\right)| \,,$$

   where

$$k_1 = \frac{\sqrt{K}}{2\pi} \left(\sum_{i=0}^{\infty} \int_{i\pi}^{(i+1)\pi} (-1)^i e^{-k_2 \rho^2} \sin\left(\sqrt{K}\rho\right) d\rho \right)^{-1} \,.$$

For $k_2 \gg K$, then k_1 may be approximated by

$$k_1 \approx \frac{i\pi^{-3/2}e^{K/4k_2}\sqrt{Kk_2}}{\text{erf}(i\sqrt{K}/2\sqrt{k_2})} .$$

Consequently, the noise average energy is given by

$$\sigma^2 \approx \frac{2i\sqrt{Kk_2}e^{K/4k_2} + \sqrt{\pi}\text{erf}(i\sqrt{K/4k_2})(2k_2 - K)}{4k_2^2\sqrt{\pi}\text{erf}(i\sqrt{K/4k_2})} .$$

In order to show that (10) is well defined, we consider the probability density function for $K = 1$, $K = 0$, $K = -1$, and the noise average energy equal to 1. Substituting these values into the previous equations, we end up with

$$p_S(\rho, \theta) = 0.31264e^{-0.8013\rho^2}|\sin(\rho)| ,$$
$$p_E(\rho, \theta) = e^{-\rho^2}\rho/\pi , \tag{17}$$
$$p_H(\rho, \theta) = 0.31572e^{-1.1492\rho^2}\sinh(\rho) .$$

As expected, we notice from (17) that the random variable θ is uniformly distributed in the three spaces, that is $p_S(\theta) = p_E(\theta) = p_H(\theta) = 1/2\pi$. Although the analytical expressions in (17) may seem different, graphically, Fig. 2(a), they seem to be the same up to the fourth decimal digit, as can be seen in Fig. 2(b). However, the most important fact can be observed in Figs. 2(c) and 2(d), where the cumulative distribution functions $P_S(\rho)$, $P_E(\rho)$ and $P_H(\rho)$ are practically the same although there is a difference from the fourth decimal digit on. We believe that this difference is related to the approximations to the error function calculations. Due to the consistence of the results obtained, we conclude that (10) is well defined.

5   Performance Analysis of M-PSK in Spaces with $K \neq 0$

Figure 3 shows the average error probability versus the signal-to-noise ratio for the 4-PSK signal constellation in four different spaces with constant sectional curvature equal to 1, 0, -1, and -2. As may be seen, the average error probability diminishes when the sectional curvature get smaller and negative. Figure 4 shows the average error probability versus the signal-to-noise ratio for the 8-PSK and 16-PSK signal constellations in spaces whose sectional curvature equals 0 and -1. As noticed for the 4-PSK, the average error probability diminishes for decreasing values of K. This fact shows that the curvature K is an important parameter to be considered in the design of signal constellations.

As mentioned previously, the average error probability of an M-PSK signal constellation in spaces with constant negative sectional curvature is upper-bounded by the average error probability of the M-PSK signal constellation in the Euclidean space. This may be explained by the fact that the P_e decreases

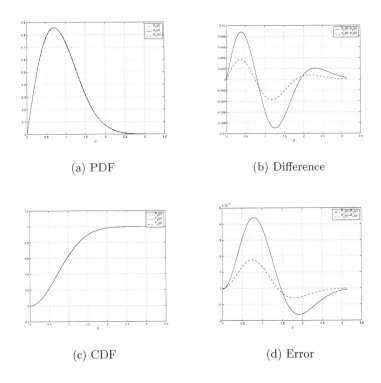

(a) PDF   (b) Difference

(c) CDF   (d) Error

Fig. 2. PDF and CDF in spaces with curvatures 1, 0, and -1

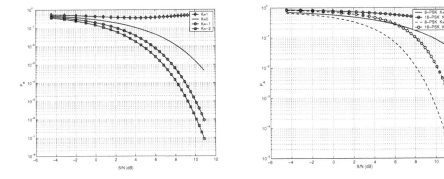

Fig. 3. $P_e \times S/N$ for 4-PSK in spaces with curvatures 1, 0, -1 and -2

Fig. 4. $P_e \times S/N$ for 8-PSK and 16-PSK in spaces with curvatures 0 and -1

when the minimum distance of the constellation, d_{min}, increases. On the other hand, the minimum distance increases when the sectional curvature decreases, for a fixed average energy of the signal constellation.

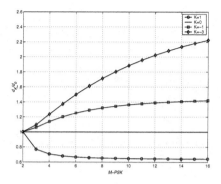

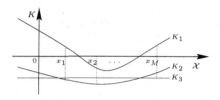

Fig. 5. $d_K/d_0 \times M$ for $E_t = (\pi/2)^2$

Fig. 6. Sectional curvature versus signal constellation $\mathcal{X}$

For instance, in Fig. 5 we notice that for an M-PSK constellation with $E_t = (\pi/2)^2$, the ratio between the minimum distance in a space with curvature K, d_K, and the minimum distance in the Euclidean space, $K = 0$, d_0, is greater than one for $K < 0$ and less than one for $K > 0$. Next, we compare the performances achieved by the same signal constellation in different Riemannian varieties. For that, we use Rauch's Theorem, [4], which intuitively states that if the gaussian curvature increases, the length decreases. Furthermore, if $K_1 < K_2$, we may conclude that the performance of a communication system with a signal constellation in a Riemannian variety with sectional curvature K_1 is better than the performance of a communication system with the same signal constellation in another Riemannian variety with sectional curvature K_2. Equivalently,

$$K_1 < K_2 \quad \Rightarrow \quad P_e(K_1) < P_e(K_2) . \tag{18}$$

From (18), we take into consideration the performance analysis of a communication system when it is subjected to the same signal constellation $\mathcal{X}$ in three different Riemannian varieties M_i, with corresponding sectional curvatures K_i, for $i = 1, 2, 3$. Figure 6 depicts the sectional curvature, K, versus the signal constellation, $\mathcal{X}$. From it, we may draw the following considerations:

- The performance of $\mathcal{X}$ in M_1 is worse than the performance of the same signal constellation in M_2 and M_3;
- We can not say anything based on (18) about the performance of $\mathcal{X}$ in M_2 being worse or better than the performance of the same signal constellation in M_3;
- In order to compare $P_e(K_2)$ and $P_e(K_3)$ we have to calculate the values of $P_{e,m}(K_2)$ and $P_{e,m}(K_3)$, for each signal x_m and we have to consider the possible values of $P(x_m)$.

When the variety does not have constant sectional curvature the determination of the symbol error probability, $P_{e,m}$, is quite difficult. However, if the variety M has non-positive sectional curvature for all of its points, then from Hadamard

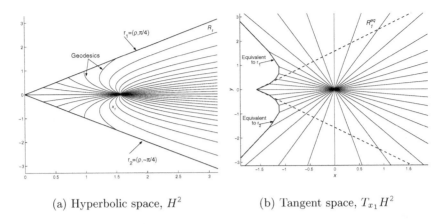

(a) Hyperbolic space, H^2 (b) Tangent space, $T_{x_1}H^2$

Fig. 7. Decision region of the signal $x_1 = (\pi/2, 0)$ in a 4-PSK signal constellation

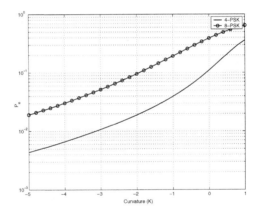

Fig. 8. Error probability versus sectional curvature

Theorem, [4], we have that the exponential mapping is a global diffeomorphism of T_pM over M. Therefore, for this particular case, a way of obtaining $P_{e,m}$ is to integrate the probability density function of the random vector $v \in T_{x_m}M$ (associated to the noise in the variety) over an equivalent decision region R_m^{eq} in $T_{x_m}M$ obtained from the decision region R_m in M. Therefore, $R_m^{eq} = \exp_{x_m}^{-1}(R_m)$.

As an example, we consider the decision region of the signal $x_1 = (\pi/2, 0)$ belonging to the 4-PSK signal constellation in the hyperbolic plane H^2, Figs. 7(a) and 7(b). Observe that R_1^{eq} contains the region delimited by the dashed line, representing the decision region of the same signal x_1, however in E^2. This is an alternative way of justifying the better performance achieved by the 4-PSK signal constellation in H^2 than that in E^2. Figure 8 shows the average error probability versus the sectional curvature of the 4-PSK and 8-PSK signal constellations for a signal-to-noise ratio equal to 4dB. As it was expected, P_e diminishes with decreasing values of K.

From the previous arguments and facts, the goal of this paper has been achieved, that is, we aimed at showing the importance of the sectional curvature in the design and performance analysis of signal constellations in Riemannian variety. Hence, the designer of a communication system has to take into consideration the sectional curvature, or equivalently, the genus of the surface where the signal constellation is on. The starting of the whole process of finding the genus or the sectional curvature comes from the embedding of the discrete memoryless channel on compact surfaces with or without borders. Therefore, surfaces with genus $g \geq 2$ or $K < 0$ are strong candidates for designing communication systems with better performance.

6   Conclusions

In this paper we have shown a dependence of the average error probability of a signal constellation in a Riemannian variety with its sectional curvature. This relationship establishes that for sectional curvatures K_1 and K_2 such that $K_1 < K_2$, then the performance a communication system with a signal constellation in a Riemannian variety M_1 with constant sectional curvature K_1 is better than the performance of a communication system with a signal constellation in a Riemannian variety M_2 with constant sectional curvature K_2.

Acknowledgement. The authors would like to thank the financial support from the Brazilian Agencies FAPESP, CNPq and CAPES.

References

1. E. Agustini, *Signal Constellations in Hyperbolic Spaces*, PhD Dissertation, IMECC-UNICAMP, Brazil, 2002, (in Portuguese).
2. E.B. da Silva, *Signal Constellations and its Performance Analysis in Hyperbolic Spaces*, PhD Dissertation, FEEC-UNICAMP, Brazil, 2000, (in Portuguese).
3. M.P. do Carmo, *Differential Geometry of Curves and Surfaces*, Prentice-Hall Inc., New Jersey, 1976.
4. M.P. do Carmo, *Geometria Riemanniana*, Impa-Projeto Euclides, 1979.
5. G.D. Forney, Jr., "Geometrically uniform codes," *IEEE Trans. Inform. Theory*, vol. IT-37, pp. 1241–1260, Sept. 1991.
6. J.D. de Lima, and R. Palazzo Jr., "Embedding Discrete Memoryless Channels on Compact and Minimal Surfaces," *IEEE Information Theory Workshop*, Bangalore, India, Oct. 20–25, 2002.
7. J.G. Proakis, *Digital Communications*, 2nd edition, McGraw-Hill, 1989.
8. R.G. Cavalcante, *Performance Analysis of Signal Constellations in Riemannian Varieties*, MS Thesis, FEEC-UNICAMP, Brazil, 2002, (in Portuguese).
9. W. Klingenberg, *A Course in Differential Geometry*, Springer, New York, 1978.

Improvements to Evaluation Codes and New Characterizations of Arf Semigroups

Maria Bras-Amorós

Universitat Politècnica de Catalunya
mbras@ma2.upc.es

Abstract. Following the ideas in [7] and [15,16], we present new improvements on evaluation codes. Some properties of the new codes will lead to new characterizations of Arf semigroups, similar to those in [3].

1 Introduction

Evaluation codes and their duals are a class of codes including the well-known families of Reed-Solomon codes, Reed-Muller codes and one-point algebraic-geometric codes.

Let $\mathbb{F}_q$ be the finite field with q elements. Given an $\mathbb{F}_q$-algebra A and a morphism of $\mathbb{F}_q$-algebras $\varphi : A \to \mathbb{F}_q^n$ the image under φ of a vector subspace of A will be a vector subspace of $\mathbb{F}_q^n$ and hence, it will be a code over $\mathbb{F}_q$. Suppose now that an order function is defined on A [10,17]. Evaluation codes are defined as the image under φ of the vector subspace of all elements in A with order bounded by a certain constant [10]. Order-prescribed evaluation codes will be defined as a generalization of evaluation codes. Indeed, they are the codes defined as the image under φ of some elements in A with a given order. An introduction to order functions and evaluation codes as well as an adaptation of the decoding algorithms for evaluation codes to order-prescribed evaluation codes is given in Sect. 2.

Feng and Rao presented in [7] what they called the *improved evaluation codes* and what we will call, for the purposes of this work, *greedy codes* correcting a given number of errors. Indeed, they are the codes defined as the image under φ of some elements in A with order in a chosen subset. This subset is chosen in function of a designed minimum distance d and they proved that these codes have minimum distance larger than or equal to d. They are in general an improvement of evaluation codes in the sense that they have dimension larger than or equal to the evaluation codes with minimum distance larger than or equal to d. More details on these codes can be found in [10] and some examples can be found in [9,8].

From a different point of view, O'Sullivan brings up a new improvement to evaluation codes. In [16,15] he restricts the error words to be corrected to the class of *generic errors*. Those are the errors whose footprint Δ_e is of the form $\{0, 1, \ldots, t-1\}$ for some t. This makes sense when we can expect that the probability of non-generic errors is very small compared to the probability

M. Fossorier, T. Hoeholdt, and A. Poli (Eds.): AAECC 2003, LNCS 2643, pp. 204–215, 2003.

of error words of weight t. This is the case of the examples worked out in [16, 15]. Using this restriction, one can construct evaluation codes with much larger dimension than that for evaluation codes correcting all errors of weight t.

Our contribution has been to mix the ideas of both Feng and Rao and O'Sullivan to define new codes. Based on the restrictions of the Berlekamp-Massey-Sakata generalized algorithm [14,10,16], we consider minimal sets for which decoding of generic errors of a given size is possible for the related order-prescribed evaluation code. In Sect. 3 we give the definitions and prove that the new codes have dimension larger than or equal to the dimension of both the Feng-Rao codes and the O'Sullivan codes.

Section 4 deals with the parameters of the new codes, based on the properties of a numerical semigroup. In particular, we give an explicit formula for the redundancy as well as an upper bound. We prove that the bound on the redundancy is tight and that equality holds if and only if the numerical semigroup is Arf. This gives a characterization of Arf numerical semigroups similar to that in [3]. On the other hand, when we wonder for which numerical semigroups the new improvements will make a difference on the related codes it turns out that the answer is all numerical semigroups but Arf semigroups.

This article is a survey of a part of my doctoral thesis, which is in preparation. Many results are just mentioned and we omitted the proofs for the sake of brevity. We refer the reader to [2].

2 Evaluation Codes

2.1 Order Functions

Definition 1. *Let A be an $\mathbb{F}_q$-algebra. An order function on A is a map $\rho : A \longrightarrow \{i \in \mathbb{Z} | i \geqslant -1\}$ satisfying the following conditions*

C0. *ρ is surjective.*
C1. *$\rho(f) = -1 \Longleftrightarrow f = 0$.*
C2. *$\rho(\xi f) = \rho(f) \; \forall \xi \in \mathbb{F}_q^*$.*
C3. *If $\rho(f) < \rho(g)$ and $h \neq 0$ then $\rho(fh) < \rho(gh)$.*
C4. *If $\rho(f) = \rho(g)$, then there exists $\xi \in \mathbb{F}_q^*$ such that $\rho(f - \xi g) < \rho(g)$.*
C5. *$\rho(f + g) \leqslant \max\{\rho(f), \rho(g)\}$ and if $\rho(f) < \rho(g)$ then equality holds.*

This definition is essentially the same given in [10,17]. Let $\mathbb{N}_0$ be the set $\{i \in \mathbb{Z} \text{ with } i \geqslant 0\}$. An operation $\oplus$ in $\mathbb{N}_0$ can be well defined by $i \oplus j = \rho(fg)$ where f and g are such that $\rho(f) = i$ and $\rho(g) = j$. Then, we can define a partial ordering $\preccurlyeq$ in $\mathbb{N}_0$ by $i \preccurlyeq j$ if and only if there exist $k \in \mathbb{N}_0$ such that $i \oplus k = j$. The integer k is denoted $j \ominus i$.

2.2 Numerical Semigroups and Weight Functions

Definition 2. *A numerical semigroup is a subset Λ of $\mathbb{N}_0$ containing 0, closed under sum and with finite complement in $\mathbb{N}_0$. For a numerical semigroup Λ we* *define the genus of Λ as the number $g = \#(\mathbb{N}_0 \setminus \Lambda)$ and the conductor of Λ as the only integer $c \in \Lambda$ such that $c - 1 \notin \Lambda$ and $c + \mathbb{N}_0 \subseteq \Lambda$. The enumeration of Λ is the unique increasing bijective map $\lambda : \mathbb{N}_0 \longrightarrow \Lambda$. We will use λ_i for $\lambda(i)$.*

Definition 3. *Let A be an $\mathbb{F}_q$-algebra and let ρ be an order function on A. We will say that it is a weight function if there exists a numerical semigroup Λ with enumeration λ such that $(\lambda \circ \rho)(fg) = (\lambda \circ \rho)(f) + (\lambda \circ \rho)(g)$ for all $f, g \neq 0$.*

Note that the condition in Definition 3 is equivalent to $\lambda_{i \oplus j} = \lambda_i + \lambda_j$ for all $i, j \in \mathbb{N}_0$. It implies moreover that $i \preccurlyeq j$ if and only if $\lambda_j - \lambda_i = \lambda_k$ for some $k \in \mathbb{N}_0$. It can be proven (see [2]) that if ρ is a weight function there exists only one numerical semigroup Λ such that its enumeration λ satisfies the condition in Definition 3. If ρ is a weight function, we will say the *numerical semigroup* of ρ, the *genus* of ρ and the *conductor* of ρ to refer to the unique numerical semigroup whose enumeration satisfies the condition in Definition 3, and its genus and conductor.

2.3   Evaluation Codes

Let A be an $\mathbb{F}_q$-algebra with an order function ρ. A subset $\{f_i, i \in \mathbb{N}_0\}$ of A such that $\rho(f_i) = i$ for all i is an $\mathbb{F}_q$-basis of A. We will call these basis ρ-*good basis* of A.

Definition 4. *Given an $\mathbb{F}_q$-algebra A, an order function ρ on A, a ρ-good $\mathbb{F}_q$-basis $\{f_i, i \in \mathbb{N}_0\}$ of A and a surjective morphism $\varphi : A \longrightarrow \mathbb{F}_q^n$, define the m-th evaluation code related to A, ρ and φ as $E_m = [\varphi(f_0), \ldots, \varphi(f_m)]_{\mathbb{F}_q}$ and its dual as $C_m = \{c \in \mathbb{F}_q^n | c \cdot \varphi(f_i) = 0 \text{ for all } i \leqslant m\}$.*

Notice that both E_m and C_m are independent of the choice of the ρ-good basis.
    For $m \in \mathbb{N}_0$ define $N_m = \{a \in \mathbb{N}_0 \mid a \preccurlyeq m\}$, $\nu_m = \#N_m$ and $d_{FR}(C_m) = \min\{\nu_i \mid i > m\}$. The parameter $d_{FR}(C_m)$ satisfies $d_{C_m} \geqslant d_{FR}(C_m)$ and it is called the *Feng-Rao bound* on the minimum distance. See [6,12,10].

Example 1. **Reed-Solomon Codes.** Let $\alpha_0, \alpha_1, \ldots, \alpha_{q-1}$ be all the elements in $\mathbb{F}_q$ and let $\mathbb{F}_q[x]_{\leqslant s}$ be the subspace of $\mathbb{F}_q[x]$ of polynomials with degree $\leqslant s$. For $k \leqslant q$ the *Reed-Solomon* code over $\mathbb{F}_q$ with dimension k, denoted $RS_q(k)$, is defined as the image of the map $ev_k : \mathbb{F}_q[x]_{\leqslant k-1} \to \mathbb{F}_q^q$, $f \mapsto (f(\alpha_0), f(\alpha_1), \ldots, f(\alpha_{q-1}))$. Notice that $\rho(f) = \deg(f)$, where $\deg(0)$ is defined to be -1, is an order function on $\mathbb{F}_q[x]$. Hence, $RS_q(k)$ is the $(k - 1)$-th evaluation code related to the algebra $A = \mathbb{F}_q[x]$, the order function $\rho = \deg$ and the morphism $\varphi : f \mapsto (f(\alpha_0), f(\alpha_1), \ldots, f(\alpha_{q-1}))$. Notice that the Vandermonde matrix $V_k(\alpha_0, \ldots, \alpha_{q-1})$ is a generating matrix for $RS_q(k)$.

Example 2. **Reed-Muller Codes.** Let $n = q^m$ and call $P_1, \ldots, P_n$ the n points in $\mathbb{F}_q^m$. The *Reed-Muller code* $RM_q(s, m)$ is defined as the image of the map $\varphi_s : \mathbb{F}_q[x_1, \ldots, x_m]_{\leqslant s} \to \mathbb{F}_q^n$, $f \mapsto (f(P_1), \ldots, f(P_n))$, where $\mathbb{F}_q[x_1, \ldots, x_m]_{\leqslant s}$

is the subspace of $\mathbb{F}_q[x_1, \ldots, x_m]$ of polynomials with total degree $\leqslant s$. Notice that Reed-Solomon codes are a particular case of Reed-Muller codes. Let $A = \mathbb{F}_q[x_1, \ldots, x_m]$ and let $\ll$ be the graded lexicographic order on monomials in A. Let f_i be the i-th monomial with respect to $\ll$. It can be shown that $\ll$ induces an order function ρ on A such that $\rho(f_i) = i$ (see [10]). Let l be such that $f_l = x_1^s$, then $f_{l+1} = x_m^{s+1}$. Consequently, $\mathbb{F}_q[x_1, \ldots, x_m]_{\leqslant s}$ is the space generated by $\{f_i \mid i \leqslant l\}$. Now, take $A = \mathbb{F}_q[x_1, \ldots, x_m]$, ρ the order function induced by $\ll$ and $\varphi : f \mapsto (f(P_1), \ldots, f(P_n))$, and obtain $E_l = RM_q(s, m)$. Notice that we get many more codes apart from the Reed-Muller codes.

Example 3. **One-Ooint Goppa Codes.** Another important example is given by the valuation of functions on a rational point P of a function field $F/\mathbb{F}_q$. For a divisor D of $F/\mathbb{F}_q$, let $\mathcal{L}(D) = \{0\} \cup \{f \in F^* | (f) + D \geqslant 0\}$. Define $A = \bigcup_{m \geqslant 0} \mathcal{L}(mP)$ and let $\Lambda = \{-v_P(f) | f \in A\} = \{-v_i | i \in \mathbb{N}_0\}$ with $-v_i < -v_{i+1}$. Define $\rho(f)$ to be -1 if $f = 0$ and i if $f \neq 0$ and $v_P(f) = v_i$. It is an order function because of the properties of v_P. The elements in Λ are called *Weierstrass non-gaps* of P, while the elements in $\mathbb{N}_0 \setminus \Lambda$ are called *Weierstrass gaps* of P. Suppose $P_1, \ldots, P_n$ are pairwise distinct rational points of $F/\mathbb{F}_q$ which are different than P and let ev be the map $A \to \mathbb{F}_q^n$ such that $f \mapsto (f(P_1), \ldots, f(P_n))$. For $m \geqslant 0$ the *one-point Goppa code* of order m related to P and $P_1, \ldots, P_n$ is defined as $C(P_1 + \ldots + P_n, mP) = ev(\mathcal{L}(mP))$. The i-th evaluation code related to A, ρ and ev is $E_i = C(P_1 + \ldots + P_n, -v_i P_0)$.

The order function $\rho(f) = \deg(f)$ in Example 1 defined on the polynomial ring $A = \mathbb{F}_q[x]$ is obviously an example of a weight function with numerical semigroup $\Lambda = \mathbb{N}_0$. It turns out that a weight function ρ on an $\mathbb{F}_q$-algebra A with semigroup $\Lambda = \mathbb{N}_0$ is possible only if $A = \mathbb{F}_q[x]$ for some $x \in A$ and ρ is the degree function [2]. The order function in Example 3 is also a weight function with numerical semigroup the one with all the Weierstrass non-gaps at P. In this case, the genus of the numerical semigroup is the same as the genus of the curve. It can be shown that the order function in Example 2 is not a weight function unless $A = \mathbb{F}_q[x_1]$ [2].

2.4   Order-Prescribed Evaluation Codes

Definition 5. *Given an $\mathbb{F}_q$-algebra A, an order function ρ on A, a ρ-good $\mathbb{F}_q$-basis $\{f_i | i \in \mathbb{N}_0\}$ of A, a surjective morphism $\varphi : A \longrightarrow \mathbb{F}_q^n$, and a subset W of $\mathbb{N}_0$, define the order-prescribed evaluation code related to A, ρ, φ and W, E_W, as the $\mathbb{F}_q$-subspace generated by $\{\varphi(f_i) | i \in W\}$ and its dual as $C_W = \{c \in \mathbb{F}_q^n | c \cdot \varphi(f_i) = 0 \text{ for all } i \in W\}$.*

Evaluation codes are a particular case of order-prescribed evaluation codes. Indeed, the m-th evaluation code C_m is the order-prescribed evaluation code C_W with W equal to $\{0, 1, \ldots, m\}$. Notice that, while in the case of evaluation codes, the choice of the ρ-good basis did not change the codes, in this case it might change them.

The interest of order-prescribed evaluation codes is that, by choosing proper subsets, we can get codes which have better parameters than evaluation codes. This is what we will do later on.

As an example of dual order-prescribed evaluation codes there are the codes introduced by Feng and Rao in [7]. Given a *designed minimum distance d* they define the *improved* evaluation code related to d as the dual order-prescribed evaluation code C_W with $W = \{i \in \mathbb{N}_0 | \nu_i < d\}$. They proved that these codes have minimum distance larger than or equal to the designed minimum distance. We will prove it by describing an algorithm that can correct up to $\lfloor \frac{d-1}{2} \rfloor$ errors for these codes. The Feng-Rao bound on the minimum distance of evaluation codes can be extended for order-prescribed evaluation codes as follows.

Definition 6. *Given A, ρ, φ and W as before, define $\delta(C_W) = \min\{\nu_i | i \notin W\}$ and $\tau(C_W) = \left\lfloor \frac{\delta(C_W)-1}{2} \right\rfloor$.*

We will see how we can adapt the Berlekamp-Massey-Sakata algorithm for evaluation codes to order-prescribed evaluation codes and the decoding algorithm will correct up to $\tau(C_W)$ errors in transmission of a word $c \in C_W$. This will prove that $d_{C_W} \geqslant \delta(C_W)$. Notice that for the Feng-Rao-improved evaluation code C_W with designed minimum distance d, we have $\delta(C_W) \geqslant d$. Then it follows that $d_{C_W} \geqslant d$.

2.5   On Decoding Algorithms

For dual evaluation codes there exists the generalized Berlekamp-Massey-Sakata decoding algorithm that corrects on each received word a number of errors up to half the Feng-Rao bound on the minimum distance. It is described in [14,10, 16]. Other historic bibliography can be found in [1,13,20,5,4].

Suppose c is the sent word, e the error word and $y = c + e$ the received word. Define the *error ideal* of e as $I_e = \{f \in A | \varphi(f) \cdot e = 0\}$ and the *footprint* of e as $\Delta_e = \mathbb{N}_0 \setminus \rho(I_e)$. Define also the set $\Sigma_e = \rho(I_e) \cap \mathbb{N}_0$. The idea of this algorithm is to find a Gröbner basis of I_e. Once we have this basis it is immediate to find the error positions and then we can get the error values from the error positions by many ways.

Define a *Gröbner subset of order i* as a set which exactly contains, for each $s \in \min_{\preccurlyeq} \rho(\{f \in A | \varphi(f_0 f) \cdot e = \varphi(f_1 f) \cdot e = \ldots \varphi(f_i f) \cdot e = 0\} \setminus \{0\})$, one $f \in A$ with $\rho(f) = s$. The interest of these subsets is that any Gröbner subset of order large enough is a Gröbner basis of the error ideal and thus, we are done. The generalized Berlekamp-Massey-Sakata algorithm will compute, for any order i, a Gröbner subset of order i.

The algorithm brings together two subalgorithms. On one hand there is the subalgorithm that generates iteratively the Gröbner subsets of order i. Let us call $\varphi(f_j) \cdot e$ the *syndrome* of order j. This subalgorithm requires all the syndromes of order up to i as well as a Gröbner subset of order $i - 1$ to give a Gröbner subset of order i. We will call this subalgorithm $\mathtt{Grob}_i$.

On the other hand, there is the subalgorithm for obtaining iteratively the required syndromes. This subalgorithm requires a Gröbner subset of order $i - 1$ as well as all the syndromes of order up to $i - 1$ to give the syndrome of order i. It moreover requires that $\nu_i > 2\#(N_i \cap \Delta_e)$. This result was introduced in [15]. We will call this subalgorithm $\mathtt{Synd}_i$.

Furthermore, if $\varphi(f_i)$ is a parity check of our code, then the syndrome of order i can be immediately computed since it is equal to $\varphi(f_i) \cdot y$. In this case we will say we applied *direct syndrome computation*.

In [14,10,16] the authors deal with dual evaluation codes. Suppose the sent word c is in C_m. They first compute all syndromes of order up to m by direct syndrome computation. Then they use iteratively $\mathtt{Grob}_i$ to get a Gröbner subset of order m. Finally they alternate $\mathtt{Synd}_i$ and $\mathtt{Grob}_i$ to get Gröbner subsets of order as large as desired.

The way in which Berlekamp-Massey-Sakata algorithm computes a basis of I_e can be adapted to dual order-prescribed evaluation codes as follows. Suppose $c \in C_W$.

- The set $\{1\}$ is a Gröbner subset of order -1. The syndrome of order -1 is 0.
- For $i = 0$ to i such that any Gröbner subset of order i is a basis of I_e,
  - If $i \in W$, use direct syndrome computation to get the syndrome of order i.
  - Else, use $\mathtt{Synd}_i$ to get the syndrome of order i.
  - Use $\mathtt{Grob}_i$ to obtain a Gröbner subset of order i.

This algorithm will work if $\mathtt{Synd}_i$ can proceed for all $i \notin W$. That is, if $\nu_i > 2\#(N_i \cap \Delta_e)$ for all $i \notin W$. This suggests the following definition that was introduced in [16] in the context of evaluation codes.

Definition 7. *We say that decoding of e is guaranteed by the code C_W if $\nu_i > 2\#(N_i \cap \Delta_e)$ for all $i \notin W$.*

Proposition 1. *If the number t of errors in e satisfies $t \leqslant \lfloor \frac{\nu_i - 1}{2} \rfloor$, then $\nu_i > 2\#(N_i \cap \Delta_e)$. In particular, if $t \leqslant \tau(C_W)$ then decoding of e is guaranteed by C_W.*

3   Improvements on Evaluation Codes and Generic Errors

Our aim in this section is to bring up codes that, for a fixed number t of errors to be corrected on each code word, have dimension as large as possible, according to the restrictions of the adapted Berlekamp-Massey-Sakata algorithm.

3.1   Greedy Codes Correcting All Errors of a Given Size

Definition 8. *Let $t \in \mathbb{N}_0$. Define*

$$m(t) = \max\{m \in \mathbb{N}_{-1} | \nu_m < 2t + 1\}, \qquad \tilde{R}(t) = \{i \in \mathbb{N}_0 | \nu_i < 2t + 1\},$$
$$R(t) = \{i \in \mathbb{N}_0 | i \leqslant m(t)\}, \qquad\qquad \tilde{r}(t) = \#\tilde{R}(t),$$
$$r(t) = \#R(t) = m(t) + 1, \qquad\qquad \tilde{C}(t) = C_{\tilde{R}(t)}.$$
$$C(t) = C_{R(t)} = C_{m(t)},$$

It is obvious that the code $C(t)$ is the largest dual evaluation code C_m with $\tau(C_m) \geq t$ while the code $\tilde{C}(t)$ is the largest dual order-prescribed evaluation code C_W with $\tau(C_W) \geq t$. Codes $\tilde{C}(t)$ were defined by Feng and Rao from a designed minimum distance. Here we adapted the definition to our purposes and defined them from a designed decoding capability.

We will have, for all $t \in \mathbb{N}_0$, $\dim \tilde{C}(t) \geq n - \tilde{r}(t)$ with equality in the case of a weight function with conductor c, if $2c \leq n$ and $1 \leq 2t + 1 \leq 2\lambda^{-1}(c) - 2$ ([19, 3]). Moreover, $\tilde{R}(t) \subseteq R(t)$ and, hence, $\dim \tilde{C}(t) \geq \dim C(t)$. Finally, $\tau(\tilde{C}(t)) = \tau(C_{\tilde{R}(t)}) \geq t$.

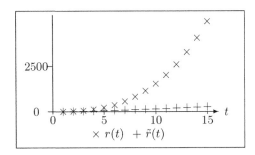

Fig. 1. Parameters $r(t)$ and $\tilde{r}(t)$ for the Reed-Muller codes on the polynomial ring with 3 variables

3.2   Generic Errors

Let us see the definition of generic error as it is given in [16]. Generic errors were also called *independent errors* in [11].

Definition 9. Let $\Delta^t = \{0, 1, \ldots, t-1\}$ and let $\Sigma^t = \{t, t+1, \ldots\}$. An error word e of weight t is said to be generic if $\Delta_e = \Delta^t$ and $\Sigma_e = \Sigma^t$.

Proposition 2. Suppose e is an error vector of weight t and suppose that I is the set of error positions of e. Define the matrix $M_e = (\varphi(f_i)_j)$ with $0 \leq i \leq t-1$ and $j \in I$. Then the following statements are equivalent:

(i) e is not generic,

(ii) There exist $\alpha_0, \alpha_1, \ldots, \alpha_{t-1}$ not all equal to zero such that $\sum_{i=0}^{t-1} \alpha_i \varphi(f_i)_j = 0$ for all $j \in I$,

(iii) $\det M_e = 0$.

In particular, if $\varphi(f_{i-1})$ is in the span of the vectors $\varphi(f_j)$ with $j < i - 1$, then any error vector of weight $t \geq i$ will be non-generic.

Example 4. **Dual Reed-Solomon Codes.** For Reed-Solomon codes, the matrix M_e will always be a square Vandermonde matrix and, hence, its determinant will always be different than zero. So, for Reed-Solomon codes, all error words will be generic.

Example 5. **Dual Reed-Muller Codes.** As in Example 2, let $n = q^m$ and call $P_1, \ldots, P_n$ the n points in $\mathbb{F}_q^m$. In this case, if the set of error positions is $I = \{i_1, \ldots, i_t\}$, statement (ii) in last proposition is equivalent to the existence of a polynomial $\sum_{i=0}^{t-1} \alpha_i f_i$ with not all α_i's equal to 0 which vanishes in $P_{i_1}, \ldots, P_{i_t}$. So, the condition of not being generic corresponds to a certain geometric distribution of the points corresponding to error positions.

For instance, suppose $m = 2$, that is, $A = \mathbb{F}_q[x_1, x_2] = \mathbb{F}_q[x, y]$ and $t \leqslant \binom{q+1}{2}$. For $t = 1$, any error vector is generic; For $t = 2$, e is not generic if and only if P_{i_1}, P_{i_2} lie in a horizontal line; For $t = 3$, e is not generic if and only if $P_{i_1}, P_{i_2}, P_{i_3}$ are collinear; For $t = 4$, e is not generic if and only if $P_{i_1}, P_{i_2}, P_{i_3}, P_{i_4}$ are in a vertical parabola; For $t = 5$, e is not generic if and only if $P_{i_1}, P_{i_2}, P_{i_3}, P_{i_4}, P_{i_5}$ are in a parabola or in an hyperbola; For $t = 6$, e is not generic if and only if $P_{i_1}, P_{i_2}, P_{i_3}, P_{i_4}, P_{i_5}, P_{i_6}$ are in any conic.

Assumption: We will make the assumption that the probability of a generic error is much higher than the probability of a non-generic error of the same weight. We have just seen that this is the case of the dual Reed-Solomon codes and the dual Reed-Muller codes of large dimension. In [15] O'Sullivan showed that non-generic errors are very improbable for the example of dual Hermitian codes.

Under this assumption, the maximal (order-prescribed) evaluation code correcting all generic errors of a given weight would fail only for a very small set of errors when applied to correct all errors of the same weight. On the other hand, the maximal (order-prescribed) evaluation code correcting all generic errors of a given weight may have dimension much larger than the maximal (order-prescribed) evaluation code correcting the whole set of errors of a given weight. This is the main idea of the improvements on evaluation codes introduced by O'Sullivan [16,15].

Proposition 3. *Condition $\nu_m > 2\#(N_m \cap \Delta^t)$ is equivalent to $m \in \Sigma^t \oplus \Sigma^t$.*

Proof. Let $X = \{s \in N_m | s \in \Delta^t, m \ominus s \in \Delta^t\}$, $Y = \{s \in N_m | s \in \Sigma^t, m \ominus s \in \Sigma^t\}$, $Z = \{s \in N_m | s \in \Delta^t, m \ominus s \in \Sigma^t\}$, $T = \{s \in N_m | s \in \Sigma^t, m \ominus s \in \Delta^t\}$. Note that $\nu_m = \#X + \#Y + \#Z + \#T$, $\#(N_m \cap \Delta^t) = \#X + \#Z$, and, since $Z = m \ominus T$, $\#Z = \#T$. Therefore, hypothesis $\nu_m > 2\#(N_m \cap \Delta^t)$ is equivalent to $\#Y > \#X$.

Now, condition $\#Y > \#X$ can be only satisfied if $Y \neq \emptyset$. But $Y \neq \emptyset$ if and only if $m \in \Sigma^t \oplus \Sigma^t$.

On the other hand, since $a \oplus b$ is stictly increasing in both a and b if they are in $\mathbb{N}_0$, $\Sigma^t \oplus \Sigma^t \cap \Delta^t \oplus \Delta^t = \emptyset$. So, if $m \in \Sigma^t \oplus \Sigma^t$ then $m \notin \Delta^t \oplus \Delta^t$. Hence, $X = \emptyset$, and condition $\#Y > \#X$ is satisfied. $\qquad \square$

3.3 Greedy Codes Correcting All Generic Errors of a Given Size

Definition 10. *Let* $t \in \mathbb{N}_0$. *Define*

$$m^*(t) = \max\{i \in \mathbb{N}_{-1} | i \notin \Sigma^t \oplus \Sigma^t\}, \quad \tilde{R}^*(t) = \{i \in \mathbb{N}_0 | i \notin \Sigma^t \oplus \Sigma^t\},$$
$$R^*(t) = \{i \in \mathbb{N}_0 | i \leqslant m^*(t)\}, \qquad \tilde{r}^*(t) = \#\tilde{R}^*(t),$$
$$r^*(t) = \#R^*(t) = m^*(t) + 1, \qquad \tilde{C}^*(t) = C_{\tilde{R}^*(t)},$$
$$C^*(t) = C_{R^*(t)} = C_{m^*(t)}, \qquad \tilde{\tau}^*(t) = \tau(\tilde{C}^*(t)).$$
$$\tau^*(t) = \tau(C^*(t)),$$

It is obvious that the code $C^*(t)$ is the largest evaluation code guaranteeing the correction of all generic errors of weight t while the code $\tilde{C}^*(t)$ is the largest order-prescribed evaluation code guaranteeing the correction of all generic errors of weight t. Notice that for all $t \in \mathbb{N}_0$, dim $(C^*(t)) \geqslant n - r^*(t)$ and dim $(\tilde{C}^*(t)) \geqslant n - \tilde{r}^*(t)$. Moreover, $R^*(t) \subseteq R(t)$, $\tilde{R}^*(t) \subseteq \tilde{R}(t)$ and $\tilde{R}^*(t) \subseteq R^*(t)$. Hence, dim $C^*(t) \geqslant$ dim $C(t)$, dim $\tilde{C}^*(t) \geqslant$ dim $\tilde{C}(t)$ and dim $\tilde{C}^*(t) \geqslant$ dim $C^*(t)$.

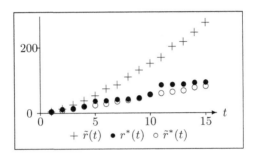

Fig. 2. Parameters $\tilde{r}(t)$, $r^*(t)$ and $\tilde{r}^*(t)$ for the Reed-Muller codes on the polynomial ring with 3 variables

It will hold that $\tilde{\tau}^*(t) \leqslant \tau^*(t) \leqslant t$. If $\tilde{\tau}^*(t) < t$ (resp. $\tau^*(t) < t$) then we say that there is *decoding beyond half the minimum distance*. By this, we mean that we will correct (generic) errors of weight t with the code $\tilde{C}^*(t)$ (resp. $C^*(t)$) although the number of errors that can be corrected on each received word by $\tilde{C}^*(t)$ (resp. $C^*(t)$) is $\tilde{\tau}^*(t)$ (resp. $\tau^*(t)$) which is strictly less than t.

4   New Characterizations of Arf Numerical Semigroups

In the case of the order function being a weight function, the definitions of $R(t), \tilde{R}(t), R^*(t)$ and $\tilde{R}^*(t)$ depend only on the numerical semigroup. The aim of this section is to study the parameters of the new codes related to the numerical semigroup.

Definition 11. *Given a numerical semigroup Λ with enumeration λ and $t \in \mathbb{N}_0$ define $\alpha(t) = \#\{h \in \mathbb{N}_0 \setminus \Lambda \mid h = \lambda_i + \lambda_j - \lambda_t$ for some $i, j \in \mathbb{N}_0, i \geqslant j \geqslant t\}$.*

Lemma 1. *Let Λ be a numerical semigroup. Then, for all $t \in \mathbb{N}_0$, $\alpha(t) = \#\{s \in \Sigma^t \oplus \Sigma^t | t \not\preccurlyeq s\}$.*

Proposition 4. *Suppose Λ is a numerical semigroup with enumeration λ. Then, for all $t \in \mathbb{N}_0$, $\tilde{r}^*(t) = \lambda_t + t - \alpha(t)$.*

Proof. For the sake of brevity we will only prove this proposition for the case in which $\lambda_t < c$. For the complete proof we refer the reader to [2]. Let $S^t = \{s \in \Sigma^t \oplus \Sigma^t | t \not\preccurlyeq s\}$, $\tilde{S}^t = \{s \in \Sigma^t \oplus \Sigma^t | t \preccurlyeq s\}$. Then, $\Sigma^t \oplus \Sigma^t = S^t \sqcup \tilde{S}^t$.

Notice that $\tilde{S}^t = t \oplus \Sigma^t$. Indeed, inclusion $\supseteq$ is obvious. For the inclusion $\subseteq$ notice that if $s \in \tilde{S}^t$ then $s = i \oplus j$ for some $i, j \geqslant t$. Moreover, $s = t \oplus k$ for some integer k. The integer k must be larger than or equal to t because $t \oplus k = i \oplus j$. So, $k \in \Sigma^t$ and $s = t \oplus k \in t \oplus \Sigma^t$.

Let $\Sigma_0 = \{t, t+1, \ldots, \lambda^{-1}(c) - 1\} = \{t, t+1, \ldots, c - g - 1\}$, $\Sigma_1 = \{\lambda^{-1}(c), \lambda^{-1}(c)+1, \ldots\} = \{c-g, c-g+1, \ldots\} = c-g+\mathbb{N}_0$. Then, $\Sigma^t = \Sigma_0 \sqcup \Sigma_1$ and $\tilde{S}^t = (t \oplus \Sigma_0) \sqcup (t \oplus \Sigma_1)$. So, $\Sigma^t \oplus \Sigma^t = S^t \sqcup (t \oplus \Sigma_0) \sqcup (t \oplus \Sigma_1)$. But, by semigroup properties, $t \oplus \Sigma_1 = \lambda_t + \Sigma_1 = \lambda_t + c - g + \mathbb{N}_0$. Hence, $\tilde{r}^*(t) = \lambda_t + c - g - \#(t \oplus \Sigma_0) - \#S^t$. Finnally, $\#(t \oplus \Sigma_0) = \#\Sigma_0 = c - g - t$, while $\#S^t = \alpha(t)$, and the proposition follows. $\qquad\square$

Corollary 1. Redundancy bound. *Suppose Λ is a numerical semigroup with enumeration λ. Then, for all $t \in \mathbb{N}_0$, $\tilde{r}^*(t) \leqslant \lambda_t + t$.*

Definition 12. *A numerical semigroup Λ with enumeration λ is called an Arf numerical semigroup if for every $i, j, k \in \mathbb{N}_0$ with $i \geqslant j \geqslant k$, it holds that $\lambda_i + \lambda_j - \lambda_k \in \Lambda$ or, equivalently, for every $i, j \in \mathbb{N}_0$ with $i \geqslant j$, it holds that $2\lambda_i - \lambda_j \in \Lambda$ [3, Definition 2.1, Proposition 2.3].*

As an example of Arf numerical semigroups there are the Weierstrass semigroups related to the codes in the second tower of Garcia-Stichtenoth attaining the Drinfeld-Vlăduţ bound.

In [3] there is a characterization of Arf numerical semigroups as the semigroups that are *stable*. That is, semigroups for which $\tilde{r}(t) = \lambda_t + t$ for any integer t. In this section we will prove two new characterizations of Arf numerical semigroups. A first one related to the parameter $\tilde{r}^*(t)$ of the new codes and a second one related to $\tilde{R}^*(t)$.

Proposition 5. *A numerical semigroup Λ is such that $\alpha(t) = 0$ for all $t \in \mathbb{N}_0$ if and only if Λ is Arf.*

Proof. If Λ is Arf then it is obvious that $\alpha(t) = 0$ for all $t \in \mathbb{N}_0$. On the other hand, suppose $\alpha(t) = 0$ for all $t \in \mathbb{N}_0$. Then, for any $i \geqslant j \geqslant k$, $\lambda_i + \lambda_j - \lambda_k \in \Lambda$ because otherwise, $\alpha(k) > 0$. Therefore, Λ is Arf. $\qquad\square$

Corollary 2. Characterization of Arf semigroups related to $\tilde{r}^*$. *A numerical semigroup Λ with enumeration λ is such that $\tilde{r}^*(t) = \lambda_t + t$ for all $t \in \mathbb{N}_0$ if and only if Λ is Arf.*

Corollary 3. *The redundancy bound given in Corollary 1 is tight.*

Lemma 2. *Let Λ be an Arf numerical semigroup. Then, for all $t \in \mathbb{N}_0$, $\Sigma^t \oplus \Sigma^t = t \oplus \Sigma^t$.*

Proof. Let λ be the enumeration of Λ. If $m \in \Sigma^t \oplus \Sigma^t$ then $m = i \oplus j$ and $\lambda_m = \lambda_i + \lambda_j$ for some $i, j \geqslant t$. By definition of Arf semigroup, $\lambda_m - \lambda_t \in \Lambda$ and so, $m \succcurlyeq t$. Now $m \ominus t$ must be in Σ^t since $m \ominus t = i \oplus j \ominus t \geqslant t \oplus t \ominus t = t$. So, $m \in t \oplus \Sigma^t$. $\qquad\square$

Proposition 6. Characterization of Arf semigroups related to $\tilde{\mathbf{R}}^*$.
Let Λ be a numerical semigroup. Then, $\tilde{R}(t) = \tilde{R}^*(t)$ (and consequently $R(t) = R^*(t)$) for all $t \in \mathbb{N}_0$ if and only if Λ is an Arf numerical semigroup.

Proof. Let λ be as usual the enumeration of Λ. By Proposition 1 and Proposition 3 we always have $\tilde{R}^*(t) \subseteq \tilde{R}(t)$ for all $t \in \mathbb{N}_0$. On the other hand, the inclusion $\tilde{R}(t) \subseteq \tilde{R}^*(t)$ is equivalent to the implication $i \in \Sigma^t \oplus \Sigma^t \implies \nu_i \geqslant 2t + 1$. So, we have to prove that this implication holds for all $t \in \mathbb{N}_0$ if and only if Λ is an Arf numerical semigroup.

Suppose Λ is such that this implication holds for all $t \in \mathbb{N}_0$. We will have that, in particular, $\nu_{t \oplus t} \geqslant 2t + 1$. Now notice that if $j \oplus k = t \oplus t$ and $j < k$, then $j < t < k$, but there are only t integers $j \in \mathbb{N}_0$ with $j < t$. So, if $\nu_{t \oplus t} = 2t + 1$ it means that for all integer j less than or equal to t, $j \preccurlyeq t \oplus t$ and hence, $2\lambda_t - \lambda_j \in \Lambda$. So, Λ must be Arf.

Now suppose Λ is Arf. Let $t \in \mathbb{N}_0$ and suppose $i \in \Sigma^t \oplus \Sigma^t$. By Lemma 2, $i = t \oplus k$ for some $k \geqslant t$. Since Λ is Arf, for all $j \leqslant t$, $\lambda_k + \lambda_t - \lambda_j \in \Lambda$ and hence, $j \preccurlyeq t \oplus k = i$. Finally, for all $j \leqslant t$, $i \ominus j \geqslant t$ with equality only if $j = k = t$. So, $\nu_i \geqslant 2t + 1$. $\qquad\square$

Notice that this proposition means that, given a numerical semigroup, the new improvements presented in Sect. 3.3 will make a difference on the related codes if and only if the numerical semigroup is not Arf.

Acknowledgements. I would like to thank Mike O'Sullivan for his dedication and for many helpful discussions.

References

1. Elwyn R. Berlekamp. *Algebraic coding theory.* McGraw-Hill Book Co., New York, 1968.
2. Maria Bras-Amorós. *On Improvements to Evaluation Codes.* PhD thesis, Universitat Politècnica de Catalunya, Barcelona, 2003. In preparation.
3. Antonio Campillo, José Ignacio Farrán, and Carlos Munuera. On the parameters of algebraic-geometry codes related to Arf semigroups. *IEEE Trans. Inform. Theory*, 46(7):2634–2638, 2000.

4. Iwan M. Duursma. Majority coset decoding. *IEEE Trans. Inform. Theory*, 39(3):1067–1070, 1993.

5. Gui Liang Feng and T.R.N. Rao. Decoding algebraic-geometric codes up to the designed minimum distance. *IEEE Trans. Inform. Theory*, 39(1):37–45, 1993.

6. Gui Liang Feng and T.R.N. Rao. A simple approach for construction of algebraic-geometric codes from affine plane curves. *IEEE Trans. Inform. Theory*, 40(4):1003–1012, 1994.

7. Gui-Liang Feng and T.R.N. Rao. Improved geometric Goppa codes. I. Basic theory. *IEEE Trans. Inform. Theory*, 41(6, part 1):1678–1693, 1995. Special issue on algebraic geometry codes.

8. Olav Geil. On codes from norm-trace curves. Submitted to Finite Fields and Their Applications, 2002.

9. Olav Geil and Tom Høholdt. On hyperbolic codes. In *Applied algebra, algebraic algorithms and error-correcting codes (Melbourne, 2001)*, volume 2227 of *Lecture Notes in Comput. Sci.*, pages 159–171. Springer, Berlin, 2001.

10. Tom Høholdt, Jacobus H. van Lint, and Ruud Pellikaan. *Algebraic Geometry codes*, pages 871–961. North-Holland, Amsterdam, 1998.

11. Helge Elbrønd Jensen, Rasmus Refslund Nielsen, and Tom Høholdt. Performance analysis of a decoding algorithm for algebraic-geometry codes. *IEEE Trans. Inform. Theory*, 45(5):1712–1717, 1999.

12. Christoph Kirfel and Ruud Pellikaan. The minimum distance of codes in an array coming from telescopic semigroups. *IEEE Trans. Inform. Theory*, 41(6, part 1):1720–1732, 1995. Special issue on algebraic geometry codes.

13. James L. Massey. Shift-register synthesis and BCH decoding. *IEEE Trans. Information Theory*, IT-15:122–127, 1969.

14. Michael E. O'Sullivan. Decoding of codes defined by a single point on a curve. *IEEE Trans. Inform. Theory*, 41(6, part 1):1709–1719, 1995. Special issue on algebraic geometry codes.

15. Michael E. O'Sullivan. Decoding of Hermitian codes: Beyond the minimum distance bound. Preprint, 2001.

16. Michael E. O'Sullivan. A generalization of the Berlekamp-Massey Sakata algorithm. Preprint, 2001.

17. Michael E. O'Sullivan. New codes for the Berlekamp-Massey-Sakata algorithm. *Finite Fields Appl.*, 7(2):293–317, 2001.

18. Ruud Pellikaan. On decoding by error location and dependent sets of error positions. *Discrete Math.*, 106/107:369–381, 1992. A collection of contributions in honour of Jack van Lint.

19. Ruud Pellikaan and Fernando Torres. On Weierstrass semigroups and the redundancy of improved geometric Goppa codes. *IEEE Trans. Inform. Theory*, 45(7):2512–2519, 1999.

20. Alexei N. Skorobogatov and Sergei G. Vlăduţ. On the decoding of algebraic-geometric codes. *IEEE Trans. Inform. Theory*, 36(5):1051–1060, 1990.

Optimal 2-Dimensional 3-Dispersion Lattices

Moshe Schwartz and Tuvi Etzion

Technion – Israel Institute of Technology
Department of Computer Science
Technion City, Haifa 32000, Israel
{moosh,etzion}@cs.technion.ac.il

Abstract. We examine 2-dimensional 3-dispersion lattice interleavers in three connectivity models: the rectangular grid with either 4 or 8 neighbors, and the hexagonal grid. We provide tight lower bounds on the interleaving degree in all cases and show lattices which achieve the bounds.

1 Introduction

In some relatively new applications, two-dimensional error-correcting codes are used. The codewords are written on the plane, and their coordinates are indexed by $\mathbb{Z}^2$. Several models of two-dimensional bursts of errors are handled in the literature. The most common burst type studied involves the rectangular grid and rectangular bursts [1,2,3,4,5]. The general two-dimensional case was studied in [6] and later in [7]. In the general case, an unrestricted burst (also called a cluster) is a connected set of points in $\mathbb{Z}^2$. The only parameter associated with such a burst is its size.

Since a burst is a connected set of points of $\mathbb{Z}^2$, we must consider several connectivity models. The simplest one is the $+$ model in which the neighbors of a given point $(x, y) \in \mathbb{Z}^2$ are, $\{(x + 1, y), (x - 1, y), (x, y + 1), (x, y - 1)\}$. A natural variation on the $+$ model is the $*$ model in which a point $(x, y) \in \mathbb{Z}^2$ has the following neighbor set, $\{(x + a, y + b) \mid a, b \in \{-1, 0, 1\}, |a| + |b| \neq 0\}$. Finally, another model of interest to us is the hexagonal model. Instead of the rectangular grid, we define the following grid: we start by tiling the plane $\mathbb{R}^2$ with regular hexagons. The vertices of the grid are the center points of the hexagons. We connect two vertices if and only if their respective hexagons are adjacent. This way, each vertex has exactly 6 neighboring vertices.

Given some connectivity model and r-points, $p_1, \ldots, p_r \in \mathbb{Z}^2$, we define $d_r(p_1, \ldots, p_r)$, also called the r-dispersion, to be the size (minus one) of the smallest burst containing all r points. The function d_2 is the known distance, while d_3 is called the *tristance*.

Bursts of errors are usually handled by interleaving several codewords together. An interleaving scheme, $\Gamma : \mathbb{Z}^2 \to \{1, 2, \ldots, m\}$ is denoted $A(t, r)$ if every burst of size t contains no more than r instances of the same integer from $\{1, 2, \ldots, m\}$. The number m of codewords needed for the interleaving, is the *interleaving degree* of Γ denoted by $\deg(\Gamma)$. If we take m codewords of an

M. Fossorier, T. Hoeholdt, and A. Poli (Eds.): AAECC 2003, LNCS 2643, pp. 216–225, 2003.

r-error-correcting code and write the i-th codeword in coordinates which are mapped by Γ to i, then a burst of size t generates no more than r errors in each of the codewords.

A simple way of creating an interleaving scheme is by taking a lattice Λ, i.e., a subspace of $\mathbb{Z}^2$, and mapping it, and each of its cosets to a unique integer. It was shown in [7] that if the $(r+1)$-dispersion of any $r+1$ points of Λ is at least t, then the interleaving scheme induced by Λ is an $A(t, r)$. Its degree is the index of Λ in $\mathbb{Z}^2$, also called the *volume* of Λ. The lattice Λ is always the span of a 2×2 matrix $\mathbf{G}$ over $\mathbb{Z}^2$ and the index of Λ in $\mathbb{Z}^2$ is also given by $|\mathbf{G}|$.

In this paper we describe optimal lattice interleavers for 2 repetitions. That is, for a given tristance d_3 we build lattices with minimal volume for which the tristance between any three of its points is at least d_3. The following three sections describe optimal lattices in each of the three connectivity models.

2   The + Model

2.1   Preliminaries

In the + model, a point $(x, y) \in \mathbb{Z}^2$ is connected to $(x+1, y), (x-1, y), (x, y+1)$, and $(x, y-1)$. We note that the distance in this model coincides with the definition of the L_1 distance between two points. Thus, for $p_i = (x_i, y_i), 1 \le i \le r$ we have

$$d_2(p_1, p_2) = |x_1 - x_2| + |y_1 - y_2| = \max_{1 \le i \le 2} x_i - \min_{1 \le i \le 2} x_i + \max_{1 \le i \le 2} y_i - \min_{1 \le i \le 2} y_i \ .$$

Lemma 1 (Theorem 2.4, [7]). *If $p_i = (x_i, y_i), 1 \le i \le 3$, are three points in $\mathbb{Z}^2$, then their tristance equals,*

$$d_3(p_1, p_2, p_3) = \max_{1 \le i \le 3} x_i - \min_{1 \le i \le 3} x_i + \max_{1 \le i \le 3} y_i - \min_{1 \le i \le 3} y_i \ .$$

In [7], Etzion and Vardy give constructions for lattice interleavers with 2 repetitions in the + model. The generator matrices for the interleavers are,

$$\mathbf{G}_{4k} = \begin{pmatrix} k & k \\ 0 & 3k \end{pmatrix} \qquad \mathbf{G}_{4k+1} = \begin{pmatrix} k & k+1 \\ 0 & 3k+2 \end{pmatrix}$$

$$\mathbf{G}_{4k+2} = \begin{pmatrix} k+1 & k \\ 1 & 3k+1 \end{pmatrix} \qquad \mathbf{G}_{4k+3} = \begin{pmatrix} k+1 & k+1 \\ 0 & 3k+2 \end{pmatrix}$$

for $k \ge 1$, and the resulting lattices are denoted Λ_{4k+i}, for $0 \le i \le 3$. It was shown ([7], Theorem 3.1) that for all $k \ge 1$ and $0 \le i \le 3$,

$$d_3(\Lambda_{4k+i}) = 4k + i \ .$$

Furthermore, the following theorem shows that Λ_{4k} and Λ_{4k+2} are optimal.

Theorem 1 ([7], Theorem 3.6). *Let Λ be any sublattice of $\mathbb{Z}^2$ with tristance $d_3(\Lambda) = t$. Set $k = \lfloor t/4 \rfloor$. Then the volume of Λ is bounded from below as follows:*

$$V(\Lambda) \geq 3k^2 \qquad\qquad\qquad if\ t \equiv 0 \pmod 4$$

$$V(\Lambda) \geq 3k^2 + \frac{3}{2}k + \frac{1}{2} \qquad\qquad if\ t \equiv 1 \pmod 4$$

$$V(\Lambda) \geq 3k^2 + 3k + 1 \qquad\qquad if\ t \equiv 2 \pmod 4$$

$$V(\Lambda) \geq 3k^2 + \frac{9}{2}k + \frac{5}{2} \qquad\qquad if\ t \equiv 3 \pmod 4$$

In the following subsection we improve on the second and fourth cases, and show that Λ_{4k+1} and Λ_{4k+3} are also optimal.

2.2   Lower Bounds

Theorem 2. *Let Λ be a sublattice of $\mathbb{Z}^2$ with $d_3(\Lambda) = 4k + 1 + 2i$, where $i \in \{0, 1\}$, then $V(\Lambda) \geq (3k + 2)(k + i)$.*

Proof. We first note that $d_2(\Lambda) \geq 2k + 1 + i$. Otherwise, let $p_0 = (0, 0)$, and $p' = (x', y')$ be two points in Λ such that $d_2(p_0, p') \leq 2k + i$, and then $d_3(p_0, p', 2p') = 4k + 2i$, so $d_3(\Lambda) \leq 4k + 2i$ which is a contradiction.

Let $p_0 = (0, 0)$, $p_1 = (x_1, y_1)$, and $p_2 = (x_2, y_2)$, for which $x_2 \geq x_1 \geq 0$, and $d_3(p_0, p_1, p_2) = d_3(\Lambda) = 4k + 1 + 2i$. We start by showing that we should only prove the case where $y_1 > y_2 \geq 0$.

If $y_1 < 0$ we take a mirror image of the lattice along the X axis and continue with the same proof. Hence we may assume that $y_1 \geq 0$. Now, if $y_2 < 0$, we move p_2 to the origin and take a mirror image of the lattice along the Y axis to achieve the required configuration, and then continue with the same proof. Therefore we may also assume that $y_2 \geq 0$. The last case is that of $y_2 \geq y_1$. In that case,

$$d_3(p_0, p_1, p_2) = d_2(p_0, p_1) + d_2(p_1, p_2) \geq 2d_2(\Lambda) \geq 4k + 2 + 2i \ ,$$

which contradicts our assumption. Thus, $y_1 > y_2 \geq 0$ is the only case left for us to handle.

We start by sharpening the inequalities. If $x_1 = x_2$ then again,

$$d_3(p_0, p_1, p_2) = d_2(p_0, p_2) + d_2(p_2, p_1) \geq 2d_2(\Lambda) \geq 4k + 2 + 2i \ ,$$

which is a contradiction. Hence $x_2 > x_1$. We now show that p_0, p_1, p_2, and $p_2 - p_1$, define a fundamental region. We actually prove a slightly stronger claim: there are no points of Λ in the rectangle $R = \{(x, y) \mid 0 < x < x_2, \ y_2 - y_1 < y < y_1\}$. Let us assume the contrary, i.e., that there exists $p = (x, y) \in \Lambda \cap R$. Now, if $y \geq 0$ then $d_3(p_0, p_2, p) = x_2 + \max\{y_2, y\} < x_2 + y_1 = 4k + 1 + 2i$, since $y, y_2 < y_1$. This is a contradiction, since $d_3(\Lambda) = 4k + 1 + 2i$. In the same manner, if $y < 0$, then $d_3(p_0, p_2, p) = x_2 + y_2 - y < x_2 + y_1 = 4k + 1 + 2i$, since $y > y_2 - y_1$, again a contradiction. Thus, p_0, p_1, p_2, and $p_2 - p_1$, define a fundamental region.

In the current configuration, $d_3(\Lambda) = 4k + 1 + 2i = x_2 + y_1$. Since one of the two summands must be strictly greater than the other, we may assume that $x_2 > y_1$, or else we exchange the X and Y axes and repeat the proof. We may therefore denote $x_2 = 2k + 1 + i + \delta$, and $y_1 = 2k + i - \delta$ for some integer $\delta \geq 0$. With the fundamental region defined above we have,

$$V(\Lambda) = \begin{vmatrix} x_2 - x_1 & y_2 - y_1 \\ x_1 & y_1 \end{vmatrix} = x_2 y_1 - x_1 y_2$$

$$= (3k + 2)(k + i) + i(i - 1) + k(k + i) - (\delta^2 + \delta + x_1 y_2)$$

$$= (3k + 2)(k + i) + k(k + i) - (\delta^2 + \delta + x_1 y_2) \qquad \text{since } i \in \{0, 1\} \ .$$

All we have to do now, is show that $\delta^2 + \delta + x_1 y_2 \leq k(k + i)$.

Using the fact that $d_2(\Lambda) \geq 2k + 1 + i$ we get the following inequalities:

$$2k + 1 + i \leq d_2(p_0, p_1) = x_1 + 2k + i - \delta \iff 0 \leq \delta \leq x_1 - 1 \qquad (1)$$
$$2k + 1 + i \leq d_2(p_1, p_2) = 4k + 1 + 2i - (x_1 + y_2) \iff x_1 + y_2 \leq 2k + i \qquad (2)$$

Two more inequalities are achieved by examining p_1, p_2, and $2p_1$. If $2x_1 \leq x_2$ then,

$$4k + 1 + 2i \leq d_3(p_1, p_2, 2p_1) \iff y_2 \leq 2k + i - x_1 - \delta \ . \qquad (3)$$

Otherwise, if $2x_1 > x_2$, then,

$$4k + 1 + 2i \leq d_3(p_1, p_2, 2p_1) \iff x_1 - y_2 \geq 2\delta + 1 \ . \qquad (4)$$

If $2x_1 \leq x_2$ then,

$$\begin{aligned}
\delta^2 + \delta + x_1 y_2 &\leq \delta^2 + \delta + x_1(2k + i - x_1 - \delta) & \text{by (3)} \\
&\leq x_1(2k + i - x_1) & \text{maximized at } \delta = 0, x_1 - 1 \text{ by (1)} \\
&\leq k(k + i) & \text{maximized at } x_1 = k, k + i \ .
\end{aligned}$$

Otherwise, $2x_1 > x_2$ and then,

$$\begin{aligned}
\delta^2 + \delta + x_1 y_2 &\leq \delta^2 + \delta + (k + \delta + 1)(k + i - \delta - 1) & \text{by (2) and (4)} \\
&= k(k + i) + (\delta + 1)(i - 1) \leq k(k + i) & \text{since } \delta \geq 0, \text{ and } i \in \{0, 1\} \ .
\end{aligned}$$

$\square$

Corollary 1. *The lattices Λ_{4k+1} and Λ_{4k+3} are optimal.*

3 The Hexagonal Model

3.1 Preliminaries

Another model of interest to us is the hexagonal model. We follow the same notations as in [8]. Instead of the rectangular grid we used up to now, we define the following graph. We start by tiling the plane $\mathbb{R}^2$ with regular hexagons. The vertices of the graph are the center points of the hexagons. We connect two vertices if and only if their respective hexagons are adjacent. This way, each vertex has exactly 6 neighboring vertices.

Since handling this grid directly is hard, we prefer an isomorphic representation of the model. This representation includes $\mathbb{Z}^2$ as the set of vertices. Each point $(x, y) \in \mathbb{Z}^2$ has the following neighboring vertices,

$$\{(x + a, y + b) \mid a, b \in \{-1, 0, 1\}, a + b \neq 0\} \enspace.$$

It may be shown that the two models are isomorphic by using the mapping $\xi : \mathbb{R}^2 \to \mathbb{Z}^2$, which is defined by $\xi(x, y) = (\frac{x}{\sqrt{3}} + \frac{y}{3}, \frac{2y}{3})$. The effect of the mapping on the neighbor set is shown in Fig. 1. From now on, by abuse of notation, we will also call the last model – the hexagonal model.

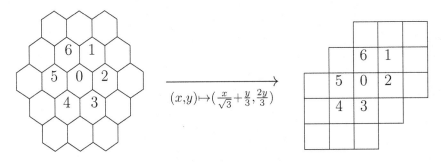

Fig. 1. The hexagonal model translation

Obviously, the distance d_2^{hex} between two points $p_i = (x_i, y_i)$, $i = 1, 2$, is

$$d_2^{\mathrm{hex}}(p_1, p_2) = \begin{cases} \max\{|x_1 - x_2|, |y_1 - y_2|\} & (x_1 - x_2)(y_1 - y_2) \geq 0 \\ |x_1 - x_2| + |y_1 - y_2| & \text{otherwise} \enspace. \end{cases}$$

Handling the tristance in the hexagonal model is a little more complicated.

Theorem 3 ([8], Theorem 6). *Let $p_i = (x_i, y_i)$, $1 \leq i \leq 3$ be points in $\mathbb{Z}^2$ for which, W.l.o.g., $x_1 \leq x_2 \leq x_3$ then,*

$$d_3^{\mathrm{hex}}(p_1, p_2, p_3) = \begin{cases} d_2^{\mathrm{hex}}(p_1, p_2) + d_2^{\mathrm{hex}}(p_2, p_3) & y_1 \leq y_2 \leq y_3 \\ d_2^{\mathrm{hex}}(p_1, \min(p_2, p_3)) + d_2^{\mathrm{hex}}(p_2, p_3) & y_1 \leq y_3 \leq y_2 \\ d_3(p_1, p_2, p_3) & y_3 \leq y_1 \leq y_2 \\ d_3(p_1, p_2, p_3) & y_3 \leq y_2 \leq y_1 \\ d_3(p_1, p_2, p_3) & y_2 \leq y_3 \leq y_1 \\ d_2^{\mathrm{hex}}(p_1, p_2) + d_2^{\mathrm{hex}}(\max(p_1, p_2), p_3) & y_2 \leq y_1 \leq y_3 \end{cases},$$

where max *(min) of two points is a component-wise* max *(min).*

An important thing to observe is that the hexagonal tristance allows scaling.

Theorem 4. *Let $p_1, p_2, p_3 \in \mathbb{Z}^2$ be three points and $k \geq 0$ an integer, then*

$$d_3^{\text{hex}}(kp_1, kp_2, kp_3) = k \cdot d_3^{\text{hex}}(p_1, p_2, p_3) .$$

Proof. The theorem simply results from Theorem 3 and the fact that the tristance in the $+$ model also allows scaling, as stated in [7]. $\qquad\square$

3.2 Constructions

For each integer $k \geq 1$ we define the lattices $\Lambda_{2k}^{\text{hex}}$ and $\Lambda_{2k+1}^{\text{hex}}$ by their respective generator matrices:

$$\mathbf{G}_{2k}^{\text{hex}} = \begin{pmatrix} k & 0 \\ 0 & k \end{pmatrix} \qquad \mathbf{G}_{2k+1}^{\text{hex}} = \begin{pmatrix} k & -1 \\ 1 & k+1 \end{pmatrix} .$$

Theorem 5.

$$\begin{aligned} d_3^{\text{hex}}(\Lambda_{2k}^{\text{hex}}) &= 2k & V(\Lambda_{2k}^{\text{hex}}) &= k^2 \\ d_3^{\text{hex}}(\Lambda_{2k+1}^{\text{hex}}) &= 2k+1 & V(\Lambda_{2k+1}^{\text{hex}}) &= k^2 + k + 1 . \end{aligned}$$

Proof. The volumes of the lattices are easily calculated by the determinants of generator matrices. Therefore, we turn to prove the minimal tristance of the lattices is as specified.

The simple case is the lattice $\Lambda_{2k}^{\text{hex}}$. This lattice is a scaling up of the trivial lattice $\mathbb{Z}^2$, by a factor of k. Since $d_3^{\text{hex}}(\mathbb{Z}^2) = 2$, we immediately get $d_3^{\text{hex}}(\Lambda_{2k}^{\text{hex}}) = 2k$.

The last case requires more care. Given three points which achieve the minimal tristance in $\Lambda_{2k+1}^{\text{hex}}$, we may always move the leftmost point to the origin. Hence, we may assume that the three points are, $p_0 = (0,0)$, $p_1 = (x_1, y_1)$, $p_2 = (x_2, y_2)$, and $x_1, x_2 \geq 0$.

We now note that both $p_1' = (k, -1)$ and $p_2' = (k+1, k)$ belong to $\Lambda_{2k+1}^{\text{hex}}$, and that $d_3^{\text{hex}}(p_0, p_1', p_2') = 2k+1$. Hence, as potential candidates for p_1 and p_2, we need to examine only points $p = (x, y)$ of $\Lambda_{2k+1}^{\text{hex}}$ for which, $x \geq 0$ and $d_2^{\text{hex}}(p_0, p) \leq 2k+1$. The only such points of $\Lambda_{2k+1}^{\text{hex}}$ are easily seen to be,

$$(k, -1), \quad (1, k+1), \quad (k+1, k), \quad (k-1, -k-2) .$$

Going over the 6 possible choices of pairs of points from the list, one may verify that $d_3^{\text{hex}}(p_0, p_1, p_2) \geq 2k+1$ in all cases. $\qquad\square$

3.3 Lower Bounds

We now show that both $\Lambda_{2k}^{\text{hex}}$ and $\Lambda_{2k+1}^{\text{hex}}$ are optimal in the sense that they have the lowest possible interleaving degree. We do so by explicitly proving $\Lambda_{2k}^{\text{hex}}$ to be optimal, and deducing that $\Lambda_{2k+1}^{\text{hex}}$ must be also optimal.

Due to lack of space, the proof that $\Lambda_{2k}^{\text{hex}}$ is optimal is omitted. The complete proof may be found in [9].

Theorem 6. *Let Λ be a sublattice of $\mathbb{Z}^2$ with $d_3^{\text{hex}}(\Lambda) = 2k$, and $d_2^{\text{hex}}(\Lambda) = k$. In that case, $V(\Lambda) \geq k^2$.*

Theorem 7. *Let Λ be a sublattice of $\mathbb{Z}^2$ with $d_3^{\text{hex}}(\Lambda) = 2k$, and $d_2^{\text{hex}}(\Lambda) > k$. In that case, $V(\Lambda) > k^2$.*

Corollary 2. *Let Λ be a sublattice of $\mathbb{Z}^2$ with $d_3^{\text{hex}}(\Lambda) = 2k$. So then, $V(\Lambda) \geq k^2$.*

We now show that $\Lambda_{2k+1}^{\text{hex}}$ is optimal also.

Theorem 8. *Let Λ be a sublattice of $\mathbb{Z}^2$ with $d_3^{\text{hex}}(\Lambda) = 2k+1$. So then, $V(\Lambda) \geq k^2 + k + 1$.*

Proof. Let us assume to the contrary, that $V(\Lambda) \leq k^2 + k$. Let Λ' be a scaling up of Λ by a factor of 2. Hence, $d_3^{\text{hex}}(\Lambda') = 2d_3^{\text{hex}}(\Lambda) = 2(2k+1)$, and $V(\Lambda') = 4V(\Lambda) \leq 4k^2 + 4k$. However, according to Corollary 2, $V(\Lambda') \geq (2k+1)^2 = 4k^2 + 4k + 1$, a contradiction. $\square$

4   The $*$ Model

4.1   Preliminaries

The $*$ model uses the rectangular grid as the previous $+$ model does, but each point $(x, y) \in \mathbb{Z}^2$ has eight neighboring points forming the set

$$\{(x + a, y + b) \in \mathbb{Z}^2 \mid a, b \in \{-1, 0, 1\}, |a| + |b| \neq 0\} \ .$$

We denote the r-dispersion in the $*$ model as d^* and in general, by affixing the $*$ to a notation we refer to its $*$ model counterpart. Etzion and Vardy [7] construct the lattices Λ_{4k}^*, Λ_{4k+1}^*, Λ_{4k+2}^*, los_{4k+3}, by providing their respective generator matrices,

$$\mathbf{G}_{4k}^* = \begin{pmatrix} k & 3k \\ 0 & 6k - 1 \end{pmatrix} \qquad \mathbf{G}_{4k+1}^* = \begin{pmatrix} k+1 & 3k+1 \\ 1 & 6k \end{pmatrix}$$

$$\mathbf{G}_{4k+2}^* = \begin{pmatrix} k+1 & 3k+1 \\ 1 & 6k+2 \end{pmatrix} \qquad \mathbf{G}_{4k+3}^* = \begin{pmatrix} k+2 & 3k+2 \\ 2 & 6k+3 \end{pmatrix} \ .$$

It was shown ([7], Theorem 7.2) that for all $k \geq 1$ and $0 \leq i \leq 3$,

$$d_3^*(\Lambda_{4k+i}^*) = 4k + i \ .$$

However, no proof is given to show that the lattices are optimal.

Our main tool for handling the $*$ model is the function φ defined in [7]. Let us denote the sublattice of $\mathbb{Z}^2$ defined as,

$$D_2 = \{(x, y) \mid x + y \equiv 0 \pmod{2}\} \ .$$

The mapping $\varphi : \mathbb{Z}^2 \rightarrow D_2$ is defined as

$$\varphi(x, y) = (x - y, x + y) \ .$$

In essence, φ rotates the plane counterclockwise by an angle of $\pi/4$ and scales it up by a factor of $\sqrt{2}$.

If Λ is a sublattice of $\mathbb{Z}^2$, then $\Lambda' = \varphi(\Lambda)$ is obviously a sublattice of D_2. By [7] Theorem 7.1,

$$d_3^*(\Lambda) = \left\lceil \frac{d_3(\Lambda')}{2} \right\rceil \ .$$

By the nature of φ, it is also easy to show that

$$V(\Lambda) = \frac{V(\Lambda')}{2} \ .$$

4.2 Lower Bounds

Theorem 9. *Let Λ be a sublattice of $\mathbb{Z}^2$ with $d_3^*(\Lambda) = 4k$, then $V(\Lambda) \geq 6k^2 - k$.*

Proof. Let us assume the contrary, i.e., that $d_3^*(\Lambda) = 4k$, and $V(\Lambda) < 6k^2 - k$. Let $\Lambda' = \varphi(\Lambda)$, so then $d_3(\Lambda')$ is either $8k - 1$ or $8k$, and $V(\Lambda') < 12k^2 - 2k$.

If $d_3(\Lambda') = 8k - 1$, then by Theorem 2, $V(\Lambda') \geq 12k^2 - 2k$. If $d_3(\Lambda') = 8k$, then by Theorem 1, $V(\Lambda') \geq 12k^2$. Either way, we have a contradiction. $\square$

Theorem 10. *Let Λ be a sublattice of $\mathbb{Z}^2$ with $d_3^*(\Lambda) = 4k + 2$, then $V(\Lambda) \geq 6k^2 + 5k + 1$.*

Proof. Let us assume the contrary, i.e., that $d_3^*(\Lambda) = 4k + 2$, and $V(\Lambda) < 6k^2 + 5k + 1$. Let $\Lambda' = \varphi(\Lambda)$, so then $d_3(\Lambda')$ is either $8k + 3$ or $8k + 4$, and $V(\Lambda') < 12k^2 + 10k + 2$.

If $d_3(\Lambda') = 8k + 3$, then by Theorem 2, $V(\Lambda') \geq 12k^2 + 10k + 2$. If $d_3(\Lambda') = 8k + 4$, then by Theorem 1, $V(\Lambda') \geq 12k^2 + 12k + 3$. Either way, we have a contradiction. $\square$

Corollary 3. *The lattices Λ_{4k}^* and Λ_{4k+2}^* are optimal.*

The two cases left require some more work. If we try to apply the method used in the last two theorems, we find that the bound we achieve is not tight. This stems from the fact that by examining $\varphi(\Lambda)$, we restrict ourselves to sublattices of D_2. We now state the equivalent theorem to Theorem 2 which refers to sublattices of D_2.

Theorem 11. *Let Λ be a sublattice of D_2 with $d_3(\Lambda) = 4k + 1$, then $V(\Lambda) \geq 3k^2 + 3k - 2$.*

Proof. The proof proceeds in a similar fashion to the proof of Theorem 2, so we will only point out the differences. The first one is the fact that in D_2, the distance between any two points is even. Hence, $d_2(\Lambda) \geq 2k + 2$.

This, in turn, changes inequalities (1) and (2) to the following:

$$2k + 2 \leq d_2(p_0, p_1) = x_1 + 2k - \delta \qquad \Longleftrightarrow \qquad 0 \leq \delta \leq x_1 - 2 \qquad (5)$$
$$2k + 2 \leq d_2(p_1, p_2) = 4k + 1 - (x_1 + y_2) \qquad \Longleftrightarrow \qquad x_1 + y_2 \leq 2k - 1 \qquad (6)$$

We now remind that $x_2 = 2k + 1 + \delta$ and $y_1 = 2k - \delta$, so x_2 and y_1 have different parity. This means that x_1 and y_2 also have different parity. We distinguish between two cases:

Case 1: $2x_1 \leq x_2$. There are two subcases according to the parity of δ.

Case 1a: δ is even. Since the parity of x_1 and y_2 is different, (3) is sharper and we get

$$y_2 \leq 2k - x_1 - \delta - 1 \ . \qquad (7)$$

Now,

$$\begin{aligned}
\delta^2 + \delta + x_1 y_2 &\leq \delta^2 + \delta + x_1(2k - x_1 - \delta - 1) &&\text{by (7)} \\
&\leq x_1(2k - x_1 - 1) &&\text{maximized at } \delta = 0 \text{ by (5)} \\
&\leq k^2 - k &&\text{maximized at } x_1 = k - 1, k \ .
\end{aligned}$$

Case 1b: δ is odd. Hence $\delta \geq 1$, so then,

$$\begin{aligned}
\delta^2 + \delta + x_1 y_2 &\leq \delta^2 + \delta + x_1(2k - x_1 - \delta) &&\text{by (3)} \\
&\leq 2 + x_1(2k - x_1 - 1) &&\text{maximized at } \delta = 1, x_1 - 2 \\
&\leq k^2 - k + 2 &&\text{maximized at } x_1 = k - 1, k \ .
\end{aligned}$$

Case 2: $2x_1 > x_2$. Then,

$$\begin{aligned}
\delta^2 + \delta + x_1 y_2 &\leq \delta^2 + \delta + (k + \delta)(k - \delta - 1) &&\text{by (6) and (4)} \\
&= k^2 - k
\end{aligned}$$

We see that in any case, $\delta^2 + \delta + x_1 y_2 \leq k^2 - k + 2$. Like in the proof of Theorem 2,

$$\begin{aligned}
V(\Lambda) &= \begin{vmatrix} x_2 - x_1 & y_2 - y_1 \\ x_1 & y_1 \end{vmatrix} \\
&= x_2 y_1 - x_1 y_2 = 4k^2 + 2k - (\delta^2 + \delta + x_1 y_2) \geq 3k^2 + 3k - 2 \ .
\end{aligned}$$

$\square$

Note that for $k = 1$, the bound of Theorem 11 is worse than the bound of Theorem 2. This does not interfere with the following theorems which do not reach that case.

Theorem 12. *Let Λ be a sublattice of $\mathbb{Z}^2$ with $d_3^*(\Lambda) = 4k + 1$ and $k \geq 1$, then $V(\Lambda) \geq 6k^2 + 3k - 1$.*

Proof. Let us assume the contrary, i.e., that $d_3^*(\Lambda) = 4k + 1$, and $V(\Lambda) < 6k^2 + 3k - 1$. Let $\Lambda' = \varphi(\Lambda)$, so then $d_3(\Lambda')$ is either $8k + 1$ or $8k + 2$, and $V(\Lambda') < 12k^2 + 6k - 2$.

Note that Λ' is a sublattice of D_2. Therefore, if $d_3(\Lambda') = 8k + 1$, then by Theorem 11, $V(\Lambda') \geq 12k^2 + 6k - 2$. If $d_3(\Lambda') = 8k + 2$, then by Theorem 1, $V(\Lambda') \geq 12k^2 + 6k + 1$. Either way, we have a contradiction. □

Theorem 13. *Let Λ be a sublattice of $\mathbb{Z}^2$ with $d_3^*(\Lambda) = 4k + 3$ and $k \geq 1$, then $V(\Lambda) \geq 6k^2 + 9k + 2$.*

Proof. Let us assume the contrary, i.e., that $d_3^*(\Lambda) = 4k + 3$, and $V(\Lambda) < 6k^2 + 9k + 2$. Let $\Lambda' = \varphi(\Lambda)$, so then $d_3(\Lambda')$ is either $8k + 5$ or $8k + 6$, and $V(\Lambda') < 12k^2 + 18k + 4$.

Note that Λ' is a sublattice of D_2. Therefore, if $d_3(\Lambda') = 8k + 5$, then by Theorem 11, $V(\Lambda') \geq 12k^2 + 18k + 4$. If $d_3(\Lambda') = 8k + 6$, then by Theorem 1, $V(\Lambda') \geq 12k^2 + 18k + 7$. Either way, we have a contradiction. □

Corollary 4. *The lattices Λ_{4k+1}^* and Λ_{4k+3}^* are optimal.*

References

1. Abdel-Ghaffar, K.A.S., McEliece, R.J., van Tilborg, H.C.A.: Two-dimensional burst identification codes and their use in burst correction. IEEE Trans. on Inform. Theory **34** (1988) 494–504
2. Blaum, M., Farrell, P.G.: Array codes for cluster-error correction. Electronic Letters **30** (1994) 1752–1753
3. Farrell, P.G.: Array codes for correcting cluster-error patterns. In: Proc. IEE Conf. Elect. Signal Processing (York, England). (1982)
4. Imai, R.M.: Two-dimensional fire codes. IEEE Trans. on Inform. Theory **19** (1973) 796–806
5. Imai, R.M.: A theory of two-dimensional cyclic codes. Inform. and Control **34** (1977) 1–21
6. Blaum, M., Bruck, J., Vardy, A.: Interleaving schemes for multidimensional cluster errors. IEEE Trans. on Inform. Theory **44** (1998) 730–743
7. Etzion, T., Vardy, A.: Two-dimensional interleaving schemes with repetitions: constructions and bounds. IEEE Trans. on Inform. Theory **48** (2002) 428–457
8. Etzion, T., Schwartz, M., Vardy, A.: Optimal tristance anticodes. manuscript in preparation, October 2002.
9. Schwartz, M., Etzion, T.: Optimal 2-dimensional 3-dispersion lattices. manuscript in preparation, August 2002.

On g-th MDS Codes and Matroids

Keisuke Shiromoto

Department of Electronics and Informatics
Ryukoku University
Seta, Otsu 520-2194, Japan
keisuke@rins.ryukoku.ac.jp

Abstract. In this paper, we give a relationship between the generalized Hamming weights for linear codes over finite fields and the rank functions of matroids. We also consider a construction of g-th MDS codes from m-paving matroids. And we determine the support weight distributions of g-th MDS codes.

1 Introduction

The closed connection between matroid theory and coding theory has been discussed by many researchers. For instance, Greene [3] gave a proof of the MacWilliams identity [7] for the Hamming weight enumerator of a linear code by using the Tutte polynomial of the corresponding matroid. Barg [2] studied the relation between the support weight enumerator of a linear code and the Tutte polynomial of the matroid. In addition, he showed the MacWilliams equation of the support weight enumerator in a simple form. In [10], Rajpal studied paving matroids and the corresponding linear codes.

The generalized Hamming weights of a linear code were introduced by Wei [13]. The weights are natural extensions of the concept of minimum Hamming weights of linear codes. Many applications of the generalized Hamming weights are well-known. They are useful in cryptography (cf. [13]), in trellis coding (cf. [5]), etc. The generalized Hamming weights have been determined for binary Hamming codes, MDS codes, Golay codes, Reed-Muller codes and their duals [13]. Similar properties were in fact considered earlier by Helleseth, Kløve and Mykkeltveit [4] for irreducible cyclic codes.

The g-th maximum distance separable (MDS) code was defined by Wei [13] as a linear code which meets the generalized Singleton bound on the g-th generalized Hamming weight. In [12], Tsfasman and Vlădut gave a construction of the codes from algebraic-geometric codes.

In this paper, we consider the generalized Hamming weights for the m-paving codes. We also look for a construction of the codes from matroid theory. Then we give some examples of the codes.

M. Fossorier, T. Hoeholdt, and A. Poli (Eds.): AAECC 2003, LNCS 2643, pp. 226–234, 2003.

2 Notation and Terminology

We begin by introducing matroids, as in [9]. A *matroid* is an ordered pair $M = (E, \mathcal{I})$ consisting of a finite set E and a collection $\mathcal{I}$ of subsets of E satisfying the following three conditions:

(I1) $\emptyset \in \mathcal{I}$.
(I2) If $I \in \mathcal{I}$ and $I' \subseteq I$, then $I' \in \mathcal{I}$.
(I3) If I_1 and I_2 are in $\mathcal{I}$ and $|I_1| < |I_2|$, then there is an element e of $I_2 - I_1$ such that $I_1 \cup e \in \mathcal{I}$.

The members of $\mathcal{I}$ are the *independent sets* of M, and a subset of E that is not in $\mathcal{I}$ is called *dependent*. A minimal dependent set in M is called a *circuit* of M, and a maximal independent set in M is called a *base* of M. For a subset X of E, we define the *rank* of X as follows:

$$r(X) := \max\{|Y| \ : \ Y \subseteq X, \ Y \in \mathcal{I}\}.$$

The *dual matroid* M^* of M is defined as the matroid, the set of bases of which is

$$\{E - B \ : \ B \text{ is a base of } M\}.$$

When we denote the rank of M^* by r^*, the following is well-known:

$$r^*(X) = |X| - r(M) + r(E - X).$$

Throughout this paper, let $\mathbb{F}_q$ be a finite field of q elements. For an $m \times n$ matrix A over $\mathbb{F}_q$, if E is the set of column labels of A and $\mathcal{I}$ is the set of subsets X of E for which the multiset of columns labelled by X is linearly independent in the vector space $\mathbb{F}_q^m$, then $M[A] := (E, \mathcal{I})$ is a matroid and is called *vector matroid* of A (cf. [9]).

For a vector $\boldsymbol{x} = (x_1, \ldots, x_n) \in \mathbb{F}_q^n$ and a subset $D \subseteq \mathbb{F}_q^n$, we define the *supports* of $\boldsymbol{x}$ and D respectively as follows:

$$\mathrm{supp}(\boldsymbol{x}) := \{i \mid x_i \neq 0\},$$
$$\mathrm{Supp}(D) := \bigcup_{\boldsymbol{x} \in D} \mathrm{supp}(\boldsymbol{x}).$$

Let C be an $[n, k]$ code over $\mathbb{F}_q$. For each g, $1 \leq g \leq k$, the g-th *generalized Hamming weight* (GHW) $d_g(C)$ is defined by Wei [13] as follows:

$$d_g(C) := \min\{|\mathrm{Supp}(D)| \ : \ D \text{ is an } [n, g] \text{ subcode of } C\}.$$

The *weight hierarchy* of C is the set of integers $\{d_g(C) \ : \ 1 \leq g \leq k\}$. The following was also proved by Wei [13]:

Monotonicity: $1 \leq d_1(C) < d_2(C) < \cdots < d_k(C) \leq n$. $\qquad(1)$

Duality: $\{d_g(C) : 1 \leq g \leq k\} = \{1, 2, \ldots, n\} - \{n + 1 - d_{g'}(C^\perp)$
$$: 1 \leq g' \leq n - k\}. \ (2)$$

Generalized Singleton Bound: $d_g(C) \leq n - k + g$. $\qquad(3)$

3 GHW and m-Paving Matroids

3.1 A Connection

First, we introduce the connection between the generalized Hamming weights of a linear code and matroid theory. It is usual, for studying the relationship between linear codes and matroids, to deal with the matroid of a generator matrix of a linear code ([2], [10], etc.). In this paper, however, we shall study the rank $n - k$ matroid $M[H]$ of a parity-check matrix H of an $[n, k]$ code C to focus on the generalized Hamming weights of C. Since $M[H]$ is determined by C (not the chosen parity-check matrix H), we shall represent $M[H] = M_C$. However, a linear code C has more information than the matroid M_C. Indeed, a matroid is the vector matroid of several linear codes. It is also clear that the dual matroid $(M_C)^*$ corresponds to the matroid $M_{C^\perp}$ of the dual code $C^\perp$.

The following is well-known [14].

Proposition 1. *Let H be a parity-check matrix for a linear code C over $\mathbb{F}_q$. Then $d_g(C) = \delta$ if and only if the following two conditions hold:*
(1) every set of $\delta - 1$ columns of H has rank $\delta - g$ or more;
(2) there exist δ columns of H with rank $\delta - g$.

The above result immediately shows the following theorem.

Theorem 2. *Let $M_C = M[H]$ be the vector matroid of a parity-check matrix H for an $[n, k]$ code C over $\mathbb{F}_q$. Then $d_g(C) = \delta$ for a g, $1 \leq g \leq k$, if and only if the following two conditions hold:*
(1) for any $(\delta - 1)$-subset X of $E(M_C)$, $r(X) \geq \delta - g$;
(2) there exists a δ-subset Y of $E(M_C)$ with $r(Y) = \delta - g$.

Proof. Let $\boldsymbol{x}_1, \boldsymbol{x}_2, \ldots, \boldsymbol{x}_{\delta-1}$ be the column vectors of H which are labelled by a $(\delta - 1)$-subset X of $E(M_C)$. If at least $\delta - g$ vectors of $\{\boldsymbol{x}_1, \boldsymbol{x}_2, \ldots, \boldsymbol{x}_{\delta-1}\}$ are linearly independent, then we have that $r(X) \geq \delta - g$, and vice versa. Let $\boldsymbol{y}_1, \boldsymbol{y}_2, \ldots, \boldsymbol{y}_\delta$ be the column vectors of H which are labelled by a δ-subset Y of $E(M_C)$. If there exists a set of $\delta - g$ vectors of $\{\boldsymbol{y}_1, \boldsymbol{y}_2, \ldots, \boldsymbol{y}_\delta\}$ which is a maximal linearly independent set, then we have that $r(Y) = \delta - g$, and vice versa. The theorem follows.   □

Example 3. Let M_C be a *uniform matroid* $U_{n-k,n}$, that is, a matroid on an n-element set E, any $(n - k)$-element subset of E of which is a base. For any $(n - k + g - 1)$-element subset X, it follows that $r(X) = n - k$ for every g, $1 \leq g \leq k$. There exists an $(n - k + g)$-element subset Y such that $r(Y) = n - k$ for every g. Therefore we have that $d_g(C) = n - k + g$ for every g. Consequently it follows that C is an MDS code.

3.2 m-Paving Matroids

An m-paving matroid was introduced by Rajpal [11] and the matroid is a generalization of a paving matroid, that is, a rank r matroid whose circuits have cardinality r or $r + 1$.

Definition 4. A rank r matroid M is *m-paving* for $m \leq r$ if all circuits of M have cardinality exceeding $r - m$.

It is not difficult to show that any uniform matroid $U_{r,n}$ is 0-paving, and any paving matroid is 1-paving. These are the only 0-paving and 1-paving matroids. In [11], Rajpal showed that if G is a generator matrix of a first-order Reed-Muller code $R(1, m)$, then the matroid $M[G]$ is a maximal binary $(m-2)$-paving matroid.

For $m \leq n - k$, we define an *m-paving code* as an $[n, k]$ code C over $\mathbb{F}_q$ such that the matroid M_C is an m-paving. From the above argument, it is clear that the dual code $R(m - 2, m)$ of a Reed-Muller code $R(1, m)$ is an $(m - 2)$-paving code.

The following result indicates the minimum Hamming weight of a paving code.

Proposition 5. [10] *For a parity-check matrix H of an $[n, k]$ code C, if $M[H]$ is a 1-paving matroid, then the minimum Hamming weight of C is $n - k$ or $n - k + 1$.*

On the generalized Hamming weights of an m-paving code, we shall prove a bound which is a generalization of the above result.

Theorem 6. *If an $[n, k]$ code C over $\mathbb{F}_q$ is an m-paving code, then*

$$d_g(C) \geq n - k + g - m \tag{4}$$

for any g, $1 \leq g \leq k$.

Proof. We assume that $d_g(C) < n - k + g - m$ for some g. Then there exists a subset $X \subseteq E(M_C)$ such that $|X| = d_g(C)$ and $r(X) = d_g(C) - g \, (< n - k - m)$. Since the matroid M_C is m-paving, it follows that any $(n-k-m)$-element subset of $E(M_C)$ is an independent set. Therefore, for any $(n - k - m)$-element subset Y of X, we have $r(Y) = n - k - m > r(X)$. A contradiction. $\square$

We remark that the bound (4) contains Proposition 5 because the bound corresponds to $d_1(C) \geq n - k$ for a 1-paving code C. Combining the above bound and the generalized Singleton bound, we note that if C is a m-paving code, then $n - k + g - m \leq d_g(C) \leq n - k + g$.

3.3 *g*-th MDS Codes

We consider a special class of linear codes defined as follows:

Definition 7. [13] *Let C be an $[n, k]$ code over $\mathbb{F}_q$. For g, C is called a g-th MDS code if $d_g(C) = n - k + g$.*

It is well-known that an MDS code is also a g-MDS code for any g and a g-MDS code is always a g'-th MDS code for any g', $g' \geq g$.

The following proposition is due to Tsfasman and Vlăduţ (Corollary 4.1 in [12]).

Proposition 8. *If C is an $[n, k, d]$ code and $r = n + 2 - k - d$, then the dual code $C^\perp$ is an r-th MDS code.*

Now we give a construction of g-th MDS codes from m-paving matroids. That also indicates a duality for g-th MDS codes. From Theorem 2, it is not difficult to prove the following lemma.

Lemma 9. *Let C be an $[n, k]$ code. Then C is a g-th MDS code if and only if $r(X) = n - k$ for any $(n - k + g - 1)$-element subset $X \subseteq E(M_C)$.*

Proof. It follows that C is a g-th MDS code if and only if the following two conditions: (1) $(n - k = r(M_C) =) r(X) \geq (n - k + g) - g$ for any $(n - k + g - 1)$-subset X of $E(M_C) = E$, that is, there exist some bases of M_C in X, and (2) there exists a $(n - k + g)$-subset Y of E with $r(Y) = (n - k + g) - g = n - k$ hold. For the above two conditions, we note that if the condition (1) hols, then the condition (2) also holds, because we can take the subset Y such that $Y = X \cup \{e\}$ for any X and any $e \in E - X$. The lemma follows. $\qquad \square$

Theorem 10. *Let C be an $[n, k]$ code over $\mathbb{F}_q$. If C is a g-paving code for $0 \leq g \leq \min\{n - k, k - 1\}$, then $C^\perp$ is a $(g + 1)$-th MDS code.*

Proof. Since M_C is a g-paving matroid, it follows that $|A| \geq (n - k) - g + 1$ for any circuit A of M_C. If we take any $(k + g)$-element subset X of $E := E(M_C) = E(M_{C^\perp})$, then we have that

$$
\begin{aligned}
r^*(X) &= |X| - r(M_C) + r(E - X) \\
&= (k + g) - (n - k) + (n - k - g) \\
&= k.
\end{aligned}
$$

From Lemma 9, it follows that $C^\perp$ is a $(g + 1)$-th MDS code. The theorem follows. $\qquad \square$

Let P, Q and R be the following binary matrices:

$$
P = \begin{pmatrix} 1\,0\,0\,0\,0\,1\,0\,0\,0 \\ 0\,1\,0\,0\,0\,1\,0\,1\,1 \\ 0\,0\,1\,0\,0\,1\,1\,0\,1 \\ 0\,0\,0\,1\,0\,1\,1\,1\,0 \\ 0\,0\,0\,0\,1\,0\,1\,1\,1 \end{pmatrix}, Q = \begin{pmatrix} 1\,0\,0\,0\,0\,0\,1\,1\,1\,1 \\ 0\,1\,0\,0\,0\,1\,0\,1\,1\,1 \\ 0\,0\,1\,0\,0\,1\,1\,0\,1\,1 \\ 0\,0\,0\,1\,0\,1\,1\,1\,0\,1 \\ 0\,0\,0\,0\,1\,1\,1\,1\,1\,0 \end{pmatrix},
$$

$$
R = \begin{pmatrix} 1\,0\,0\,0\,0\,0\,1\,1\,1\,1\,1\,0\,1\,0\,0\,1 \\ 0\,1\,0\,0\,0\,1\,0\,1\,1\,1\,0\,1\,0\,1\,0\,1 \\ 0\,0\,1\,0\,0\,1\,1\,0\,1\,0\,1\,1\,0\,0\,1\,1 \\ 0\,0\,0\,1\,0\,1\,1\,1\,0\,0\,0\,0\,1\,1\,1\,1 \\ 0\,0\,0\,0\,1\,0\,0\,0\,1\,1\,1\,1\,1\,1\,1 \end{pmatrix}.
$$

In [11], it is shown that the matroids $M[P]$, $M[Q]$ and $M[R]$ are the only binary maximal 2-paving matroids of rank 5. From Theorem 10, we have the following result.

Corollary 11. *The binary codes whose generator matrices are P, Q and R are the binary third MDS codes.*

Remark 12. We can also give a proof of Proposition 8 by using Theorem 10. That means Theorem 10 is a generalization of the proposition. Let C be an $[n, k, d]$ code C and set $g = n + 2 - k - d$. We set $g' = n + 1 - k - d$. Since the minimun Hamming weight d corresponds to the minimum order of curcuits in M_C, we have that $|A| \geq (n - k) - g' + 1 = d$ for any curcuit A. So C is a g'-paving code. Therefore it follows, from Theorem 10, that $C^\perp$ is a $(g' + 1)$-th MDS code.

Now we give a characterization of a generator matrix of a g-th MDS code.

Corollary 13. *Let G be a generator matrix of an $[n, k]$ code C over $\mathbb{F}_q$. For a g, $1 \leq g \leq \min\{n - k, k - 1\}$, C is a g-th MDS code if and only if the matroid $M[G]$ is a $(g - 1)$-paving matroid.*

Proof. If $M[G]$ is a $(g - 1)$-paving matroid, then $C^\perp$ is a $(g - 1)$-paving code and so it follows that C is a g-th MDS code from Theorem 10.

Conversely, we assume that C is a g-th MDS code. We take a subset $X \subseteq E(M[G]) = E$ such that $|X| = n - k + g - 1$. From Lemma 9, we have that $r^*_{M[G]}(X) = n - k$. Then it follows that

$$
\begin{aligned}
r_{M[G]}(E - X) &= |E - M| - r^*_{M[G]}(E) + r^*_{M[G]}(X) \\
&= (k - g + 1) - (n - k) + (n - k) \\
&= k - (g - 1).
\end{aligned}
$$

Therefore, for any circuit T of $M[G]$, we have that $|T| > k - (g - 1)$. $\square$

3.4 Support Weight Enumerators

We study the connection between the support weight enumerator of a linear code and the Tutte polynimial of a matroid. We now consider the support weight enumerators of linear codes over finite fields.

Definition 14. *Let C be an $[n, k]$ code over $\mathbb{F}_q$. For a g, $0 \leq g \leq k$, the g-th support weight enumerator $A^g_C(z)$ of C is defined as follows:*

$$
A^g_C(z) := \sum_{i=0}^{n} A^g_i z^i,
$$

where

$$
A^g_i := |\{D \ : \ D \text{ is an } [n, g] \text{ subcode of } C \text{ and } |\mathrm{Supp}(D)| = i\}|.
$$

We note that $W_C(z) := 1 + (q - 1)A^1_C(z)$ is the Hamming weight enumerator of a linear code C over $\mathbb{F}_q$ (cf. [2], [6]).

Definition 15. For a matroid $M = (E, \mathcal{I})$, the *Tutte polynomial* $T(M; x, y)$ of M is defined by

$$T(M; x, y) := \sum_{S \subseteq E} (x-1)^{r(E)-r(S)}(y-1)^{|S|-r(S)}.$$

For nonnegative integers a, b and a prime power q, we set that

$$[a]_b := \prod_{i=0}^{b-1}(q^a - q^i)$$

$$\langle a \rangle := [a]_a = \prod_{i=0}^{a-1}(q^a - q^i)$$

$$\begin{bmatrix} a \\ b \end{bmatrix} := \frac{[a]_b}{\langle b \rangle},$$

where $\begin{bmatrix} a \\ b \end{bmatrix}$ is the Gaussian binomial coefficient, that is, the number of b-dimensional subspaces of an a-dimensional vector space over $\mathbb{F}_q$.

Let C be an $[n, k]$ code over $\mathbb{F}_q$ having generator matrix G. The following proposition was proved by Barg [2].

Proposition 16.

$$\sum_{t=0}^{g} [g]_t A_C^t(z) = (1-z)^k z^{n-k} T\left(M[G]; \frac{1 + (q^g - 1)z}{1-z}, \frac{1}{z}\right), \tag{5}$$

where $0 \leq g \leq k$.

Using this proposition, Barg gave the simple proof of the MacWilliams type identity for the support weight enumerators.

The following lemma is essential.

Lemma 17. *If C is a g-th MDS code, then*

$$\sum_{i=n-k+g}^{n-s} \binom{n-i}{s} A_i^g = \frac{\binom{n}{s}}{\langle g \rangle} \sum_{t=0}^{g} (-1)^{g-t} q^{\binom{g-t}{2}+t(k-s)} \begin{bmatrix} g \\ t \end{bmatrix}, \tag{6}$$

for any s, $s = 0, 1, \ldots, k - g$.

Proof. We denote the right-hand side in the equation (5) by $\Psi^g_{M[G]}(z)$. Then we have that

$$\Psi^g_{M[G]}(z) = (1-z)^k z^{n-k} \sum_{S \subseteq E} \left(\frac{1 + (q^g - 1)z}{1-z} - 1\right)^{k-r(S)} \left(\frac{1}{z} - 1\right)^{|S|-r(S)}$$

$$= (1-z)^k z^{n-k} \sum_{S \subseteq E} q^{g(k-r(S))} z^{k-|S|}(1-z)^{|S|-k}$$

$$= \sum_{S \subseteq E} q^{g(k-r(S))} z^{n-|S|}(1-z)^{|S|}.$$

From Theorem 10, $M[G]$ is a $(g-1)$-paving matroid and so we have that

$$\Psi^g_{M[G]}(z) = \sum_{j=0}^{k-(g-1)} \binom{n}{j} q^{g(k-j)} z^{n-j} (1-z)^j$$

$$+ \sum_{\substack{S \subseteq E \\ |S| \geq k-(g-1)+1}} q^{g(k-r(S))} z^{n-|S|} (1-z)^{|S|}.$$

From the equation (5), we have that

$$\sum_{t=0}^{g} \begin{bmatrix} g \\ t \end{bmatrix} \langle t \rangle A_C^t(z) = \Psi^g_{M[G]}(z).$$

Using the Gauss inversion formula (see 3.38 Collorary in [1]), it follows that

$$\langle g \rangle A_C^g(z) = \sum_{t=0}^{g} (-1)^{g-t} q^{\binom{g-t}{2}} \begin{bmatrix} g \\ t \end{bmatrix} \Psi^t_{M[G]}(z).$$

Replacing z by z^{-1} and then multiplying by $\frac{z^n}{\langle g \rangle}$ in the above equation, we have that

$$\sum_{i=0}^{n} A_i^g z^{n-i} = \frac{1}{\langle g \rangle} \sum_{t=0}^{g} (-1)^{g-t} q^{\binom{g-t}{2}} \begin{bmatrix} g \\ t \end{bmatrix} \left\{ \sum_{j=0}^{k-g+1} \binom{n}{j} q^{g(k-j)} (z-1)^j \right.$$

$$\left. + \sum_{\substack{S \subseteq E \\ |S| \geq k-g+2}} q^{g(k-r(S))} (z-1)^{|S|} \right\}.$$

Differentiating this equation s times, $s = 0, 1, \ldots, k-g$, and substituting $z = 1$, we have that

$$\sum_{i=0}^{n-s} \frac{(n-i)!}{(n-i-s)!} A_i^g = \frac{1}{\langle g \rangle} \sum_{t=0}^{g} (-1)^{g-t} q^{\binom{g-t}{2}} \begin{bmatrix} g \\ t \end{bmatrix} q^{t(k-s)} \binom{n}{s} s!.$$

The lemma follows.   □

Theorem 18. *Let C be an $[n, k]$ code over $\mathbb{F}_q$ having generator matrix G. If C is a g-th MDS code, then*

$$A^g_{n-k+g} = \frac{\binom{n}{k-g}}{\langle g \rangle} \sum_{t=0}^{g} (-1)^{g-t} q^{\binom{g-t}{2}+tg} \begin{bmatrix} g \\ t \end{bmatrix}$$

and

$$A^g_{n-k+g+1} = \frac{\binom{n}{k-g-1}}{\langle g \rangle} \sum_{t=0}^{g} (-1)^{g-t} q^{\binom{g-t}{2}+tg} \begin{bmatrix} g \\ t \end{bmatrix} (q^t - n + k - g - 1).$$

Proof. By substituting $s = k - g$ and $s = k - g - 1$ in the equation (6), the theorem follows.   □

Corollary 19. *If there exists an $[n, k]$ g-th MDS code with $k \geq g + 1$, then*

$$\frac{\sum_{t=0}^{g}(-1)^{g-t}q^{\binom{g-t}{2}+t(g+1)}\begin{bmatrix}g\\t\end{bmatrix}}{\sum_{t=0}^{g}(-1)^{g-t}q^{\binom{g-t}{2}+tg}\begin{bmatrix}g\\t\end{bmatrix}} \geq n - k + g + 1.$$

Proof. Since the number $A_{n-k+g+1}^{g}$ must be nonnegative, we have the collorary.
                                                                            □

Example 20. Let C_P, C_Q, C_R be the binary third MDS codes having generator matrices P, Q, R in Corollary 11, respectively. From Theorem 18, we have easily that for C_P, $A_7^3 = 36$ and $A_8^3 = 70$, for C_Q, $A_8^3 = 45$ and $A_9^3 = 60$ and for C_R, $A_{14}^3 = 120$ and $A_{15}^3 = 0$.

Acknowledgments. The author would like to thank the referees for their useful comments.

References

1. M. Aigner, *Combinatorial Theory*, Springer-Verlag, Berlin-New York (1980).
2. A. Barg, The matroid of supports of a linear code, *Applicable Algebra in Engineering, Communication and Computing*, **8** (1997) pp. 165–172.
3. C. Greene, Weight enumeration and the geometry of linear codes, *Studies in Applied Mathematics* **55** (1976) pp. 119–128.
4. T. Helleseth, T. Kløve, and J. Mykkeltveit, The weight distribution of irreducible cyclic codes with block lengths $n_1((q^l - 1)/N)$, *Discrete Mathematics* **18** (1977) pp. 179–211.
5. T. Kasami, T. Tanaka, T. Fujiwara, and S. Lin, On the optimum bit orders with respect to the state complexity of trellis diagrams for binary linear codes, *IEEE Trans. Inform. Theory* **39** (1993) pp. 242–245.
6. T. Kløve, Support weight distribution of linear codes. *Discrete Mathematics* **106/107** (1992) pp. 311–316.
7. F.J. MacWilliams, A theorem on the distribution of weights in a systematic code, *Bell System Tech. J* **42** (1963) pp.79–94.
8. F.J. MacWilliams and N.J.A. Sloane, *The Theory of Error-Correcting Codes*, North-Holland, Amsterdam 1977.
9. J.G. Oxley, *Matroid Theory*, Oxford University Press, Oxford, 1992.
10. S. Rajpal, On paving matroids and a generalization of MDS codes, *Discrete Applied Mathematics* **60** (1995) pp. 343–347.
11. S. Rajpal, On binary k-paving matroids and Reed-Muller codes, *Discrete Mathematics* **190** (1998) pp. 191–200.
12. M.A. Tsfasman and S.G. Vlădut, Geometric approach to higher weights, *IEEE Trans. Inform. Theory* **41** (1995) pp. 1564–1588.
13. V.K. Wei, Generalized Hamming weights for linear codes, *IEEE Trans. Inform. Theory* **37** (1991) pp. 1412–1418.
14. V.K. Wei, Generalized Hamming weights; Fundamental open problems in coding theory, *Arithmetic, geometry and coding theory* (Luminy, 1993) pp. 269–281.
15. D.J.A. Welsh, *Matroid Theory*, Academic Press, London, 1976.

On the Minimum Distance of Some Families of $\mathbb{Z}_{2^k}$-Linear Codes

F. Galand

INRIA Rocquencourt, Projet CODES, Le Chesnay, France
and GREYC, University of Caen, Caen, France
Fabien.Galand@inria.fr

Abstract. With the help of a computer, we obtain the minimum distance of some codes belonging to two families of $\mathbb{Z}_{2^k}$-linear codes: the first is the generalized Kerdock codes which aren't as good as the best linear codes and the second is the Hensel lift of quadratic residue codes. In the latter, we found new codes with same minimum distances as the best linear codes of same length and same cardinality. We give a construction of binary codes starting with a $\mathbb{Z}_{2^k}$-linear code and adding cosets to it, increasing its cardinality and keeping the same minimum distance. This construction allows to derive a non trivial upper bound on cardinalities of $\mathbb{Z}_{2^k}$-linear Codes.

1 Introduction

The article of Hammons *et al.*, [1], gives a construction of Kerdock codes as images by the Gray map of linear codes over the ring of integers modulo 4, $\mathbb{Z}_4$ (see also [2]). Such codes are called $\mathbb{Z}_4$-linear. The point is that the Gray map is a translation-invariant isometric mapping from $\mathbb{Z}_4$ with Lee metric, to $GF(2)^2$ with Hamming metric.

One can define a generalization of the Lee metric, called homogeneous metric (see [3]) over the ring of integers modulo 2^k, $\mathbb{Z}_{2^k}$, by the weight function

$$w_L(a) = \begin{cases} 0 & \text{if } a = 0, \\ 2^{k-2} & \text{if } a \neq 2^{k-1}, \\ 2^{k-1} & \text{if } a = 2^{k-1}. \end{cases}$$

The nice feature of this metric is that we know a translation-invariant isometric mapping (see [4]) from $\mathbb{Z}_{2^k}$ to a subset of $GF(2)^{2^{k-1}}$, namely the Reed-Muller code of length 2^{k-1} and order 1, $RM(1, k-1)$. This mapping is called the generalized Gray map, Ψ and is defined by

$$\Psi(a) : (y_1, \ldots, y_{k-1}) \mapsto a_k + \sum_{i=1}^{k-1} a_i y_i \ ,$$

where $a = \sum a_i 2^i \in \mathbb{Z}_{2^k}$ and $a_i \in GF(2)$. With this mapping, binary codes can be constructed from codes over $\mathbb{Z}_{2^k}$ by taking their images with the generalized

M. Fossorier, T. Hoeholdt, and A. Poli (Eds.): AAECC 2003, LNCS 2643, pp. 235–243, 2003.

Gray map. The binary images have same cardinalities as codes over $\mathbb{Z}_{2^k}$ and their minimum Hamming distances are the minimum homogeneous distances of the codes over $\mathbb{Z}_{2^k}$. Moreover, if the code over $\mathbb{Z}_{2^k}$ is linear, its binary image C is distance-invariant, that is all its cosets of the form $c + C$, with $c \in C$ have the same distance distribution. We call $\mathbb{Z}_{2^k}$-linear the codes equal to images by the generalized Gray map of linear codes over $\mathbb{Z}_{2^k}$.

With the generalized Gray map came a natural generalization of Kerdock codes, which can be better than the best linear codes. In Sect. 3, we address the issue of their minimum distance with the help of a computer. We show that these codes are not good for short lengths.

The Hensel lift to $\mathbb{Z}_4$ of quadratic residue codes is known to give good binary codes [5,6,7,8]. We study their Hensel lift to $\mathbb{Z}_{2^k}$, still using a computer. We obtain new results for several codes. In particular we find a $\mathbb{Z}_{16}$-linear code whose parameters are equal to those of the best known linear code.

Finally, in [8], Duursma *et al.* use the union of 2 cosets of a $\mathbb{Z}_8$-linear code to construct a bigger code with same length and minimum distance. We generalize this construction to $\mathbb{Z}_{2^k}$, the number of cosets which are used used depending on k and on the parameters of the $\mathbb{Z}_{2^k}$-linear code. This construction yields a non trivial upper bound on the cardinalities of $\mathbb{Z}_{2^k}$-linear codes.

2   Background

2.1   Galois Rings

We recall basic facts about Galois rings ([9,1]) since we will need them to define generalized Kerdock codes in §3.

Galois rings are finite rings isomorphic to quotient rings $\mathbb{Z}_{p^k}[X]/(P)$ where p is a prime and P is a unitary polynomial such that $P \pmod{p}$ is an irreducible polynomial with coefficients in $\mathrm{GF}(p)$. We denote $\mathrm{GR}(p^k, m)$ the Galois ring isomorphic to $\mathbb{Z}_{p^k}[X]/(P)$ where P has degree m. The set of roots of $X^{p^m-1} - 1$ is a cyclic multiplicative group of order $p^m - 1$. Adding 0 to this group, we get the Teichmuller set, $\mathcal{T} = \{0, \zeta, \ldots, \zeta^{p^m-1}\}$ with ζ a generator of the cyclic group. If we choose for P a polynomial such that $P \pmod{p}$ is primitive, then we can take $\zeta = X \pmod{P}$. The quotient ring $\mathrm{GR}(p^k, m)/(p)$ is simply the field $\mathrm{GF}(p^m)$ and the Teichmuller set is a set of representatives of the equivalence classes of $\mathrm{GR}(p^k, m)$ modulo p.

As with finite fields, the automorphism group of $\mathrm{GR}(p^k, m)$ is cyclic of order m, and generated by a particular element σ, the Frobenius map, which coincides with the power function exponent p on the Teichmuller set and with identity on the subring $\mathbb{Z}_{p^k}$. Using the additive representation of the elements of $\mathrm{GR}(p^k, m)$ (the elements are polynomials over $\mathbb{Z}_{p^k}$ in ζ of degree strictly less than m), σ is defined by

$$\sigma(\alpha) = \sum_{i=0}^{m-1} a_i \zeta^{ip} ,$$

for every $\alpha = \sum a_i \zeta^i$. Over the ring $\mathrm{GR}(p^k, m)$ one can define a trace mapping, $\mathrm{Tr}\colon \mathrm{GR}(p^k, m) \to \mathbb{Z}_{p^k}$, which is the sum over all the automorphisms,

$$\mathrm{Tr}(\alpha) = \sum_{j=0}^{m-1} \sigma^j(\alpha).$$

This mapping is a linear form over $\mathrm{GR}(p^k, m)$. Moreover, every linear form over $\mathrm{GR}(p^k, m)$ has the form $x \mapsto \mathrm{Tr}(\alpha x)$ for some $\alpha \in \mathrm{GR}(p^k, m)$.

2.2 Hensel Lift and Cyclic Codes

The Hensel lift can be viewed as a toolbox for constructing cyclic codes over $\mathbb{Z}_{2^k}$ from binary cyclic codes. We focus on cyclic codes of odd length n over $\mathbb{Z}_{2^k}$ which are free modules. These codes are ideals of $\mathbb{Z}_{2^k}[X]/(X^n - 1)$ generated by a polynomial $g^{(k)}$ dividing $X^n - 1$. Such polynomials can be obtained by Hensel lifting from binary polynomials dividing $X^n - 1$, more precisely, if f divides $X^n - 1$ in $\mathrm{GF}(2)$, then there exists a unique unitary polynomial $f^{(k)}$ of $\mathbb{Z}_{2^k}[X]$ dividing $X^n - 1$ such that $f^{(k)} = f \pmod 2$.

In the sequel a cyclic code over $\mathbb{Z}_{2^k}$ obtained this way will be referred to as Hensel lift of the binary code (for properties of (cyclic) codes over $\mathbb{Z}_{p^k}$ see [10, 11]).

3   Generalized Kerdock Codes and Their Parameters

The Kerdock code of length 2^{m+1} (m odd) is known to be the image by the Gray map of affine functions over $\mathrm{GR}(4, m)$ (viewed as a $\mathbb{Z}_4$-module) restricted to the Teichmuller set (cf. [1]). This led to define generalized Kerdock codes $\mathcal{K}(k, m)$ as the image by his generalized Gray map of affine functions over $\mathrm{GR}(2^k, m)$ (viewed as a $\mathbb{Z}_{2^k}$-module) restricted to the Teichmuller set, write $\mathrm{AF} = \{x \mapsto \mathrm{Tr}(\alpha x) + b \mid \alpha \in \mathrm{GR}(2^k, m), b \in \mathbb{Z}_{2^k}\}$ the set of affine functions,

$$\mathbf{K}(k, m) = \left\{ \left(\mathbf{c}(0), \mathbf{c}(1), \mathbf{c}(\zeta), \ldots, \mathbf{c}\left(\zeta^{2^m-2}\right) \right) \mid \mathbf{c} \in \mathrm{AF} \right\}$$

the set of 2^m-tuples corresponding to restrictions of affine functions to the Teichmuller set, then $\mathcal{K}(k, m) = \Psi(\mathbf{K}(k, m))$ (see [4]). Thus, these codes are of lengths 2^{k+m-1} and cardinality $2^{k(m+1)}$. We have the following lower bound on the minimum distance δ

Proposition 1 (Carlet [4, §IV Cor. 1]).

$$\delta \geq 2^{m+k-2} - 2^{k + \lceil \frac{m}{2^{k-1}} \rceil - 4} \cdot \left\lfloor 2^{\frac{m}{2} + 2 - \lceil \frac{m}{2^{k-1}} \rceil} (2^{k-1} - 1) \right\rfloor. \tag{1}$$

Cyclic Structure of Generalized Kerdock Codes. We proved that generalized Kerdock codes are related to cyclic codes over $\mathbb{Z}_{2^k}$ (recall that over $\mathbb{Z}_{2^k}$ we can define duality with respect to the usual inner product).

Proposition 2. *The code* $\mathbf{K}(k, m)$ *is the dual of an extended cyclic code over* $\mathbb{Z}_{2^k}$. *Moreover, if* $k \leq m$, *then* $\mathbf{K}(k, m)$ *is an extended cyclic code over* $\mathbb{Z}_{2^k}$.

Proof. Due to length constraint we refer to [12]. □

Generalized Preparata Codes. Kerdock codes ($k = 2$) are related to Preparata codes, their $\mathbb{Z}_4$-duals: the Gray image of dual of $\mathbf{K}(2, m)$ is a Preparata code (cf. [1]) in the sense it has the same parameters and weight distribution. A natural generalization of Preparata codes is

Definition 1. *The generalized Preparata code,* $\mathcal{P}(k, m)$, *is the generalized Gray image of the dual of* $\mathbf{K}(k, m)$.

Minimum distance. Carlet asked for the accuracy of the lower bound (1), particularly in the case $k = 3$, since for this value of k, it is close to the minimum distance of the dual of 3-error correcting BCH code of the same length (parameters of this dual are $[2^{m+2} - 1, 3m + 6, 2^{m+1} - 2^{(m+3)/2}]$). With the help of a computer we computed some minimum distances of generalized Kerdock codes. The results are in Table 1 (notation $\{n, k, d\}$ means length n, cardinality 2^k and minimum distance d). In fact, when possible we computed the whole weight distribution of the code $\mathbf{K}(k, m)$, and we used the MacWilliams identities (over $\mathbb{Z}_{2^k}$) to get the weight distribution of its dual, leading to minimum distance of generalized Preparata Codes (see Table 2). Comparing this with Brouwer's table of the best linear codes show (see Table 3 and [13]), except for the case $k = m = 3$, generalized Kerdock and Preparata codes are not as good as the best linear codes for the parameters we were able to compute. Moreover, the bound seems tight, so it is very likely that duals of BCH codes are better than generalized Kerdock codes and we check it is actually true for the parameters we computed: they are shorter, bigger and with larger minimum distance.

4   Hensel Lift of Binary Quadratic Residue Codes

The idea was to lift good binary cyclic codes to $\mathbb{Z}_{2^k}$, we choose binary quadratic residue codes, QR(n), since they fulfill this criteria. Codes obtained by Hensel lift of their generating polynomial to $\mathbb{Z}_4$ had been studied for several lengths, see [5] ($n = 17, 23$), [6] ($n = 31, 47$), [7] ($n = 31$). The codes of [5] are as good as the best linear codes and those of [6] are better (for same length and cardinality).

Duursma *et al.* lifted QR(23) to $\mathbb{Z}_8$ and thus got a code better than previously known codes (linear or not). It is the only example of a good code obtained by Hensel lift to $\mathbb{Z}_{2^k}$ with $k > 2$. But the authors didn't stop at this point, they used a union of 2 cosets to construct a bigger code. We will go back to this in the next section.

Table 1. Parameters of $\mathcal{K}(k,m)$, $\left\{2^{(m+k-1)}, k(m+1), \delta\right\}$.

$m\backslash k$	2	3	4	5
3	$\{16, 8, 6\}$	$\{32, 12, 10\}$	$\{64, 16, 20\}$	$\{128, 20, 40\}$
4	$\{32, 10, 12\}$	$\{64, 15, 20\}$	$\{128, 20, 40\}$	$\{256, 25, 80\}$
5	$\{64, 12, 28\}$	$\{128, 18, 44\}$	$\{256, 24, 88\}$	$\{512, 30, 176\}$
6	$\{128, 14, 56\}$	$\{256, 21, 96\}$	$\{512, 28, 192\}$	$\{1024, 35, 384\}$
7	$\{256, 16, 120\}$	$\{512, 24, 212\}$	$\{1024, 32, 424\}$	
8	$\{512, 18, 240\}$	$\{1024, 27, 440\}$		
9	$\{1024, 20, 496\}$	$\{2048, 30, 928\}$		
10	$\{2048, 22, 992\}$	$\{4096, 33, 1888\}$		

$m\backslash k$	6	7	8
3	$\{256, 24, 80\}$	$\{512, 28, 160\}$	$\{1024, 32, 320\}$
4	$\{512, 30, 160\}$		

Table 2. Parameters of $\mathcal{P}(k,m)$, $\left\{2^{(m+k-1)}, k\left(2^m - (m+1)\right), \delta\right\}$.

$m\backslash k$	2	3	4	5
3	$\{16, 8, 6\}$	$\{32, 12, 10\}$	$\{64, 16, 20\}$	$\{128, 20, 40\}$
4	$\{32, 22, 4\}$	$\{64, 33, 8\}$	$\{128, 44, 16\}$	$\{256, 55, 32\}$
5	$\{64, 52, 6\}$	$\{128, 78, 10\}$	$\{256, 104, 20\}$	$\{512, 130, 40\}$
6	$\{128, 114, 4\}$	$\{256, 171, 8\}$	$\{512, 228, 16\}$	
7	$\{256, 240, 6\}$	$\{512, 360, 10\}$	$\{1024, 480, 20\}$	
8	$\{512, 494, 4\}$	$\{1024, 741, 8\}$		
9	$\{1024, 1004, 6\}$	$\{2048, 1506, 10\}$		
10	$\{2048, 2026, 4\}$			

$m\backslash k$	6	7
3	$\{256, 24, 80\}$	$\{512, 28, 160\}$
4	$\{512, 66, 64\}$	

Table 3. Extract from (Binary) Brouwer's table: bound on the best minimum distance for linear codes of the same length and the same cardinality than $\mathcal{K}(k,m)$ (left) and $\mathcal{P}(k,m)$ (right).

$m\backslash k$	2	3	4	5	6
3	5	10	24	48-53	100-114
4	12	24	48-53	100-114	
5	25-26	48-54	100-114		
6	56-57	112-116			
7	113-120				

$m\backslash k$	2	3	4	5	6
3	5	10	24	48-53	100-114
4	5	12-14	28-38	68-96	
5	5	16-22	46-71		
6	5	24-34			
7	5				

Table 4. Bound (1) and true minimum distance δ of $\mathcal{K}(3,m)$.

m	3	4	5	6	7	8	9	10
δ	10	20	44	96	212	440	928	1888
bound (Prop. 1)	0	8	32	80	190	416	892	1856
error (%)	-	60	27.3	16.7	10.4	5.5	3.9	1.7

Following this work, we lifted generating polynomial of QR(n) for $n =$ $17, 23, 31, 47$ to $\mathbb{Z}_8$ and $\mathbb{Z}_{16}$, extended the resulting codes by a parity-check symbol, then computed their minimum distance with computer assistance. Results are shown in Table 5, bold (resp. italic) is for codes better than (resp. as good as) the best linear ones and * is only an upper bound on minimum distance. Note that extended lift to $\mathbb{Z}_8$ of QR for length 31 and 47 and extended lift to $\mathbb{Z}_{16}$ of QR(31) reach the parameters of best linear codes.

Table 5. Parameters of generalized Gray image of extended Hensel lift of QR(n) (left) and minimum distance of the best known linear codes of same length and size (right).

n	$\mathbb{Z}_4$	$\mathbb{Z}_8$	$\mathbb{Z}_{16}$
17	$\{36, 18, 8\}$	$\{72, 27, 16\}$	$\{144, 36, 32\}$
23	$\{48, 24, 12\}$	$\{96, 36, \mathbf{24}\}$	$\{192, 48, 48\}$
31	$\{64, 32, \mathbf{14}\}$	$\{128, 48, 28\}$	$\{256, 64, 56^*\}$
47	$\{96, 48, \mathbf{18}\}$	$\{192, 72, 36\}$	

n	$\mathbb{Z}_4$	$\mathbb{Z}_8$	$\mathbb{Z}_{16}$
17	8	19	38
23	12	**20**	48
31	12	28	62
47	**16**	36	

*: only an upper bound

5 Constructing Binary Codes Using $\mathbb{Z}_{2^k}$-Linear Codes

By definition of the generalized Gray map Ψ, the images of codes of length n over $\mathbb{Z}_{2^k}$ are subsets of $\mathrm{RM}(1, k-1)^n$.

Let $\mathcal{C} \subset \mathrm{RM}(r, m)^n$ be a binary code of minimum distance d. If we can find cosets of $\mathcal{C}$ at distance at least d from each other then, we can perform the union of these cosets, getting a code with same length and same minimum distance as $\mathcal{C}$, but with larger cardinality. We study below a construction of such cosets relying on non binary codes and possibility to find cosets of $\mathrm{RM}(r, m)$ far enough from each other.

Namely we are searching for a set $T \subset \left(\mathrm{GF}(2)^{2^m}\right)^n$ such that the sum of any two elements is at Hamming distance at least d from the set $\mathrm{RM}(r, m)^n$.

Theorem 1. *Let $\mathcal{C} \subset \mathrm{RM}(r, m)^n$ be a binary code of minimum distance d. Let S be a subset of $\mathrm{GF}(2)^{2^m}$ such that*

$$\forall z \in \mathrm{RM}(r, m), \ \forall (x, y) \in S \times S, x \neq y \quad \mathrm{w}(x + y + z) \geq d_S \ ,$$

and $T \subset S^n$ with

$$\forall (\mathbf{u}, \mathbf{v}) \in T \times T, \mathbf{u} \neq \mathbf{v}, \quad |\{i : u_i \neq v_i\}| \geq d_T \ .$$

If the inequality $d_T \cdot d_S \geq d$ holds, then the cosets $\mathbf{t} + \mathcal{C}$, $\mathbf{t} \in T$, are pairwise disjoints and the code $\mathcal{G}$ of length $n \cdot 2^m$ defined by

$$\mathcal{G} = \bigcup_{\mathbf{t} \in T} (\mathbf{t} + \mathcal{C})$$

is of cardinality $|T| \cdot |\mathcal{C}|$ and minimum distance d.

Proof. Let $\mathbf{a} = (a_1, \ldots, a_n), \mathbf{b} = (b_1, \ldots, b_n)$ be two distinct codewords of $\mathcal{C}$, with a_i, b_i in $\mathrm{RM}(r, m) \subset \mathrm{GF}(2)^{2^m}$ and $\mathbf{u} = (u_1, \ldots, u_n), \mathbf{v} = (v_1, \ldots, v_n)$ with u_i, v_i in S. Then,

$$\mathrm{w}(\mathbf{a} + \mathbf{u} + \mathbf{b} + \mathbf{v}) = \sum_{i=0}^{n} \mathrm{w}(a_i + u_i + b_i + v_i) \ .$$

We have the two following cases:

1. $\mathbf{u} = \mathbf{v}$, so $\mathbf{a} + \mathbf{u}$ and $\mathbf{b} + \mathbf{v}$ are in the same coset of $\mathcal{C}$, thus

$$\mathrm{w}(\mathbf{a} + \mathbf{u} + \mathbf{b} + \mathbf{v}) = \mathrm{w}(\mathbf{a} + \mathbf{b}) \geq d \ ;$$

2. $\mathbf{u} \neq \mathbf{v}$, so $\mathbf{a} + \mathbf{u}$ and $\mathbf{b} + \mathbf{v}$ are in different cosets, and we have

$$\mathrm{w}(\mathbf{a} + \mathbf{u} + \mathbf{b} + \mathbf{v}) \geq \sum_{u_i \neq v_i} \mathrm{w}(a_i + u_i + b_i + v_i) \ .$$

By linearity of $\mathrm{RM}(r, m)$, $a_i + b_i$ is in $\mathrm{RM}(r, m)$ and the hypothesis on S leads to

$$\mathrm{w}(\mathbf{a} + \mathbf{u} + \mathbf{b} + \mathbf{v}) \geq |\{i : u_i \neq v_i\}| \cdot d_S \ .$$

Since we have $|\{i : u_i \neq v_i\}| \geq d_T$ and $d_T \cdot d_S \geq d$, we have $\mathrm{w}(\mathbf{a} + \mathbf{u} + \mathbf{b} + \mathbf{v}) \geq d$ and two different cosets are at distance at least d, in particular they are disjoint.

$\square$

Now, we have to find sets S and T fulfilling the required hypothesis to get an effective construction. In view of Theorem 1, it is natural to take a concatenated code [14] for T and regard S as the inner code.

We start with S, using the well known construction of the Reed-Muller code $\mathrm{RM}(r + a, m)$ from $\mathrm{RM}(r, m)$ ([15, Chp. 13 §3]). Recall that $\mathrm{RM}(r + a, m)$ is the union of 2^l different cosets of $\mathrm{RM}(r, m)$, with $l = \binom{m}{r+a} + \binom{m}{r+a-1} + \cdots + \binom{m}{r+1}$. We take for S a set of coset representatives, so $|S| = 2^l$ and it is easy to see $d_S = 2^{m-r-a}$, which is the minimum distance of $\mathrm{RM}(r + a, m)$. To construct T, we use as an outer code, a code C of length n and minimum distance $d_T \geq d/2^{m-r-a}$ over $(GF)(2^l)$, we choose a one to one mapping $\varphi \colon \mathrm{GF}(2^l) \to S$ and we define T by

$$T = \varphi(C) = \left\{ (\varphi(c_1), \ldots, \varphi(c_n)) \ \middle| \ (c_1, \ldots, c_n) \in C \right\} \ .$$

Thus, we can apply Theorem 1. Note that the code $\mathcal{G}$ we obtain this way is a subset of $\mathrm{RM}(r + a, m)^n$, and so we can iterate the construction.

Example 1. For $\mathbb{Z}_8$ we have $r = 1, m = 2$, and $S = \{0000, 1000\}$. If the $\mathbb{Z}_8$-linear code is $\{96, 36, 24\}$, we have only one possibility for C: the trivial binary code of length 24. This was used in [8].

Bound on Cardinality of $\mathbb{Z}_{2^k}$-linear codes. Since we can apply our construction, one can think intuitively that the subsets of $C \subset \mathrm{RM}(1, k-1)^n$ aren't good codes, in particular for $\mathbb{Z}_{2^k}$-linear codes, $k > 2$. This can be formalized and yields to a bound on the cardinality of codes which are subsets of $\mathrm{RM}(r, m)^n$.

Corollary 1. *Let $C \subset \mathrm{RM}(r, m)^n$ be of minimum distance d. Then, we have the following bound on the cardinality of C :*

$$|C| \le \frac{A_{n \cdot 2^m}^d(2)}{A_n^{\lceil 2^{r+a-m} \cdot d \rceil}(2^l)} \quad,$$

where $A_n^d(q)$ denote the maximum cardinality for a code of length n and minimum distance less or equal to d over $\mathrm{GF}(q)$ and $l = \sum_{i=1}^a \binom{m}{r+i}$.

Proof. Choose for C a code of length n, minimum distance $\lceil 2^{r+a-m \cdot d} \rceil$ and maximal cardinality $A_n^{\lceil 2^{r+a-m \cdot d} \rceil}(2^l)$ over $\mathrm{GF}(2^l)$. Using for S a set of coset representatives of $\mathrm{RM}(r, m)$ in $\mathrm{RM}(r + a, m)$, and applying Theorem 1 the code $\mathcal{G}$ is of length $n \cdot 2^m$ and minimum distance d. Therefor, $|\mathcal{G}| \le A_{n \cdot 2^m}^d(2)$, since $|\mathcal{G}| = |C| \cdot |C|$, we have the bound

$$|C| \le \frac{A_{n \cdot 2^m}^d(2)}{A_n^{\lceil 2^{r+a-m} \cdot d \rceil}(2^l)} \quad.$$

$\square$

Using the fact we can iterate the construction leads to the following

Corollary 2. *Let $C \subset \mathrm{RM}(r, m)^n$ be of minimum distance d, and $s_0 = r < s_1 < \cdots < s_t \le m$. Then*

$$|C| \le \frac{A_{n \cdot 2^m}^d(2)}{\prod_{j=1}^t A_n^{\lceil 2^{s_j-m} \cdot d \rceil}(2^{l_j})},$$

with $l_j = \sum_{i=s_{j-1}+1}^{s_j} \binom{m}{i}$.

Remark 1. We use for S a set of cosets representatives of $\mathrm{RM}(r, m)$ in $\mathrm{RM}(r + a, m)$ but, there exist other possibilities. For instance, when m is even and $r = a = 1$ we can use a set of cosets representatives of $\mathrm{RM}(1, m)$ in the Kerdock code of length 2^m. It leads to a smaller S (only 2^m elements instead of $2^{m(m-1)/2}$) but d_S is nearly twice as big ($2^{m-1} - 2^{(m-2)/2}$ instead of 2^{m-2}). Of course, this yields different bounds.

References

1. Hammons, R., Kumar, P., Calderbank, A., Sloane, N., Solé, P.: Kerdock, Preparata, Goethals and other codes are linear over $\mathbb{Z}_4$. IEEE Transactions on Information Theory **IT-40** (1994) 301–319
2. Nechaev, A.: Kerdock codes in a cyclic form. Discrete Mathematics and Applications **1** (1991) 365–384 (Translated from russian, *Diskretnaya Matematika*, vol. 1, n° 4, 1989, pp. 123–139).
3. Constantinescu, I., Heise, W.: A metric for codes over residue class rings. Problems of Information Transmission **33** (1997) 208–213 (Translated from russian, *Problemy Peredachi Informatsii*, vol. 33, No. 3, 1997, pp. 22–28).
4. Carlet, C.: $\mathbb{Z}_{2^k}$-linear codes. IEEE Transactions on Information Theory **IT-44** (1998) 1543–1547
5. Bonnecaze, A., Solé, P., Calderbank, A.: Quaternary quadratic residue codes and unimodular lattices. IEEE Transactions on Information Theory **IT-41** (1995) 366–377
6. Pless, V., Qian, Z.: Cyclic codes and quadratic residue codes over $\mathbb{Z}_4$. IEEE Transactions on Information Theory **IT-42** (1996) 1594–1600
7. Calderbank, A., McGuire, G., Kumar, P., Helleseth, T.: Cyclic codes over $\mathbb{Z}_4$, locator polynomials and Newton's identities. IEEE Transactions on Information Theory **IT-42** (1996) 217–226
8. Duursma, I., Greferath, M., Litsyn, S., Schmidt, S.: A $\mathbb{Z}_8$-linear lift of the binary Golay code and a non-linear binary $(96, 2^{37}, 24)$ code. IEEE Transactions on Information Theory **IT-47** (2001) 1596–1598
9. MacDonald, B.: Finite Rings with Identity. Dekker (1974)
10. Calderbank, A., Sloane., N.: Modular and p-adic cyclic codes. Designs, Codes and Cryptography **6** (1995) 21–35
11. Kanwar, P., López-Permouth, S.: Cyclic codes over the integers modulo p^m. Finite Fields and Their Applications (1997) 334–352
12. Galand, F.: Codes $\mathbf{Z}_{2^k}$-lineaires. Technical report, (INRIA) To appear.
13. Brouwer, A.: Bounds on the Size of Linear Codes. In: Handbook of Coding Theory. Volume 1. North-Holland (1998) 295–461.
    (http://www.win.tue.nl/~aeb/voorlincod.html).
14. Dumer, I.: Concatened Codes and Their Multilevel Generalizations. In: Handbook of Coding Theory. Vol. 2. North-Holland (1998) 1911–1988
15. MacWilliams, F., Sloane, N.: The Theory of Error-Correcting Codes. 3rd ed. North-Holland (1996)

Quasicyclic Codes of Index ℓ over F_q Viewed as $F_q[x]$-Submodules of $F_{q^\ell}[x]/\langle x^m - 1\rangle$

Kristine Lally

Department of Mathematics, RMIT University, Melbourne, Australia
kristine.lally@rmit.edu.au

Abstract. Quasicyclic codes of length $n = m\ell$ and index ℓ over the finite field F_q are linear codes invariant under cyclic shifts by ℓ places. They are shown to be isomorphic to the $F_q[x]/\langle x^m - 1\rangle$-submodules of $F_{q^\ell}[x]/\langle x^m - 1\rangle$ where the defining property in this setting is closure under multiplication by x with reduction modulo $x^m - 1$. Using this representation, the dimension of a 1-generator code can be determined straightforwardly from the chosen generator, and improved lower bounds on minimum distance are developed. A special case of multi-generator codes, for which the dimension can be algebraically recovered from the generating set is described. Every possible dimension of a quasicyclic code can be obtained in some such special form. Lower bounds on minimum distance are also given for all multi-generator quasicyclic codes.

1 Introduction

An $[n, k]$ linear block code C over F_q is a quasicyclic code if every cyclic shift of a codeword by ℓ places, for some integer ℓ, $1 \leq \ell \leq n$, results in another codeword. The smallest such ℓ is called the index of C and in this case $n = m\ell$ for some multiple m. Quasicyclic codes are of course a natural generalization of cyclic codes ($\ell = 1$), and have closely linked algebraic structure.

Early studies on quasicyclic codes include [1] and [7]. Links with convolutional codes have been established in [13]. In [8] it was shown that quasicyclic codes meet a modified version of the Gilbert-Varshamov bound, and thus, unlike BCH cyclic codes, are known to contain asymptotically good codes.

A quasicyclic code is conventionally viewed as the rowspace of a block matrix consisting of rows of $m \times m$ circulant matrices. To date most of the literature on quasicyclic codes has been largely concerned with the 1-generator case, that is, when only one row of circulants is present in this generator matrix. Construction techniques, such as those described in [4,5,15], have resulted in the discovery of a large number of binary quasicyclic codes which meet the best possible values of minimum distance of any linear code of the same length and dimension. However despite being well reputed as a very good class of codes, good algebraic lower bounds on the minimum distance have not previously been developed and good codes have largely been found by intensive computational searching techniques.

The algebraic structure of quasicyclic codes has been studied by Séguin and others [2,12], and Ling and Solé [11]. A quasicyclic code with generator matrix in

M. Fossorier, T. Hoeholdt, and A. Poli (Eds.): AAECC 2003, LNCS 2643, pp. 244–253, 2003.

conventional 'row-circulant' form can be viewed as an R-submodule of R^ℓ, where $R = F_q[x]/\langle x^m - 1 \rangle$. In this paper this conventional representation is not used. We introduce a new module representation of a quasicyclic code which follows naturally from their definition. Our main result is an algebraic lower bound on minimum distance which can be read directly from the choice of generators.

2   Module Structure

Let C be a quasicyclic code of length $m\ell$ and index ℓ over F_q. Let T be the cyclic shift operator acting on the vectors in $F_q^{m\ell}$ and let T^i be defined recursively by $\mathrm{T}^i v = T(T^{i-1} v)$ for $i = 1, \ldots,$ where $\mathrm{T}^0 = 1$ is the identity mapping on $F_q^{m\ell}$. The code C over F_q is a linear subspace of $F_q^{m\ell}$ which is invariant under T^ℓ.

As described in [2], if $a \in F_q[x]$ then $F_q^{m\ell}$ can be viewed as an $F_q[x]$-module by defining $a(x)\mathbf{f} = a(\mathrm{T}^\ell)\mathbf{f}$ for $\mathbf{f} \in F_q^{m\ell}$. With this definition of $F_q[x]$-multiplication a quasicyclic code of block length $m\ell$ and index ℓ is an $F_q[x]$-submodule of $F_q^{m\ell}$. Since the ideal $I = \langle x^m - 1 \rangle$ is the annihilator of $F_q^{m\ell}$, we can also view $F_q^{m\ell}$ as a module over the quotient ring $F_q[x]/I$ by defining $(a + I)\mathbf{f} = a\mathbf{f}$, for $\mathbf{f} \in F_q^{m\ell}$. Similarly quasicyclic codes are just the $F_q[x]/I$-submodules of $F_q^{m\ell}$.

Let C be the quasicyclic code generated by $\mathbf{f}_1, \mathbf{f}_2, \ldots, \mathbf{f}_t \in F_q^{m\ell}$ as an $F_q[x]/I$-submodule of $F_q^{m\ell}$. Then $C = \langle \mathbf{f}_1, \mathbf{f}_2, \ldots, \mathbf{f}_t \rangle = \{a_1 \mathbf{f}_1 + a_2 \mathbf{f}_2 + \cdots + a_t \mathbf{f}_t \mid a_i \in F_q[x]/I, i = 1, \ldots, t\} \subseteq F_q^{m\ell}$, and as a vector space over F_q, is spanned by the set $\{\mathbf{f}_1, x\mathbf{f}_1, \ldots, x^{m-1}\mathbf{f}_1, \ \mathbf{f}_2, x\mathbf{f}_2, \ldots, x^{m-1}\mathbf{f}_2, \ \ldots, \ \mathbf{f}_t, x\mathbf{f}_t, \ldots, x^{m-1}\mathbf{f}_t\} = \{\mathbf{f}_1, \mathrm{T}^\ell \mathbf{f}_1, \ldots, \mathrm{T}^{(m-1)\ell}\mathbf{f}_1, \ \mathbf{f}_2, \mathrm{T}^\ell \mathbf{f}_2, \ldots, \mathrm{T}^{(m-1)\ell}\mathbf{f}_2, \ \ldots, \ \mathbf{f}_t, \mathrm{T}^\ell \mathbf{f}_t, \ldots, \mathrm{T}^{(m-1)\ell}\mathbf{f}_t\}$. A quasicyclic code C in $F_q^{m\ell}$ with a cyclic basis $\mathbf{c}, \mathrm{T}^\ell \mathbf{c}, \mathrm{T}^{2\ell}\mathbf{c}, \ldots, \mathrm{T}^{(k-1)\ell}\mathbf{c}$, for some $\mathbf{c} \in F_q^{m\ell}$ and $k \leq m$, corresponds to a 1-generator $F_q[x]/I$-submodule of $F_q^{m\ell}$ where $C = \langle \mathbf{c} \rangle = \{a\mathbf{c} \mid a \in F_q[x]/I\}$.

Let us now consider $\mathbf{v} = (v_{(0,0)}, \ldots, v_{(0,\ell-1)}, \ldots, v_{(m-1,0)}, \ldots, v_{(m-1,\ell-1)}) \in F_q^{m\ell}$ as a sequence of m blocks of length ℓ, and associate with each ℓ-tuple $\mathbf{v}_j = (v_{(j,0)}, v_{(j,1)}, \ldots, v_{(j,\ell-1)}) \in F_q^\ell, j = 0, 1, \ldots, m-1$, the element $v_j = v_{(j,0)} + v_{(j,1)}\alpha + \cdots + v_{(j,\ell-1)}\alpha^{\ell-1}$ in F_{q^ℓ}, where $\{1, \alpha, \alpha^2, \ldots, \alpha^{\ell-1}\}$ is some fixed choice of basis of F_{q^ℓ} as a vector space over F_q. Indicating the block positions with increasing powers of x, the vector $\mathbf{v} \in F_q^{m\ell}$ can then be associated with the polynomial $v_0 + v_1 x + \cdots + v_{m-1} x^{m-1} \in F_{q^\ell}[x]$. An $F_q[x]/I$-module isomorphism between $F_q^{m\ell}$ and $F_{q^\ell}[x]/I$, which preserves the weight structure of the submodules, is given by $\Phi(\mathbf{v}) = v(x) + I = v_0 + v_1 x + \cdots + v_{m-1} x^{m-1} + I \in F_{q^\ell}[x]/I$, for $\mathbf{v} \in F_q^{m\ell}$, where $v_j = v_{(j,0)} + v_{(j,1)}\alpha + \cdots + v_{(j,\ell-1)}\alpha^{\ell-1} \in F_{q^\ell}, j = 0, 1, \ldots, m-1$. In the latter module, scalar multiplication by elements of $F_q[x]/I$ now means coset multiplication, that is, $(f+I)(v+I) = fv + I \in F_{q^\ell}[x]/I$ for $f + I \in F_q[x]/I$. The action of the cyclic shift operator T^ℓ in this setting is equivalent to multiplication by x modulo $x^m - 1$. It follows that, with respect to a fixed F_q-basis of F_{q^ℓ}, there is a one-to-one correspondence between $F_q[x]/I$-submodules of $F_{q^\ell}[x]/I$ and quasicyclic codes of index ℓ and length $m\ell$ over F_q.

Henceforth we use vector notation $\mathbf{v} = (v_{(0,0)}, \ldots, v_{(0,\ell-1)}, \ldots \ldots, v_{(m-1,\ell-1)}) = (\mathbf{v}_0, \mathbf{v}_1, \ldots, \mathbf{v}_{m-1})$ and polynomial notation $v(x) + I = v_0 + v_1 x + \cdots +$

$v_{m-1}x^{m-1} + I$ freely, and vary between the two as is appropriate. We also often drop the coset notation and write $v(x)$ for $v(x) + I$ with the understanding that $v(x)$ is the unique polynomial of degree less than m in $v(x) + I$, and multiplication is performed mod $x^m - 1$.

Theorem 1. $F_{q^\ell}[x]/I$ *is generated by the set* $\{1, \alpha, \alpha^2, \ldots, \alpha^{\ell-1}\}$ *as a module over* $F_q[x]/I$, *and is the direct sum,* $F_{q^\ell}[x]/I = \langle 1 \rangle \oplus \langle \alpha \rangle \oplus \cdots \oplus \langle \alpha^{\ell-1} \rangle$, *of cyclic submodules.*

A basis for $F_{q^\ell}[x]/I$ as a vector space over F_q is therefore the set $\{\alpha^i x^j, i = 0, 1, \ldots \ell - 1, j = 0, 1, \ldots, m - 1\}$.

Corollary 1. $F_{q^\ell}[x]/I$ *has a cyclic basis as a vector space over* F_q *if and only if* $\ell = 1$.

When $\ell = 1$ an $F_q[x]/I$-submodule of $F_{q^\ell}[x]/I$ is of course the well-known case of a cyclic code viewed as an ideal in $F_q[x]/I$. Since $F_q[x]/I$ is a principal ideal ring, a unique monic generator polynomial can be found which generates the code. We also note that when $\ell > 1$, a quasicyclic code C, as an $F_q[x]/I$-submodule of $F_{q^\ell}[x]/I$, is a subcode of the cyclic code $\widetilde{C}$ over F_{q^ℓ} generated as an $F_{q^\ell}[x]/I$-submodule of $F_{q^\ell}[x]/I$ (that is, an ideal in $F_{q^\ell}[x]/I$) by the same set of generators. Similarly a unique monic generator polynomial can be found in $F_{q^\ell}[x]/I$ which generates the code $\widetilde{C}$.

3   1-Generator Quasicyclic Codes

A 1-generator quasicyclic code C is an $F_q[x]/I$-submodule of $F_{q^\ell}[x]/I$ generated by a single element, that is, $C = \langle f \rangle = \{af \mid a \in F_q[x]/I\}$ for some $f \in F_{q^\ell}[x]/I$, and as a vector space over F_q, the code C is spanned by the set $\{f, xf, \ldots, x^{m-1}f\} \bmod x^m - 1$ where $f = f_0 + f_1 x + \cdots + f_{m-1}x^{m-1} \in F_{q^\ell}[x]/I$ and each coefficient $f_j = f_{(j,0)} + f_{(j,1)}\alpha + \cdots + f_{(j,\ell-1)}\alpha^{\ell-1} \in F_{q^\ell}$ is an F_q-linear combination of the basis elements $\{1, \alpha, \alpha^2, \ldots, \alpha^{\ell-1}\}$ of F_{q^ℓ}. Let $a, b \in F_{q^\ell}[x]$. We denote by $\gcd_q(a, b)$ the largest monic divisor of $\gcd(a, b) \in F_{q^\ell}[x]$ which lies in $F_q[x]$, where, as is conventional, we take the gcd as the unique monic representative in all cases. The dimension of the code C can be read straightforwardly from the generating polynomial as follows. We note that formulae for dimension of a 1-generator code have been previously given by van Tilborg [15], and Séguin *et al.* [12], in the context of the conventional 'row-circulant' generator matrix.

Theorem 2. *Let C be a 1-generator quasicyclic code of length $m\ell$ and index ℓ over F_q generated by $f \in F_{q^\ell}[x]/I$. The dimension of the code C is $k' = m - \deg(g')$ where $g' = \gcd_q(f, x^m - 1) \in F_q[x]$.*

Proof. Let $g = \gcd(f, x^m - 1) \in F_{q^\ell}[x]$ and $g' = \gcd_q(f, x^m - 1) \in F_q[x]$. We write $x^m - 1 = g'h'$ with $\deg(g') = m - k'$, so $\deg(h') = k'$. Every codeword $c + I$ can be written in the form $af + I$ for some $a \in F_q[x]/I$. Dividing a by h' in $F_q[x]$ to obtain $a = qh' + r$ with $\deg(r) < \deg(h')$, we have $c + I = af + I = (qh' + r)f + I = rf + I$.

Hence $c+I$ can be expressed in the form $rf+I$ with $r \in F_q[x]/I$ and $\deg(r) < k'$, and so $c+I$ is an F_q-linear combination of the set $\{f+I, xf+I, \ldots, x^{k'-1}f+I\}$.

This set is also linearly independent over F_q. For if $rf+I = 0+I$ then $x^m - 1$ divides the product rf as polynomials in $F_{q^\ell}[x]$. This implies that $(x^m - 1)/g$ divides r. Since r is a polynomial in $F_q[x]$ it follows that $(x^m - 1)/g' = h'$ must also divide r. Now $\deg(r) < \deg(h') = k'$ and so $r = 0$.

This result mimics closely the analogous result for cyclic codes. A generator matrix for the code C, comprising of linearly independent rows, can now be constructed directly from the generating polynomial $f \in F_{q^\ell}[x]/I$ as follows. The set $\{f, xf, \ldots, x^{k'-1}f\} \bmod x^m - 1$ forms a basis for the submodule C as a vector space over F_q. The $k' \times m\ell$ matrix over F_q of the form

$$
G = \begin{pmatrix}
f_0 & f_1 & \cdots\cdots & f_{m-1} \\
f_{m-1} & f_0 & \cdots\cdots & f_{m-2} \\
\vdots & \vdots & \ddots & \vdots \\
f_{((m-k'+1))} & f_{((m-k'+2))} & \cdots\cdots & f_{((m-k'))}
\end{pmatrix} \tag{1}
$$

is therefore a generator matrix for the code $C \subseteq F_q^{m\ell}$, where the double parentheses in the indices denote reduction modulo m, and each $f_j = f_{(j,0)} + f_{(j,1)}\alpha + \cdots + f_{(j,\ell-1)}\alpha^{\ell-1} \in F_{q^\ell}$ represents the vector $\mathbf{f}_j = (f_{(j,0)}, f_{(j,1)}, \ldots, f_{(j,\ell-1)}) \in F_q^\ell$, $j = 0, 1, \ldots, m-1$. The cyclic nature of the code is immediately evident from this form of generator matrix, the rows of which form the first k' rows of a single circulant matrix over F_{q^ℓ}. This observation prompts the following simple result.

Lemma 1. *A 1-generator quasicyclic code C of length $m\ell$ and index ℓ over F_q generated by $f \in F_{q^\ell}[x]/I$ has lower bound on minimum distance given by*

$$d_{\min}(C) \geq d_{\min}(\widetilde{C})$$

where $\widetilde{C}$ is the cyclic code of length m over F_{q^ℓ} with generator polynomial $g = \gcd(f, x^m - 1) \in F_{q^\ell}[x]/I$.

Proof. The cyclic code $\widetilde{C} = \langle f \rangle = \langle g \rangle$ is an ideal in $F_{q^\ell}[x]/I$ and thus closed under multiplication by $F_{q^\ell}[x]/I$ which contains $F_q[x]/I$ and the result follows.

When $\gcd(m, q) = 1$ we can of course apply the well-known BCH lower bound on minimum distance to the cyclic code $\widetilde{C}$. If $\gcd(m, q) = 1$ there exists some integer s such that $m \mid ((q^\ell)^s - 1)$ and $m \nmid ((q^\ell)^{s-1} - 1)$, and so $F_{q^{\ell s}}$ is the smallest extension field of F_{q^ℓ} which contains a primitive m^{th}-root of unity β. The code $\widetilde{C}$ has BCH designed minimum distance $\delta + 1$, where δ is the largest number of consecutive powers of β in $F_{q^{\ell s}}$ which are roots of the generator polynomial $g \in F_{q^\ell}[x]$. Henceforth we denote the size of this consecutive set by $\#CR_{q^\ell}(g)$ where the appropriate extension field of F_{q^ℓ} is implied.

We recall that dimension of the code C is $k' = m - \deg(\gcd_q(f, x^m - 1))$, where $\gcd_q(f, x^m - 1) \in F_q[x]$, whereas the BCH lower bound depends on

$g = \gcd(f, x^m - 1) \in F_{q^\ell}[x]$. Let $J_i = \{i, iq^\ell, iq^{2\ell}, \dots\}$ be the ith cyclotomic coset modulo m over F_{q^ℓ}, and $J = \bigcup_{i=0}^{m-1} J_i$ be the set of all such cyclotomic cosets. Similarly let $J_i' = \{i, iq, iq^2, \dots\}$ be the ith cyclotomic coset modulo m over F_q, and $J' = \bigcup_{i=0}^{m-1} J_i'$. It is easy to see that a coset $J_i' \in J'$ may be the union of more than one coset in J. Hence a coset in $J \setminus J'$ can be chosen such that the corresponding minimal polynomial lies in $F_{q^\ell}[x]$, and thus could contribute to the polynomial $\gcd(f, x^m - 1) \in F_{q^\ell}[x]$. Note however that such a minimal polynomial does not lie in $F_q[x]$ and therefore may not contribute to the polynomial $\gcd_q(f, x^m - 1)$. It follows that $\gcd(f, x^m - 1)$ can have roots in $F_{q^{\ell s}}$ which increase the BCH designed minimum distance but do not result in a decrease in the dimension of C.

Viewing a quasicyclic code as a subcode of a cyclic code over an extension field is not however new. Similar BCH-type lower bounds on minimum distance have been presented by Tanner [14], and more recently by Dey and Rajan [3], using discrete Fourier transform techniques.

We also note that as ℓ gets larger the bound given in Lemma 1 becomes less useful. Each non-zero coefficient $c_j \in F_{q^\ell}$ of a codeword $c = c_0 + c_1 x + \cdots + c_{m-1} x^{m-1} \in F_{q^\ell}[x]/I$ adds only a count of 1 to the weight as an element of the cyclic code $\widetilde{C}$, whereas in C each c_j can contribute up to weight ℓ to the total weight of the length $m\ell$ codeword. We now present an improved lower bound on minimum distance which follows naturally from our module representation.

A codeword $c = c_0 + c_1 x + \cdots + c_{m-1} x^{m-1} \in F_{q^\ell}[x]/I$ is an F_q-linear combination of the basis vectors $\{f, xf, \dots, x^{k'-1} f\} \bmod x^m - 1$, and from the circulant nature of the generating matrix given in (1), we see that each coefficient $c_j \in F_{q^\ell}, 0 \le j \le m - 1$, is an F_q-linear combination of (a k'-subset of) the coefficients $f_0, f_1, \dots, f_{m-1} \in F_{q^\ell}$ of the generating polynomial $f \in F_{q^\ell}[x]/I$. It follows that the c_j can be viewed as codewords $\mathbf{c}_j \in F_q^\ell$ in the block linear code of length ℓ generated by $\{\mathbf{f}_0, \mathbf{f}_1, \dots, \mathbf{f}_{m-1}\} \subseteq F_q^\ell$.

Theorem 3. *A quasicyclic code C of length $m\ell$ and index ℓ over F_q with generator $f = f_0 + f_1 x + \cdots + f_{m-1} x^{m-1} \in F_{q^\ell}[x]/I$ has lower bound on minimum distance given by*

$$d_{\min}(C) \ge d_{\min}(\widetilde{C}) d_{\min}(B)$$

where $\widetilde{C}$ is the cyclic code of length m over F_{q^ℓ} with generator polynomial $g = \gcd(f, x^m - 1) \in F_{q^\ell}[x]/I$, and B is the linear block code of length ℓ generated by $\{\mathbf{f}_0, \mathbf{f}_1, \dots, \mathbf{f}_{m-1}\} \subseteq F_q^\ell$, where each $\mathbf{f}_j$ is the vector equivalent of the coefficients $f_i \in F_{q^\ell}$ with respect to a fixed choice of F_q-basis $\{1, \alpha, \alpha^2, \dots, \alpha^{\ell-1}\}$ of F_{q^ℓ}.

Proof. Each $c = c_0 + c_1 x + \cdots + c_{m-1} x^{m-1} \in C$ is an element of the cyclic code $\widetilde{C}$ over F_{q^ℓ} generated by g, and thus has at least $d_{\min}(\widetilde{C})$ non-zero coefficients. In turn each coefficient $c_i \in F_{q^\ell}$ when viewed as a vector $\mathbf{c}_i \in F_q^\ell$ is a codeword in the linear block code generated by $\{\mathbf{f}_0, \mathbf{f}_1, \dots, \mathbf{f}_{m-1}\} \subseteq F_q^\ell$.

When $\gcd(m, q) = 1$, a method for constructing 1-generator quasicyclic codes based on this result, proceeds as follows. Every generator polynomial $f = f_0 +$

$f_1 x + \cdots + f_{m-1} x^{m-1} \in F_{q^\ell}[x]/I$ is of the form $f = eg = e\rho g'$, where $g = \gcd(f, x^m - 1) \in F_{q^\ell}[x]$ and $g' = \gcd_q(f, x^m - 1) \in F_q[x]$. First we choose g' a divisor of $x^m - 1$ over $F_q[x]$, to determine the dimension k'. Then we choose ρ, a divisor of $(x^m - 1)/g'$ over $F_{q^\ell}[x]$, which increases $\#CR_{q^\ell}(g) = \#CR_{q^\ell}(\rho g')$ and thus the BCH designed minimum distance of $\widetilde{C}$. Finally we choose $e \in F_{q^\ell}[x]$ with $\deg(e) < m - \deg(g)$, satisfying $\gcd(e, (x^m - 1)/g) = 1$, so that the coefficients of $f = eg$ generate a code B of length ℓ with high minimum distance. For a given choice of dimension, the final search for suitable polynomial e is, in principle, the most computationally intensive step in the procedure. The complexity of this step is limited by the degree of e and the fact that we only need to find the minimum distance of a code B, a fraction of the length of the required quasicyclic code C. We now give an example to illustrate this technique.

Example 1. Let $q = 2, m = 9$ and $\ell = 3$. F_{2^6} is the smallest field extension of F_{2^3} which contains a primitive 9th-root of unity $\zeta = \beta^7$, where β is a primitive element in F_{2^6}, (taken here as a root of the primitive polynomial $x^6 + x + 1$). The cyclotomic cosets modulo 9 over F_{2^3} are $\{0\}, \{1, 8\}, \{2, 7\}, \{3, 6\}, \{4, 5\}$ and over F_2 are $\{0\}, \{1, 2, 4, 8, 7, 5\}, \{3, 6\}$. Let $g' = (x + \zeta^3)(x + \zeta^6) = 1 + x + x^2$ and $\rho = (x + \zeta^4)(x + \zeta^5) = 1 + \alpha^2 x + x^2$, where $\alpha = \beta^{27}$ is a primitive element of $F_{2^3} \subseteq F_{2^6}$. Then the quasicyclic code C has dimension $k' = 7$, and the cyclic code $\widetilde{C}$ generated by $g = (1 + x + x^2)(1 + \alpha^2 x + x^2) \in F_{2^3}[x]/I$ has BCH designed minimum distance $\#CR_{2^3}(g) + 1 = 5$. Choosing $e = \alpha^4 \in F_{2^3}[x]$ we obtain $f = \alpha^4 g = \alpha^4 + \alpha^3 x + \alpha^6 x^2 + \alpha^3 x^3 + \alpha^4$. The set of non-zero coefficients of f expressed in vector form (with respect to the basis $\{1, \alpha, \alpha^2\}$) is $\{f_0 = f_4 = (0, 1, 1), f_1 = f_3 = (1, 1, 0), f_2 = (1, 0, 1)\}$ and generates a code B of length 3 with minimum distance 2. A lower bound on the minimum distance of C is therefore 10. We note that the weight of f is 10 and so C has parameters $[27, 7, 10]$.

In the above example we see from the spanning set of $\{\mathbf{f}_0, \mathbf{f}_1, \ldots, \mathbf{f}_{m-1}\}$ that the code B of length ℓ is also a cyclic code. This 'double-cyclic' structure can be imposed in general on the form of a generator $f = f_0 + f_1 x + \cdots + f_{m-1} x^{m-1} \in F_{q^\ell}[x]/I$. If we consider each $f_j = f_{(j,0)} + f_{(j,1)} \alpha + \cdots + f_{(j,\ell-1)} \alpha^{\ell-1}$ in F_{q^ℓ} simultaneously as a polynomial with indeterminate α in $F_q[\alpha]/\langle \alpha^\ell - 1 \rangle$, we can apply the following BCH-type lower bound on minimum distance.

Corollary 2. *Let* $\gcd(q, m\ell) = 1$. *If* C *is quasicyclic code of length* $m\ell$ *and index* ℓ *over* F_q *with generator* $f = f_0 + f_1 x + \cdots + f_{m-1} x^{m-1} \in F_{q^\ell}[x]/I$ *where* $\{f_0, f_1, \ldots, f_{m-1}\}$ *are codewords of the cyclic code* $B \trianglelefteq F_q[\alpha]/\langle \alpha^\ell - 1 \rangle$, *then* C *has lower bound on minimum distance given by*

$$d_{\min}(C) \geq \left(\#CR_{q^\ell}(g) + 1 \right) \left(\#CR_q(b) + 1 \right)$$

where $g = \gcd(f, x^m - 1) \in F_{q^\ell}[x]$ *and* $b = \gcd(f_0, f_1, \ldots, f_{m-1}, \alpha^\ell - 1) \in F_q[\alpha]$.

If $m = q^\ell - 1$ then the splitting field of $x^m - 1$ over F_{q^ℓ} is F_{q^ℓ}. In this case the code $\widetilde{C}$ is a Reed-Solomon code over F_{q^ℓ}. The cyclotomic cosets modulo m over F_{q^ℓ} each contain only one element, giving us maximum flexibility in our choice of polynomial $\rho \in F_{q^\ell}[x]$. We illustrate this in the following example.

Example 2. Let $q = 2, m = 7$ and $\ell = 3$. F_{2^3} contains a primitive 7th-root of unity α (taken here as a root of the primitive polynomial $x^3 + x + 1$). The cyclotomic cosets modulo 7 over F_{2^3} are $\{0\}, \{1\}, \{2\}, \{3\}, \{4\}, \{5\}, \{6\}$ and over F_2 are $\{0\}, \{1, 2, 4\}, \{3, 6, 5\}$. Let $g' = x + 1$. Then C has dimension $k' = 6$. If $\rho = (x + \alpha)(x + \alpha^6)$ then the cyclic code $\widetilde{C}$ generated by $g = (x + 1)(x + \alpha)(x + \alpha^6) \in F_{2^3}[x]/I$ has BCH designed minimum distance $\#CR_{2^3}(g) + 1 = 4$. Choosing $e = 1$ we obtain $f = (1 + \alpha^2) + (1 + \alpha)x + (1 + \alpha)x^2 + (1 + \alpha^2)x^3$. The code B generated by $\{1 + \alpha^2, 1 + \alpha\}$ is the cyclic code in $F_2[\alpha]/\langle \alpha^3 - 1 \rangle$ with generator polynomial $b = 1 + \alpha \in F_2[\alpha]$ and BCH designed minimum distance $\#CR_2(b) + 1 = 2$. A lower bound on the minimum distance of C is therefore 8. The weight of the generator f is 8 and so we have a $[21, 6, 8]$ code. This is in fact the largest possible minimum distance for any binary linear code of the same length and dimension. A generator matrix for the code is given by

$$G = \begin{pmatrix} 101\ 110\ 110\ 101\ 000\ 000\ 000 \\ 000\ 101\ 110\ 110\ 101\ 000\ 000 \\ 000\ 000\ 101\ 110\ 110\ 101\ 000 \\ 000\ 000\ 000\ 101\ 110\ 110\ 101 \\ 101\ 000\ 000\ 000\ 101\ 110\ 110 \\ 110\ 101\ 000\ 000\ 000\ 101\ 110 \end{pmatrix}$$

4   Multi-generator Quasicyclic Codes

Previous studies of quasicyclic codes have been mainly concerned with the 1-generator case. Little is known about quasicyclic codes generated by an arbitrary set of generators. Methods proposed for determining dimension of such arbitrary multi-generator codes involve lengthy reduction procedures, such as that described in [9]. Here we introduce sufficient conditions on a generating set of elements in $F_{q^\ell}[x]/I$ which allow us to determine dimension straightforwardly from this choice of generating set.

A quasicyclic code C generated by the t generators $f_1, f_2, \ldots, f_t \in F_{q^\ell}[x]/I$, is the set $C = \{a_1 f_1 + a_2 f_2 + \cdots + a_t f_t \mid a_i \in F_q[x]/I, i = 1, 2, \ldots, t\} \subseteq F_{q^\ell}[x]/I$. Each generator $f_i = f_{i,0} + f_{i,1}x + \cdots + f_{i,m-1}x^{m-1} \in F_{q^\ell}[x]/I, 1 \le i \le t$, can be written in the form $f_i = e_i g_i = e_i \rho_i g'_i$ where $g_i = \gcd(f_i, x^m - 1) \in F_{q^\ell}[x]$ and $g'_i = \gcd_q(f_i, x^m - 1) \in F_q[x]$. We denote the product $e_i \rho_i \in F_{q^\ell}[x]$ by m_i and also write $f_i = m_i g'_i, 1 \le i \le t$, where $\deg(m_i) < m - \deg(g'_i) = k'_i$, as required.

Theorem 4. *Let C be a t-generator quasicyclic code of length $m\ell$ and index ℓ over F_q with generating set $\{f_i = m_i g'_i, i = 1, \ldots, t\} \subseteq F_{q^\ell}[x]/I$ where $g'_i = \gcd_q(f_i, x^m - 1) \in F_q[x]$ and $f_i/g'_i = m_i = m_{i,0} + m_{i,1}x + \cdots + m_{i,k'_i-1}x^{k'_i-1} \in F_{q^\ell}[x]$. If the set $\{m_{i,0}, m_{i,1}, \ldots, m_{i,k'_i-1}, i = 1, \ldots, t \mid m_{i,j} \ne 0\} \subseteq F_{q^\ell}$ of non-zero coefficients of the $m_i, 1 \le i \le t$, (each expressed as an F_q-linear combination of the basis elements $\{1, \alpha, \alpha^2, \ldots, \alpha^{\ell-1}\}$ of F_{q^ℓ}) are an F_q-linearly independent set in F_{q^ℓ}, then the dimension of the code C is $k = \sum_{i=1}^t k'_i = \sum_{i=1}^t (m - \deg(g'_i))$ where $g'_i = \gcd_q(f_i, x^m - 1) \in F_q[x]$.*

Proof. Every codeword $c + I$ can be written in the form $a_1 f_1 + a_2 f_2 + \cdots + a_t f_t + I$ for some $a_i \in F_q[x]/I$ with $\deg(a_i) < m - \deg(g_i') = k_i'$. If $c + I = a_1 f_1 + a_2 f_2 + \cdots + a_t f_t + I = 0 + I$ then $\sum_{i=1}^{t} a_i (m_i g_i') = \sum_{i=1}^{t} (a_i g_i')(m_{i,0} + m_{i,1} x + \cdots + m_{i,k_i'-1} x^{k_i'-1}) = \sum_{i=1}^{t} \sum_{j=0}^{k_i'-1} (x^j a_i g_i') m_{i,j} = 0 \bmod x^m - 1$. Let $b_i^j = x^j a_i g_i' \bmod x^m - 1 \in F_q[x], 1 \leq i \leq t, 1 \leq j \leq k_i' - 1$. Since the $m_{i,j}$ are constant values in F_{q^ℓ} it follows that $\sum_{i=1}^{t} \sum_{j=0}^{k_i'-1} b_i^j m_{i,j} = 0$. For each i there exists $j, 1 \leq j \leq k_i' - 1$ such that $m_{i,j} \neq 0$, since $m_i \neq 0$. If the set $\{m_{i,0}, m_{i,1}, \ldots, m_{i,k_i'-1}, i = 1, \ldots, t \mid m_{i,j} \neq 0\} \subseteq F_{q^\ell}$ are F_q-linearly independent in F_{q^ℓ}, then equating the various coefficients of powers of x, it follows that for each i there exists $j, 1 \leq j \leq k_i' - 1$, such that $b_i^j = x^j a_i g_i' \bmod x^m - 1 = 0$. Then $a_i g_i' = 0 \bmod x^m - 1, 1 \leq i \leq t$, and since $\deg(a_i g_i') < m$ we have $a_i g_i' = 0$. This implies that $a_i = 0$ for each $i, 1 \leq i \leq t$, and so the set $\{f_1, x f_1, \ldots, x^{k_1'-1} f_1, \ldots, f_t, \ldots, x^{k_t'-1} f_t\} \bmod x^m - 1$ forms an F_q-basis for the code C.

Since each term $m_i, 1 \leq i \leq t$, in the generating set described above, must have at least one non-zero coefficient and there are at most ℓ possible linearly independent elements in F_{q^ℓ} to choose from, it follows that a submodule can have at most ℓ generators in this restricted form. We know from [9] that every quasicyclic code has dimension equal to $\sum_{i=1}^{\ell} (m - \deg(d_i))$ for some set $\{d_i \in F_q[x], i = 1, \ldots, \ell\}$, each a divisor of $x^m - 1$, and so it is easy to see that every possible dimension permissible by the degrees of the irreducible factors of $x^m - 1$ can be achieved by some such generating set (for example in a code with generating set $\{d_1, \alpha d_2, \ldots, \alpha^{\ell-1} d_\ell\} \bmod x^m - 1$).

A generator matrix for the code $C \subseteq F_q^{m\ell}$ with generating set of the form described in Theorem 4 is the $\left(\sum_{i=1}^{t} k_i'\right) \times m\ell$ matrix over F_q consisting of a column of $t \leq \ell$ truncated circulants over F_{q^ℓ}, the top row of each these circulant being the coefficients of the generator $f_i = f_{i,0} + f_{i,1} x + \cdots + f_{i,m-1} x^{m-1} \in F_{q^\ell}[x]/I, i = 1, \ldots, t$, respectively, where each coefficient $f_{i,j} = f_{i,(j,0)} + f_{i,(j,1)} \alpha + \cdots + f_{i,(j,\ell-1)} \alpha^{\ell-1} \in F_{q^\ell}$ represents the vector $\mathbf{f}_{i,j} = (f_{i,(j,0)}, f_{i,(j,1)}, \ldots, f_{i,(j,\ell-1)}) \in F_q^\ell, j = 0, 1, \ldots, m - 1$.

We now present a lower bound on minimum distance that holds not only for the special case of the multi-generator codes described above, but for all codes with any arbitrary set of generators.

Theorem 5. *Let C be a t-generator quasicyclic code of length $m\ell$ and index ℓ over F_q with generating set $\{f_i = f_{i,0} + f_{i,1} x + \cdots + f_{i,m-1} x^{m-1}, i = 1, \ldots, t\} \subseteq F_{q^\ell}[x]/I$. Then C has lower bound on minimum distance given by*

$$d_{\min}(C) \geq d_{\min}(\widetilde{C}) d_{\min}(B)$$

where $\widetilde{C}$ is the cyclic code of length m over F_{q^ℓ} with generator polynomial $\gcd(f_1, f_2, \ldots, f_t, x^m - 1) \in F_{q^\ell}[x]/I$, and B is the linear block code of length ℓ generated by $\{\mathbf{f}_{i,j}, i = 1, \ldots, t, j = 0, 1, \ldots, m - 1\} \subseteq F_q^\ell$ where each $\mathbf{f}_{i,j}$ is the vector equivalent of the coefficients $f_{i,j} \in F_{q^\ell}$ with respect to a fixed choice of F_q-basis $\{1, \alpha, \alpha^2, \ldots, \alpha^{\ell-1}\}$ of F_{q^ℓ}.

Example 3. Let $q = 2, m = 7, \ell = 3$ and α be defined as described in Example 2. Let C be the 2-generator quasicyclic code with generators $f_1 = m_1 g_1' = ((1 + \alpha + \alpha^2)x^2 + (\alpha^2)x^5)(x+1) = (1+\alpha+\alpha^2)x^2 + (1+\alpha+\alpha^2)x^3 + \alpha^2 x^5 + \alpha^2 x^6$ and $f_2 = m_2 g_2' = ((\alpha + \alpha^2)x^3)(x+1) = (\alpha + \alpha^2)x^3 + (\alpha + \alpha^2)x^4$ where $g_i' = \gcd_q(f_i, x^m - 1) = x + 1 \in F_q[x], i = 1, 2$. The non-zero coefficients $m_{1,2} = 1 + \alpha + \alpha^2, m_{1,5} = \alpha^2$ and $m_{2,3} = \alpha + \alpha^2$ are linearly independent over F_2, and so C has dimension $k = k_1' + k_2' = 6 + 6 = 12$. A generator matrix for the code is given by

$$G = \begin{pmatrix}
000\ 000\ 111\ 111\ 000\ 001\ 001 \\
001\ 000\ 000\ 111\ 111\ 000\ 001 \\
001\ 001\ 000\ 000\ 111\ 111\ 000 \\
000\ 001\ 001\ 000\ 000\ 111\ 111 \\
111\ 000\ 001\ 001\ 000\ 000\ 111 \\
111\ 111\ 000\ 001\ 001\ 000\ 000 \\
\\
000\ 000\ 000\ 011\ 011\ 000\ 000 \\
000\ 000\ 000\ 000\ 011\ 011\ 000 \\
000\ 000\ 000\ 000\ 000\ 011\ 011 \\
011\ 000\ 000\ 000\ 000\ 000\ 011 \\
011\ 011\ 000\ 000\ 000\ 000\ 000 \\
000\ 011\ 011\ 000\ 000\ 000\ 000
\end{pmatrix}$$

Furthermore we find that $\gcd(f_1, x^m - 1) = (x + \alpha)(x + 1)$ and $\gcd(f_2, x^m - 1) = (x + 1)$. Hence the cyclic code $\widetilde{C}$ has generator polynomial $g = \gcd(f_1, f_2, x^m - 1) = (x + 1) \in F_{2^3}[x]/I$ and BCH designed minimum distance $\#CR_{2^3}(g) + 1 = 2$. The code B generated by $\{1 + \alpha + \alpha^2, \alpha + \alpha^2, \alpha^2\}$ has minimum distance 1. A lower bound on minimum distance of C is therefore 2. A quick search through all codewords shows that the actual minimum distance is in fact 4 and so the code has parameters $[21, 12, 4]$.

5   Decoding

We now briefly consider some simple observations on decoding in the light of this new module representation. We note that a good general decoding algorithm for quasicyclic codes is as yet undeveloped. Decoding techniques for 1-generator rate-$1/\ell$ quasicyclic codes have been described by Karlin in [7], and Heijnen and van Tilborg in [6], and for codes of any rate admissible to a 1-generator code by this author in [10]. Here we discuss decoding for the more general case of any arbitrary set of generators.

In the context of the terminology used in Theorem 5, since C is a subcode of the cyclic code $\widetilde{C}$, a codeword $c = c_0 + c_1 x + \cdots + c_{m-1}x^{m-1} \in C$ can be decoded, using well-known cyclic decoding techniques (such as the Berlekamp-Massey or Peterson-Gorenstein-Zierler algorithms), as an element of the cyclic code $\widetilde{C}$, and will be correctly decoded if at most $r = \lfloor (d_{\min}(\widetilde{C}) - 1)/2 \rfloor$ of the coefficients $\{c_0, c_1, \ldots, c_{m-1}\} \subseteq F_{q^\ell}$, each representing a vector of length ℓ, have been received in error. We remark that quasicyclic codes are therefore suitable when multiple burst errors, (each burst being at most length ℓ), are likely to occur in a codeword. Alternatively, if such a decoding approach should fail, we can of course decode each vector $\mathbf{c}_j \in F_q^\ell, j = 0, 1, \ldots, m - 1$, as an element of the same block linear code B of length ℓ with error correcting capability $s = \lfloor (d_{\min}(B) - 1)/2 \rfloor$ in each vector. This latter scenario may be more suitable when up to ms random errors are likely to occur, but no more than s in any block of length ℓ.

6 Conclusion

Viewing a quasicyclic code of index ℓ and length $m\ell$ over F_q as an $F_q[x]/I$-submodule of $F_{q^\ell}[x]/I$ is a natural and intuitive representation of these codes, and provides the framework in which to develop results on dimension and minimum distance for quasicyclic codes analogous to those known for cyclic codes.

References

1. C.L. Chen, W.W. Peterson and E.J. Weldon, Jr., "Some results on quasi-cyclic codes", *Inform. Contr.*, 15 pp. 407–423, 1969.
2. J. Conan and G. Séguin, "Structural properties and enumeration of quasicyclic codes", *Appl. Alg. in Eng. Comm. Comput.* 4, pp. 25–39, 1993.
3. B.K. Dey and B.S. Rajan, "F_q-linear cyclic codes over F_{q^m} : DFT characterization", AAECC-14, LNCS 2227, Eds S. Boztas, I.E. Shparlinski, Springer, pp. 67–76, 2001.
4. T.A. Gulliver and V.K. Bhargava, "Some best rate $1/p$ and rate $(1-p)/p$ systematic quasi-cyclic codes", *IEEE Trans. Inform. Theory*, 37, pp. 552–555, 1991.
5. T.A. Gulliver and V.K. Bhargava, "Nine good rate $(m-1)/pm$ quasi-cyclic codes", *IEEE Trans. Inform. Theory*, 38, pp. 1366–1369, 1992.
6. P. Heijnen, H. van Tilborg and S. Weijs, "On quasi-cyclic codes", *Proc. ISIT 1998, Cambridge, MA, USA, August 1998*, pp. 293.
7. M. Karlin, "Decoding of circulant codes", *IEEE Trans. Inform. Theory*, 16, pp. 797–802, 1970.
8. T. Kasami, "A Gilbert-Varshamov bound for quasi-cyclic codes of rate 1/2", *IEEE Trans. Inform. Theory*, 20, pp. 679, 1974.
9. K. Lally and P. Fitzpatrick, "Algebraic structure of quasicyclic codes", *Discrete Appl. Math.*, 111, no. 1–2, pp. 157–175, 2001.
10. K. Lally, "Quasicyclic codes – some practical issues", *Proc. ISIT 2002, Lausanne, Switzerland, July 2002*, pp. 174.
11. S. Ling and P. Solé, "On the algebraic structure of quasi-cyclic codes I: finite fields", *IEEE Trans. Inform. Theory*, 47, pp. 2751–2759, 2001.
12. G.E. Séguin and G. Drolet, "The theory of 1-generator quasi-cyclic codes", Dept. of Electrical and Computer Engineering, Royal Military College of Canada, Kingston, Ontario, June 1990.
13. G. Solomon and H.C.A. van Tilborg, "A connection between block and convolutional codes", *SIAM J. Appl. Math.* 37, 1979, 358–369.
14. R.M. Tanner, "A transform theory for a class of group invariant codes", *IEEE Trans. Inform. Theory*, 34, pp. 752–775, 1988.
15. H.C.A. van Tilborg, "On quasi-cyclic codes with rate $1/m$", *IEEE Trans. Inform. Theory*, 24, pp. 628–630, 1978.

Fast Decomposition of Polynomials with Known Galois Group

Andreas Enge and François Morain

École polytechnique (LIX, CNRS/FRE 2653) and INRIA Futurs
91128 Palaiseau cedex, France
enge@lix.polytechnique.fr
morain@lix.polytechnique.fr

Abstract. Let $f(X)$ be a separable polynomial with coefficients in a field K, generating a field extension M/K. If this extension is Galois with a solvable automorphism group, then the equation $f(X) = 0$ can be solved by radicals. The first step of the solution consists of splitting the extension M/K into intermediate fields. Such computations are classical, and we explain how fast polynomial arithmetic can be used to speed up the process. Moreover, we extend the algorithms to a more general case of extensions that are no longer Galois. Numerical examples are provided, including results obtained with our implementation for Hilbert class fields of imaginary quadratic fields.

1 Introduction

Let M_0/K_0 be a finite separable field extension. Our aim is to describe the extension by a tower of intermediate fields. Clearly, it is sufficient to show how to obtain one intermediate field L_0 and to describe the extensions L_0/K_0 and M_0/L_0, since this process can be iterated. If M_0/K_0 is defined by some irreducible polynomial $f \in K_0[X]$, then the decomposition of the field extension into two relative extensions naturally leads to a factorisation of f.

In the case that M_0/K_0 is Galois with solvable automorphism group, such a factorisation of f is an important step towards a solution of the equation by radicals. This can also be seen as a particular case of the subfield computations of algorithms presented in [2,14,15].

The algorithms needed to perform these computations are explained in [11, 12] and applied to the practical solution of equations arising in the theory of class fields. Our work extends this approach to polynomials f which are not Galois themselves, but whose Galois groups split nicely into one part corresponding to the simple extension described by f and another one describing the Galois closure. We also explain how to carry out the computations directly for M_0/K_0 without working in the Galois closure, which results in a significant speed-up. The necessary theory is developed in Sect. 2.

Relying on the Hecke representation of algebraic numbers reviewed in Sect. 3, we show that fast polynomial arithmetic can be applied to compute such a representation and thus to solve the problem of determining intermediate fields,

M. Fossorier, T. Hoeholdt, and A. Poli (Eds.): AAECC 2003, LNCS 2643, pp. 254–264, 2003.
© Springer-Verlag Berlin Heidelberg 2003

cf. Sect. 4. We present and analyse the complete algorithm in Sect. 6. All these ideas are explained on numerical examples and again the case of class fields is detailed. This has important consequences in the implementation of the primality proving algorithm ECPP, see Sects. 7 and 8.

2   Galois Theory

Let M be the Galois closure of M_0 of Galois group $\hat{G} = \text{Gal}(M/K_0)$. Let C be the subgroup of $\hat{G}$ fixing M_0 elementwise, that is, $M_0 = \text{Fix}(C)$. We assume that C possesses a normal complement G in $\hat{G}$, so that in fact $\hat{G} = G \rtimes C$. If K denotes the fixed field of G, then M/K is Galois of group G, and K/K_0 is Galois of group $C|_K$. Let H be a normal subgroup of G and $L = \text{Fix}(H)$, so that M/L is Galois of group H and L/K is Galois of group G/H. Denote by L_0 the fixed field of $\langle H, C \rangle$. Then $K_0 \subseteq L_0 \subseteq M_0$.

From now on, we assume that C normalises H, that is, $chc^{-1} \in H$ for any $c \in C$, $h \in H$. Then, any element ch of $\langle H, C \rangle$ with $c \in C$ and $h \in H$ can be written as $h'c$ with $h' = chc^{-1} \in H$, and $\langle H, C \rangle$ is in fact the semidirect product $H \rtimes C$. Consequently, L/L_0 is Galois of group $C|_L$.

In particular, $[M_0 : L_0] = [M : L]$ and $[L_0 : K_0] = [L : K]$. We are thus in a situation where the tower of extensions $M_0/L_0/K_0$, although not Galois itself, bears strong similarities with the Galois tower $M/L/K$. The situation is summarised in the following figure.

We now consider some irreducible polynomial $f \in K_0[X]$ which has a root $x \in M_0$. As $M = K \cdot M_0 = K \cdot K_0(x) = K(x)$ and $[M : K] = [M_0 : K_0]$, the polynomial f remains irreducible over K. Denoting the image of a field element under a Galois automorphism by adding a superscript, we thus obtain

$$f(X) = \prod_{g \in G} (X - x^g).$$

Conversely, let M/K be defined by an irreducible polynomial $f \in K[X]$ which has a root $x \in M_0$. Then $f = \prod_{g \in G}(X - x^g)$, and in fact $f \in K_0[X]$, as it is stable under C: For $c \in C$, we have

$$f^c(X) = \prod_{g \in G} (X - x^{cg}) = \prod_{g \in G} (X - x^{(cgc^{-1})c})$$

$$= \prod_{g' \in G} (X - x^{g'c}) \text{ as } C \text{ normalises } G$$

$$= \prod_{g \in G} (X - x^g) \text{ since } x \in M_0 \text{ is stable under } C$$

$$= f$$

Taking into account that $[M : K] = [M_0 : K_0]$, we find that $M_0 = K_0(x)$. Hence, the polynomials generating M_0/K_0 are exactly the polynomials generating M/K which additionally have a root in M_0.

To construct M_0/L_0, let us define for $\overline{g} \in G/H$ and $g \in G$ a representative of $\overline{g}$,

$$U_{\overline{g}}(X) = \prod_{h \in H} (X - x^{hg}).$$

The polynomial $U_{\overline{g}}$ does not depend on the chosen representative g and is invariant under the action of H. Thus, $U_{\overline{g}} \in L[X]$. Precisely, f factors as $\prod_{\overline{g} \in G/H} U_{\overline{g}}$. Any of the $U_{\overline{g}}$ generates M/L, and G resp. G/H act by permutation on the $U_{\overline{g}}$:

$$U_{\overline{g}}^{g'} = U_{\overline{g'g}} \text{ for } g' \in G.$$

Let us compute the action of C on the $U_{\overline{g}}$. For $c \in C$,

$$U_{\overline{g}}^c = \prod_{h \in H} (X - x^{chg}) = \prod_{h \in H} \left(X - x^{(chc^{-1})(cgc^{-1})c} \right) = U_{\overline{cgc^{-1}}} \qquad (1)$$

as C normalises H and G and x is stable under C.

So C also acts by permutation on the $U_{\overline{g}}$, and more precisely by conjugation of the index. In particular, $U_{\overline{1}} = (X - x) \prod_{h \in H \setminus \{1\}} (X - x^h)$ is stable under C. It is thus a polynomial in $L_0[X]$ of degree $|H| = [M_0 : L_0]$ with root x such that $M_0 = L_0(x)$. Hence, $U_{\overline{1}}$ is irreducible and generates M_0/L_0. Write

$$U_{\overline{g}} = \sum_{k=0}^{|H|} (-1)^{|H|-k} \vartheta_{\overline{g}}^{(k)} X^k \text{ with } \vartheta_{\overline{g}}^{(k)} \in L, \vartheta_{\overline{1}}^{(k)} \in L_0.$$

To construct L_0/K_0, we define polynomials V_k whose roots are the Galois images of $\vartheta_{\overline{1}}^{(k)}$ under G/H, that is, $V_k = \prod_{\overline{g} \in G/H} \left(Y - \vartheta_{\overline{g}}^{(k)} \right)$. Notice that the actions of G and C on the $\vartheta_{\overline{g}}^{(k)}$ are the same as on the $U_{\overline{g}}$. So as G and C permute the $\vartheta_{\overline{g}}^{(k)}$, the polynomials V_k lie in $K_0[Y]$. Furthermore, the $\vartheta_{\overline{1}}^{(k)}$ are elements of L_0. Thus, if some V_{k_0} is irreducible, then it generates L_0/K_0. We assume henceforth that this is the case; otherwise, one has to apply a translation to the roots of f. We have thus split $M_0 = K_0[X]/(f)$ into a tower of two field extensions $L_0 = K_0[Y]/(V_{k_0})$ and $M_0 = L_0[X]/(U_{\overline{1}})$.

Notice that if M_0/K_0 is Galois itself, then $C = 1$, the fields with index 0 coincide with the indexless fields, and we recover the approach of [11].

3   Hecke Representation

It remains to explicitly write down the coefficients of the polynomial $U_{\overline{1}}$ as elements of $K_0[Y]/(V_{k_0})$, assuming V_{k_0} is irreducible. In [11], it was suggested to use a power basis representation for algebraic numbers. When dealing with algebraic integers, this is already not satisfying from a theoretical point of view since integral power bases do not always exist. In practice, this leads to tedious computations based on the integrality defect. As already observed in [12], the following *Hecke representation* provides an elegant solution, which is furthermore well suited for computations. This representation is based on the following lemma [13].

Lemma 1. *Let L/K be a separable field extension of degree n with primitive element $\alpha = \alpha_0$. Denote by $(\alpha_i)_{0 \leq i < n}$ the conjugates of α and by $f = \prod_{i=0}^{n-1}(Y - \alpha_i)$ its minimal polynomial. Then any element $\vartheta \in L$ may be written as*

$$\vartheta = \frac{g_\vartheta(\alpha)}{f'(\alpha)} \text{ for some } g_\vartheta \in K[Y].$$

Moreover, if $\mathcal{O}_K$ is an integrally closed subring of K and α and ϑ are integral over $\mathcal{O}_K$, then $f, g_\vartheta \in \mathcal{O}_K[Y]$.

Proof. Let $(\vartheta_i)_{0 \leq i < n}$ be the conjugates of ϑ. Put

$$g_\vartheta(Y) = \sum_{i=0}^{n-1} \vartheta_i \frac{f(Y)}{Y - \alpha_i}. \tag{2}$$

By Galois theory, we have $g_\vartheta \in K[Y]$. Furthermore, $g_\vartheta(\alpha) = \vartheta f'(\alpha)$. If α and ϑ are integral over $\mathcal{O}_K$, then so are the coefficients of f and g_ϑ, whence f, $g_\vartheta \in \mathcal{O}_K[Y]$ since $\mathcal{O}_K$ is integrally closed.

Formula (2) is the key to writing the coefficients $\vartheta_{\overline{1}}^{(k)}$ of $U_{\overline{1}}$ in terms of the generating element $\vartheta_{\overline{1}}^{(k_0)}$ of L_0/K_0. We postpone the formulation of the decomposition algorithm to Sect. 6, as the analysis of the algorithm requires to have a closer look at the underlying polynomial arithmetic.

4   Fast Polynomial Arithmetic

The problem of the fast computation of the Hecke representation is very closely related to that of multiple polynomial interpolation over the same set of abscissae, but with varying ordinates. In this section, we thus recall fast algorithms on polynomials, within the framework of [8, Chapt. 10].

Let $(u_0, u_1, \ldots, u_{n-1})$ be n elements of a field K. The subproduct tree associated to them is a binary tree built from subproducts of the u_i obtained by a divide and conquer approach. To begin with, assume first that $n = 2^k$. We define for $0 \leq i \leq k = \log_2 n$ and $0 \leq j < 2^{k-i}$:

$$M_{i,j} = \prod_{\ell=0}^{2^i-1}(X - u_{j \cdot 2^i + \ell}).$$

The $M_{0,j}$ form the leaves of a binary tree, and the parent of $M_{i,j}$ and $M_{i,j+1}$ for j even is given by $M_{i+1,j/2}$. When n is not a power of 2, with $2^{k-1} < n < 2^k$, we build the tree with 2^k leaves, replacing the missing $M_{i,j}$ by 1. The resulting procedure is named BUILDSUBPRODUCTTREE.

If the underlying polynomial arithmetic uses Karatsuba multiplication and counting only multiplications in K, it is shown in [3] that the strategy of adding leaves with content 1 rightmost in the tree is optimal. The cost of building the tree is then bounded above by 3^k multiplications in K, or otherwise said, it is in $O(M_K(n))$, where $M_K(n)$ denotes the cost of multiplying two polynomials over K with n coefficients. If the polynomial multiplication uses the FFT, then the cost of creating the tree is no more than $M_K(n) \log n$, cf. [8].

We take the opportunity to rewrite [8, Algorithm 10.9], which is the heart of the computation of the Hecke representation, in a version more suited to implementation and taking care of the case where n is not a power of 2. Note the pruning of the tree in that case.

The tree M is conveniently stored in an array T of size $2^{k+1} - 1$ in such a way that `T[2^(k-i)-1+j]` contains $M_{i,j}$.

procedure LCFLM(u, c, k, T, s)
INPUT: $u_0, \ldots, u_{n-1}, c_0, \ldots, c_{n-1}$ with $2^{k-1} < n \leq 2^k$; T containing the subproduct tree; $s = 2^{k-i} - 1 + j$ s.t. $0 \leq i < k$, $0 \leq j < 2^{k-i}$.

OUTPUT: $\displaystyle\sum_{\ell=0}^{2^i-1} c_{j \cdot 2^i + \ell} \frac{M_{i,j}}{X - u_{j \cdot 2^i + \ell}} \in K[X]$.

1. **if** $i = 0$, i.e., $s \geq 2^k - 1$, **then return** $c_j = c_{s-2^k+1}$.
2. **if** $T[2s + 2] = 1$ **then return** LCFLM$(u, c, k, T, 2s + 1)$;
3. $r_0 :=$ LCFLM$(u, c, k, T, 2s + 1)$;
4. $r_1 :=$ LCFLM$(u, c, k, T, 2s + 2)$;
5. **return** $r_0 T[2s + 2] + r_1 T[s + 1]$.

and the launching procedure is then simply

procedure LINEARCOMBINATIONFORLINEARMODULI(u, c, n, T)
1. **return** LCFLM$(u, c, \lceil \log_2 n \rceil, T, 0)$.

The cost analysis of LINEARCOMBINATIONFORLINEARMODULI is closely related to that of BUILDSUBPRODUCTTREE, which is easily seen by rewriting the latter in a recursive style, cf. [3]. It is easily shown that $L(n) = 2B(n)$. Moreover, the analysis of [3] carries over and shows that in the Karatsuba model, our strategy of filling the subproduct tree with 1 in the rightmost leaves remains optimal also for LINEARCOMBINATIONFORLINEARMODULI.

5   Tricks in the Non-Galois Case

In order to determine Hecke representations in the non-Galois setting of Sect. 2, the procedures BUILDSUBPRODUCTTREE and LINEARCOMBINATIONFORLINEARMODULI can be applied verbatim. Then all computations take place in K, and the final result will be in K_0. It is desirable, however, to work directly in K_0, where in general the basic arithmetic is less costly. To this goal, notice that with the notations of Sects. 2 and 4, while the values u_i and c_i are elements of L, they nevertheless form complete orbits under $C|_L = \mathrm{Gal}(L/L_0)$. This is precisely what causes the result to be defined over K_0 instead of K. We may then modify our procedures by computing first the results for the orbits of C, which will be elements of $K_0[X]$, and filling the leaves of the trees with these results. Otherwise said (and not taking the complications into account coming from the fact that not all integers are powers of 2), we stop our recursion at a higher level.

To do so, we need control of the orbits. Precisely, we require a function that takes as input the $u_0, \ldots, u_{n-1}$ and $c_0, \ldots, c_{n-1}$ and the action of C on them. It should return the number n' of joint orbits of the u_i and c_i, their cardinalities n_j and the orbits $(U_j, (c_{j,0}, \ldots, c_{j,n_j-1}))$ with $U_j = (u_{j,0}, \ldots, u_{j,n_j-1})$ for $0 \le j < n'$. Notice that the orbits will in general not all have the same size (in particular, one of them is trivial, but not all of them are, unless we are in the Galois case $M = M_0$). In the setting of Sect. 2, the procedure can be realised symbolically making use of the group structure of $\mathrm{Gal}(M/K_0) = G \rtimes C$ and of the action of the Galois group on the u_i and c_i. Another approach is possible, for instance, when C is generated by the complex conjugation, in which case the orbits can be obtained by simple inspection of the u_i and c_i, see Sect. 7.

The two procedures now generalise as follows. The function ORBITALBUILD-SUBPRODUCTTREE is a trivial modification of BUILDSUBPRODUCTTREE, replacing n by n' and k by k' s.t. $2^{k'-1} < n' \le 2^{k'}$, and filling the leaves with $M_{0,j} := \prod_{r=0}^{n_j-1}(X - u_{j,r}) \in K_0[X]$. In LCFLM, we need to replace the input c_j corresponding to an element u_j by the interpolation polynomial C_j corresponding to an orbit U_j, letting

$$C_j = \sum_{r=0}^{n_j-1} c_{j,r} \frac{M_{0,j}}{X - u_{j,r}} \in K_0[X].$$

The C_j may be obtained by applying the original BUILDSUBPRODUCTTREE and LINEARCOMBINATIONFORLINEARMODULI; in many cases with small C, however, they are trivial to write down, cf. Sect. 7. The remainder of the function ORBITALLINEARCOMBINATIONFORLINEARMODULI is straightforward.

6 Algorithm

6.1 Description

We use the notations of Sect. 2. Let $m = |H|$ and $n = |G/H|$. We assume that a root $x \in M_0$ of f is identified, that $H = \{h_0 = 1, h_1, \ldots, h_{m-1}\}$ and a system of representatives $\{g_0 = 1, g_1, \ldots, g_{n-1}\}$ of G/H are given, and that the action of G on x is explicitly known. This allows to create an ordered list of conjugates
$$(x_0 = x, x_1, \ldots, x_{mn-1}) = \left(x^{h_0 g_0}, x^{h_1 g_0}, \ldots, x^{h_{m-1} g_0}, x^{h_0 g_1}, \ldots, x^{h_{m-1} g_{n-1}}\right),$$
i.e., $x_{i+mj} = x^{h_i g_j}$. In the non-Galois case, where C is non-trivial, we furthermore need an enumeration of the elements of C and must be able to determine their action on the x_i.

procedure DECOMPOSITION($f(X)$, $\mathcal{R}$, m, n, C)

INPUT: $\mathcal{R} = (x_0, \dots, x_{mn-1})$ a list of the roots of f ordered as explained above.

OUTPUT: a defining polynomial $V_{k_0}(Y)$ of L_0/K_0 and a factor $U_0(X, Y)$ of f defining M_0/L_0.

1. **for** $i = 0..n-1$ **do** compute

$$U_i(X) = U_{\bar{g}_i}(X) = \prod_{j=0}^{m-1} (X - x_{mi+j}) = \sum_{k=0}^{m} (-1)^{m-k} \vartheta_i^{(k)} X^k$$

2. **do** compute

$$V_k(Y) = \prod_{i=0}^{n-1} \left(Y - \vartheta_i^{(k)}\right) \in K_0[Y]$$

   for $k = 0..m-1$ **until** one of them, $V_{k_0}(Y)$, is irreducible over K. If none of them is irreducible, declare failure.

3. **for** $k = 0..m-1$ **do** compute the polynomials $g_k(Y) = g_{\vartheta_0^{(k)}}(Y)$ defining the Hecke representations of the coefficients of $U_0 = U_{\bar{1}}$ according to (2).

4. **return** $V_{k_0}(Y)$ and

$$U_0(X, Y) = X^m + \frac{1}{V'_{k_0}(Y)} \sum_{k=0}^{m-1} (-1)^{m-k} g_k(Y) X^k.$$

6.2   Analysis

Step 1: The computation of the $U_i(X)$ costs n applications of BUILDSUBPRODUCTTREE for polynomials of degree m. Notice that if we can compute the action of C on the coefficients of the U_i, then we may use (1) to compute only some of the U_i and deduce the others via the action of C. In the best case, this may reduce the complexity by a factor of about $|C|$, cf. the examples of Sect. 7.

Step 2: The computation of the $V_k(Y)$ costs at most m applications of BUILDSUBPRODUCTTREE for polynomials of degree n in the basic case. If C is nontrivial, then we need to identify the orbits of the $\vartheta_i^{(k)}$ and then apply at most m times ORBITALBUILDSUBPRODUCTTREE on polynomials whose degrees are defined by the number of orbits.

   As $V_k(Y)$ is a power of an irreducible polynomial over K (the minimal polynomial of $\vartheta_0^{(k)}$) we may check for irreducibility by verifying that $\gcd(V_k, V'_k) = 1$. In many cases, it is possible to perform a sufficient simpler test. Over the rationals, for instance, we may reduce modulo small prime numbers and check for irreducibility.

Step 3: Having kept the subproduct tree leading to V_{k_0} in Step 2, we need $m - 1$ calls to LINEARCOMBINATIONFORLINEARMODULI on a polynomial of degree n. (Notice that we already know $g_{k_0} = Y V'_{k_0}(Y)$.) For non-trivial C, one may use ORBITALLINEARCOMBINATIONFORLINEARMODULI on the orbits already identified in Step 2.

Denote by $B(n)$ the cost of BUILDSUBPRODUCTTREE and by $L(n)$ the cost of LINEARCOMBINATIONFORLINEARMODULI. In the basic Galois case and computing all possible V_k, the total cost of the algorithm then becomes

$$nB(m) + mB(n) + (m-1)L(n) \approx nB(m) + 3mB(n),$$

where $B(n) = O(M_K(n))$ in the Karatsuba and $B(n) = O(\log n M_K(n))$ in the FFT case.

As a possible improvement to the algorithm, one may want to compute all the V_k in Step 2 and choose the best one according to some optimality criterion, for instance the heights of the resulting polynomials, the sizes of their roots, etc. If the process is to be iterated with V_{k_0} in the role of f, it may pay off to carefully arrange the subproduct tree in Step 2, so that the subproduct trees of the U_i of the next iteration appear as branches and need not be recomputed; see the example in Sect. 7.

7   Application: Class Fields of Imaginary Quadratic Fields

Class fields of imaginary quadratic fields are used in a crucial way in the primality proving algorithm ECPP, which motivated our research. During the algorithm, one needs to find a root of the polynomial defining the class field modulo a prime. A fast way of doing so is obtained by first decomposing the polynomial (independently of the prime) over a tower of algebraic field extensions, then reducing and solving a series of lower degree equations in the finite prime field. See [1,11,4,5] for more details and a description of the general approach. This large class of fields fits into the framework developed above.

With the notations of the preceding sections, we put $K_0 = \mathbb{Q}$ and $K = \mathbb{Q}(\sqrt{-D})$, an imaginary quadratic field of discriminant $-D < 0$. The field M is the Hilbert class field, $G \sim \mathrm{Cl}(-D)$ is the ideal class group and its cardinality $h(-D)$ the class number of K. The group C is generated by the complex conjugation; dealing with this action is perhaps the simplest possible case if we use a floating point representation of the roots.

The polynomials $f(X)$ in which we are interested are minimal polynomials $H_D[u](X)$ of class invariants u, such as the j-invariant, or (generalised) Weber functions, for which the explicit Galois action is known cf. [4,6]. We restrict the presentation to the cases where $H_D[u](X)$ is in fact defined over $\mathbb{Q}$, so that our theory applies.

Let us consider the case of $D = 95$ for Weber's function $\mathfrak{f}$. We choose the defining polynomial

$$H_{4\cdot95}[\mathfrak{f}/\sqrt{2}](X) = X^8 - 2X^7 - 2X^6 + X^5 + 2X^4 - X^3 + X - 1,$$

whose roots in the right order for $m = 2$ and $n = 4$ are given by

i	x_i	comment	i	x_i	comment
0	2.533	real	4	-0.9146	real
1	$-0.8287 - 0.6954i$		5	$0.2107 + 0.6954i$	$= \overline{x_3}$
2	$0.8090 + 0.2097i$		6	$0.8090 - 0.2097i$	$= \overline{x_2}$
3	$0.2107 - 0.6954i$		7	$-0.8287 + 0.6954i$	$= \overline{x_1}$

Using our theoretical knowledge, we are able to spot the complex conjugates without actually computing them. We compute the polynomials U_i as follows:

i	U_i	comment
0	$(X - x_0)(X - x_4) = X^2 - \vartheta_{0,1}X + \vartheta_{0,0} = X^2 - 1.618X - 2.317$	real
1	$(X - x_1)(X - x_5) = X^2 - \vartheta_{1,1}X + \vartheta_{1,0} = X^2 + 0.6180X + 0.3090 - 0.7229i$	
2	$(X - x_2)(X - x_6) = X^2 - \vartheta_{2,1}X + \vartheta_{2,0} = X^2 - 1.618X + 0.6985$	real
3	$(X - x_3)(X - x_7) = X^2 - \vartheta_{3,1}X + \vartheta_{3,0}$	$= \overline{U_1(X)}$

Since we have the information on the complex conjugates, we do not need to compute U_3, which saves some time. We can also derive the conjugacy properties of the V, namely that $\vartheta_{3,j} = \overline{\vartheta_{1,j}}$, and $\vartheta_{0,j}$ and $\vartheta_{2,j}$ are real.

To compute V_0, we can precompute the trees for the U of the next iteration, namely

$$U_0^*(X) = (X + 2.317)(X - 0.6985),$$
$$U_1^*(X) = (X - (0.3090 - 0.7229i))\overline{(X - (0.3090 - 0.7229i))}$$

and deduce

$$V_0(Y) = U_0^*(Y)U_1^*(Y) = Y^4 + Y^3 - 2Y^2 + 2Y - 1.$$

For the sake of completeness, note that

$$V_1(Y) = Y^4 - 2Y^3 - Y^2 + 2Y + 1 = \left(Y^2 - Y - 1\right)^2.$$

We finally compute

$$U_0(X, Y) = X^2 + \frac{1}{V_0'(Y)}\left((-2Y^3 + Y^2 - 3Y - 1)X - Y^3 + 4Y^2 - 6Y + 4\right).$$

The next and last iteration yields $V_0^*(Y_1) = Y_1^2 + Y_1 - 1$ and

$$U_0^*(Y, Y_1) = Y^2 + \frac{1}{V_0^*(Y_1)'}((Y_1 - 2)Y - Y_1 + 2),$$

so that we replaced the resolution of a degree 8 equation by the solution of $V_0^*(Y_1) = 0$, followed by that of $U_0^*(Y, Y_1) = 0$ and finally $U_0(X, Y) = 0$.

8   Implementation

We have implemented our algorithms in C. As we represent the roots by complex numbers with high precision, we rely on the packages gmp, mpfr and mpc [9,10,7]

Table 1.

D	$h(-D)$	$\mathrm{Cl}(-D)$	prec	u	time
2519	64	$C(64)$	250	0.1	0.1
10295	128	$C(32) \times C(4)$	508	0.3	0.8
39431	256	$C(128) \times C(2)$	1095	1.6	5.4
132599	512	$C(512)$	2320	12.4	19.7
328319	1024	$C(1024)$	4571	86.5	138.4
1333631	2048	$C(2048)$	10130	777.6	1256.8
4599839	4096	$C(2048) \times C(2)$	21601	6586.9	9023.7
361919	$31 \cdot 29$	$C(899)$	4327	62.1	82.2

for the basic arithmetic. The fast polynomial arithmetic as described in Sect. 4 is implemented in the Karatsuba model. We linked the field decomposition code with our programs for computing class fields and ultimately with the second author's current implementation of ECPP. In Table 1, we provide some timings for class polynomials $H_D[\mathfrak{f}/\sqrt{2}]$. We give the precision of the computations (in bits), the time in seconds needed for computing complex approximations of the roots to this precision, and the total running time. All examples have been obtained on a Pentium III with 450 MHz.

Let us examine the last example in more detail. We use $m = 31$ and $n = 29$. We first compute the 29 polynomials U of degree 31. There are 1 real polynomial and 14 pairs of complex conjugate ones. The real one takes 0.16 seconds and each pair of complex ones takes 0.32 seconds. Constructing the 31 polynomials V of degree 29 takes 13 seconds; reconstructing the algebraic factor $U(X, Y)$ requires 30 interpolations, for a total of 7.15 seconds.

The time needed in the Hecke phase for $D = 132599$ of degree 512 illustrates the effect of fast multiplication. Reconstructing one polynomial $U(X, Y)$ of degree 256 takes 2.98 seconds, whereas one of degree 128 takes 0.99 seconds: this is exactly the factor 3 predicted theoretically in the Karatsuba model.

Acknowledgements. We thank G. Hanrot and P. Zimmermann for rapidly correcting the (minor) bugs we found in `mpfr`. The second author is on leave from the French Department of Defense, Délégation Générale pour l'Armement.

References

1. A.O.L. Atkin and F. Morain. Elliptic curves and primality proving. *Math. Comp.*, 61(203):29–68, July 1993.
2. J.D. Dixon. Computing subfields in algebraic number fields. *J. Austral. Math. Soc. Ser. A*, 49:434–448, 1990.
3. A. Enge, P. Gaudry, G. Hanrot, and P. Zimmermann. Reconstructing a polynomial from its roots. In preparation, 2002.
4. A. Enge and F. Morain. Further investigations of the generalised Weber functions. Preprint, July 2001.

5. A. Enge and F. Morain. Comparing invariants for class fields of imaginary quadratic fields. In C. Fieker and D. R. Kohel, editors, *ANTS IV – Algorithmic Number Theory*, volume 2369 of *Lecture Notes in Comput. Sci.*, pages 252–266. Springer-Verlag, 2002.

6. A. Enge and R. Schertz. Constructing elliptic curves from modular curves of positive genus. Preprint, 2001.

7. A. Enge and P. Zimmermann. `mpc` – a library for multiprecision complex arithmetic with exact rounding. Version 0.4.1, available from `http://www.lix.polytechnique.fr/Labo/Andreas.Enge`.

8. J. von zur Gathen and J. Gerhard. *Modern Computer Algebra*. Cambridge University Press, 1999.

9. T. Granlund et. al. `gmp` – gnu multiprecision library. Version 4.1.2, available from `http://www.swox.com/gmp`.

10. G. Hanrot, V. Lefèvre, and P. Zimmermann et. al. `mpfr` – a library for multiple-precision floating-point computations with exact rounding. Version contained in [9]. Available from `http://www.mpfr.org`.

11. G. Hanrot and F. Morain. Solvability by radicals from an algorithmic point of view. In B. Mourrain, editor, *Symbolic and algebraic computation*, pages 175–182. ACM, 2001. Proceedings ISSAC'2001, London, Ontario.

12. G. Hanrot and F. Morain. Solvability by radicals from a practical algorithmic point of view. Submitted. Available from `http://www.lix.polytechnique.fr/Labo/Francois.Morain`, November 2001.

13. E. Hecke. *Vorlesungen über die Theorie der algebraischen Zahlen*. Chelsea Publishing Company, 2nd ed., 1970.

14. J. Klüners. On computing subfields. A detailed description of the algorithm. *J. Théor. Nombres Bordeaux*, 10:243–271, 1998.

15. J. Klüners and M. Pohst. On computing subfields. *J. Symbolic Comput.*, 24:385–397, 1997.

Author Index

Abdel-Ghaffar, Khaled   98

Billet, Olivier   34
Bouganis, Thanasis   169, 180
Bras-Amorós, Maria   204
Brier, Eric   43

Castro, Francis N.   129
Cavalcante, Rodrigo Gusmão   191
Cohen, Gérard   79
Coles, Drue   180

Ding, Cunsheng   24
Djurdjevic, Ivana   98

Encheva, Sylvia   79
Enge, Andreas   254
Etzion, Tuvi   216

Galand, Fabien   235
Gashkov, Igor B.   108

Hatano, Yasuo   61
Horadam, K.J.   150

Imai, Hideki   87

Janwa, H.   119
Joye, Marc   34, 43

Kaneko, Toshinobu   61
Kazarin, L.S.   108
Klappenecker, Andreas   139

Lally, Kristine   244
Lange, Tanja   51
Lauder, Alan G.B.   18
Lin, Shu   98

Morain, François   254
Moreno, Oscar   129
Mueller-Quade, Joern   87

Nascimento, Anderson C.A.   87
Niederreiter, Harald   6

Otsuka, Akira   87

Palazzo, Reginaldo Jr.   191

Ramirez-Alzola, Domingo   158
Rötteler, Martin   139

Salomaa, Arto   24
Schaathun, Hans Georg   71
Schwartz, Moshe   216
Shiromoto, Keisuke   226
Shparlinski, Igor E.   6
Sidelnikov, V.M.   108
Solé, Patrick   24
Stern, Jacques   1

Tanaka, Hidema   61
Tian, Xiaojian   24

Winterhof, Arne   51

Xu, Jun   98

Lecture Notes in Computer Science

For information about Vols. 1–2561

please contact your bookseller or Springer-Verlag

Vol. 2562: V. Dahl, P. Wadler (Eds.), Practical Aspects of Declarative Languages. Proceedings, 2003. X, 315 pages. 2002.

Vol. 2563: Y. Manolopoulos, S, Evripidou, A.C. Kakas (Eds.), Advances in Informatics. Proceedings, 2001. XI, 498 pages. 2002.

Vol. 2565: J.M.L.M. Palma, J. Dongarra, V. Hernández, A. Augusto Sousa (Eds.), High Performance Computing for Computational Science – VECPAR 2002. Proceedings, 2002. XVII, 732 pages. 2003.

Vol. 2566: T.Æ. Mogensen, D.A. Schmidt, I.H. Sudborough (Eds.), The Essence of Computation. XIV, 473 pages. 2002.

Vol. 2567: Y.G. Desmedt (Ed.), Public Key Cryptography – PKC 2003. Proceedings, 2003. XI, 365 pages. 2002.

Vol. 2568: M. Hagiya, A. Ohuchi (Eds.), DNA Computing. Proceedings, 2002. X, 338 pages. 2003.

Vol. 2569: D. Gollmann, G. Karjoth, M. Waidner (Eds.), Computer Security – ESORICS 2002. Proceedings, 2002. XIII, 648 pages. 2002. (Subseries LNAI).

Vol. 2570: M. Jünger, G. Reinelt, G. Rinaldi (Eds.), Combinatorial Optimization – Eureka, You Shrink!. Proceedings, 2001. X, 209 pages. 2003.

Vol. 2571: S.K. Das, S. Bhattacharya (Eds.), Distributed Computing. Proceedings, 2002. XIV, 354 pages. 2002.

Vol. 2572: D. Calvanese, M. Lenzerini, R. Motwani (Eds.), Database Theory – ICDT 2003. Proceedings, 2003. XI, 455 pages. 2002.

Vol. 2574: M.-S. Chen, P.K. Chrysanthis, M. Sloman, A. Zaslavsky (Eds.), Mobile Data Management. Proceedings, 2003. XII, 414 pages. 2003.

Vol. 2575: L.D. Zuck, P.C. Attie, A. Cortesi, S. Mukhopadhyay (Eds.), Verification, Model Checking, and Abstract Interpretation. Proceedings, 2003. XI, 325 pages. 2003.

Vol. 2576: S. Cimato, C. Galdi, G. Persiano (Eds.), Security in Communication Networks. Proceedings, 2002. IX, 365 pages. 2003.

Vol. 2577: P. Petta, R. Tolksdorf, F. Zambonelli (Eds.), Engineering Societies in the Agents World III. Proceedings, 2002. X, 285 pages. 2003. (Subseries LNAI).

Vol. 2578: F.A.P. Petitcolas (Ed.), Information Hiding. Proceedings, 2002. IX, 427 pages. 2003.

Vol. 2580: H. Erdogmus, T. Weng (Eds.), COTS-Based Software Systems. Proceedings, 2003. XVIII, 261 pages. 2003.

Vol. 2581: J.S. Sichman, F. Bousquet, P. Davidsson (Eds.), Multi-Agent-Based Simulation II. Proceedings, 2002. X, 195 pages. 2003. (Subseries LNAI).

Vol. 2582: L. Bertossi, G.O.H. Katona, K.-D. Schewe, B. Thalheim (Eds.), Semantics in Databases. Proceedings, 2001. IX, 229 pages. 2003.

Vol. 2583: S. Matwin, C. Sammut (Eds.), Inductive Logic Programming. Proceedings, 2002. X, 351 pages. 2003. (Subseries LNAI).

Vol. 2584: A. Schiper, A.A. Shvartsman, H. Weatherspoon, B.Y. Zhao (Eds.), Future Directions in Distributed Computing. X, 219 pages. 2003.

Vol. 2585: F. Giunchiglia, J. Odell, G. Weiß (Eds.), Agent-Oriented Software Engineering III. Proceedings, 2002. X, 229 pages. 2003.

Vol. 2586: M. Klusch, S. Bergamaschi, P. Edwards, P. Petta (Eds.), Intelligent Information Agents. VI, 275 pages. 2003. (Subseries LNAI).

Vol. 2587: P.J. Lee, C.H. Lim (Eds.), Information Security and Cryptology – ICISC 2002. Proceedings, 2002. XI, 536 pages. 2003.

Vol. 2588: A. Gelbukh (Ed.), Computational Linguistics and Intelligent Text Processing. Proceedings, 2003. XV, 648 pages. 2003.

Vol. 2589: E. Börger, A. Gargantini, E. Riccobene (Eds.), Abstract State Machines 2003. Proceedings, 2003. XI, 427 pages. 2003.

Vol. 2590: S. Bressan, A.B. Chaudhri, M.L. Lee, J.X. Yu, Z. Lacroix (Eds.), Efficiency and Effectiveness of XML Tools and Techniques and Data Integration over the Web. Proceedings, 2002. X, 259 pages. 2003.

Vol. 2591: M. Aksit, M. Mezini, R. Unland (Eds.), Objects, Components, Architectures, Services, and Applications for a Networked World. Proceedings, 2002. XI, 431 pages. 2003.

Vol. 2592: R. Kowalczyk, J.P. Müller, H. Tianfield, R. Unland (Eds.), Agent Technologies, Infrastructures, Tools, and Applications for E-Services. Proceedings, 2002. XVII, 371 pages. 2003. (Subseries LNAI).

Vol. 2593: A.B. Chaudhri, M. Jeckle, E. Rahm, R. Unland (Eds.), Web, Web-Services, and Database Systems. Proceedings, 2002. XI, 311 pages. 2003.

Vol. 2594: A. Asperti, B. Buchberger, J.H. Davenport (Eds.), Mathematical Knowledge Management. Proceedings, 2003. X, 225 pages. 2003.

Vol. 2595: K. Nyberg, H. Heys (Eds.), Selected Areas in Cryptography. Proceedings, 2002. XI, 405 pages. 2003.

Vol. 2596: A. Coen-Porisini, A. van der Hoek (Eds.), Software Engineering and Middleware. Proceedings, 2002. XII, 239 pages. 2003.

Vol. 2597: G. Păun, G. Rozenberg, A. Salomaa, C. Zandron (Eds.), Membrane Computing. Proceedings, 2002. VIII, 423 pages. 2003.

Vol. 2598: R. Klein, H.-W. Six, L. Wegner (Eds.), Computer Science in Perspective. X, 357 pages. 2003.

Vol. 2599: E. Sherratt (Ed.), Telecommunications and beyond: The Broader Applicability of SDL and MSC. Proceedings, 2002. X, 253 pages. 2003.

Vol. 2600: S. Mendelson, A.J. Smola, Advanced Lectures on Machine Learning. Proceedings, 2002. IX, 259 pages. 2003. (Subseries LNAI).

Vol. 2601: M. Ajmone Marsan, G. Corazza, M. Listanti, A. Roveri (Eds.) Quality of Service in Multiservice IP Networks. Proceedings, 2003. XV, 759 pages. 2003.

Vol. 2602: C. Priami (Ed.), Computational Methods in Systems Biology. Proceedings, 2003. IX, 214 pages. 2003.

Vol. 2603: A. Garcia, C. Lucena, F. Zambonelli, A. Omicini, J. Castro (Eds.), Software Engineering for Large-Scale Multi-Agent Systems. XIV, 285 pages. 2003.

Vol. 2604: N. Guelfi, E. Astesiano, G. Reggio (Eds.), Scientific Engineering for Distributed Java Applications. Proceedings, 2002. X, 205 pages. 2003.

Vol. 2606: A.M. Tyrrell, P.C. Haddow, J. Torresen (Eds.), Evolvable Systems: From Biology to Hardware. Proceedings, 2003. XIV, 468 pages. 2003.

Vol. 2607: H. Alt, M. Habib (Eds.), STACS 2003. Proceedings, 2003. XVII, 700 pages. 2003.

Vol. 2609: M. Okada, B. Pierce, A. Scedrov, H. Tokuda, A. Yonezawa (Eds.), Software Security – Theories and Systems. Proceedings, 2002. XI, 471 pages. 2003.

Vol. 2610: C. Ryan, T. Soule, M. Keijzer, E. Tsang, R. Poli, E. Costa (Eds.), Genetic Programming. Proceedings, 2003. XII, 486 pages. 2003.

Vol. 2611: S. Cagnoni, J.J. Romero Cardalda, D.W. Corne, J. Gottlieb, A. Guillot, E. Hart, C.G. Johnson, E. Marchiori, J.-A. Meyer, M. Middendorf, G.R. Raidl (Eds.), Applications of Evolutionary Computing. Proceedings, 2003. XXI, 708 pages. 2003.

Vol. 2612: M. Joye (Ed.), Topics in Cryptology – CT-RSA 2003. Proceedings, 2003. XI, 417 pages. 2003.

Vol. 2613: F.A.P. Petitcolas, H.J. Kim (Eds.), Digital Watermarking. Proceedings, 2002. XI, 265 pages. 2003.

Vol. 2614: R. Laddaga, P. Robertson, H. Shrobe (Eds.), Self-Adaptive Software: Applications. Proceedings, 2001. VIII, 291 pages. 2003.

Vol. 2615: N. Carbonell, C. Stephanidis (Eds.), Universal Access. Proceedings, 2002. XIV, 534 pages. 2003.

Vol. 2616: T. Asano, R. Klette, C. Ronse (Eds.), Geometry, Morphology, and Computational Imaging. Proceedings, 2002. X, 437 pages. 2003.

Vol. 2617: H.A. Reijers (Eds.), Design and Control of Workflow Processes. Proceedings, 2002. XV, 624 pages. 2003.

Vol. 2618: P. Degano (Ed.), Programming Languages and Systems. Proceedings, 2003. XV, 415 pages. 2003.

Vol. 2619: H. Garavel, J. Hatcliff (Eds.), Tools and Algorithms for the Construction and Analysis of Systems. Proceedings, 2003. XVI, 604 pages. 2003.

Vol. 2620: A.D. Gordon (Ed.), Foundations of Software Science and Computation Structures. Proceedings, 2003. XII, 441 pages. 2003.

Vol. 2621: M. Pezzè (Ed.), Fundamental Approaches to Software Engineering. Proceedings, 2003. XIV, 403 pages. 2003.

Vol. 2622: G. Hedin (Ed.), Compiler Construction. Proceedings, 2003. XII, 335 pages. 2003.

Vol. 2623: O. Maler, A. Pnueli (Eds.), Hybrid Systems: Computation and Control. Proceedings, 2003. XII, 558 pages. 2003.

Vol. 2625: U. Meyer, P. Sanders, J. Sibeyn (Eds.), Algorithms for Memory Hierarchies. Proceedings, 2003. XVIII, 428 pages. 2003.

Vol. 2626: J.L. Crowley, J.H. Piater, M. Vincze, L. Paletta (Eds.), Computer Vision Systems. Proceedings, 2003. XIII, 546 pages. 2003.

Vol. 2627: B. O'Sullivan (Ed.), Recent Advances in Constraints. Proceedings, 2002. X, 201 pages. 2003. (Subseries LNAI).

Vol. 2628: T. Fahringer, B. Scholz, Advanced Symbolic Analysis for Compilers. XII, 129 pages. 2003.

Vol. 2631: R. Falcone, S. Barber, L. Korba, M. Singh (Eds.), Trust, Reputation, and Security: Theories and Practice. Proceedings, 2002. X, 235 pages. 2003. (Subseries LNAI).

Vol. 2632: C.M. Fonseca, P.J. Fleming, E. Zitzler, K. Deb, L. Thiele (Eds.), Evolutionary Multi-Criterion Optimization. Proceedings, 2003. XV, 812 pages. 2003.

Vol. 2633: F. Sebastiani (Ed.), Advances in Information Retrieval. Proceedings, 2003. XIII, 546 pages. 2003.

Vol. 2634: F. Zhao, L. Guibas (Eds.), Information Processing in Sensor Networks. Proceedings, 2003. XII, 692 pages. 2003.

Vol. 2636: E. Alonso, D, Kudenko, D. Kazakov (Eds.), Adaptive Agents and Multi-Agent Systems. XIV, 323 pages. 2003. (Subseries LNAI).

Vol. 2637: K.-Y. Whang, J. Jeon, K. Shim, J. Srivastava (Eds.), Advances in Knowledge Discovery and Data Mining. Proceedings, 2003. XVIII, 610 pages. 2003. (Subseries LNAI).

Vol. 2639: G. Wang, Q. Liu, Y. Yao, A. Skowron (Eds.), Rough Sets, Fuzzy Sets, Data Mining, and Granular Computing. Proceedings, 2003. XVII, 741 pages. 2003. (Subseries LNAI).

Vol. 2642: X. Zhou, Y. Zhang, M.E. Orlowska (Eds.), Web Technologies and Applications. Proceedings, 2003. XIII, 608 pages. 2003.

Vol. 2643: M. Fossorier, T. Høholdt, A. Poli (Eds.), Applied Algebra, Algebraic Algorithms and Error-Correcting Codes. Proceedings, 2003. X, 256 pages. 2003.

Vol. 2644: D. Hogrefe, A. Wiles (Eds.), Testing of Communicating Systems. Proceedings, 2003. XII, 311 pages. 2003.

Vol. 2646: H. Geuvers, F, Wiedijk (Eds.), Types for Proofs and Programs. Proceedings, 2002. VIII, 331 pages. 2003.

Vol. 2648: T. Ball, S.K. Rajamani (Eds.), Model Checking Software. Proceedings, 2003. VIII, 241 pages. 2003.

Vol. 2649: B. Westfechtel, A. van der Hoek (Eds.), Software Configuration Management. Proceedings, 2003. VIII, 241 pages. 2003.

Vol. 2656: E. Biham (Ed.), Advances in Cryptology – EUROCRPYT 2003. Proceedings, 2003. XIV, 649 pages. 2003.

Vol. 2663: E. Menasalvas, J. Segovia, P.S. Szczepaniak (Eds.), Advances in Web Intelligence. Proceedings, 2003. XII, 350 pages. 2003. (Subseries LNAI).